Scientific Instruments between East and West

Scientific Instruments and Collections

VOLUME 7

The titles published in this series are listed at *brill.com/sico*

Scientific Instruments between East and West

Edited by

Neil Brown
Silke Ackermann
Feza Günergun

TUTA SUB AEGIDE PALLAS · 1683 ·

BRILL

LEIDEN | BOSTON

Cover illustration: Portrait of "Feyzi Bey the engineer" with the sextant (*Muhadarat-ı Feyzi*, Istanbul University Rare Books Library, MS T6833, fol. 1v, detail)

The Library of Congress Cataloging-in-Publication Data is available online at http://catalog.loc.gov

Typeface for the Latin, Greek, and Cyrillic scripts: "Brill". See and download: brill.com/brill-typeface.

ISSN 1877-7066
ISBN 978-90-04-41283-5 (hardback)
ISBN 978-90-04-41284-2 (e-book)

This book is printed on acid-free paper and produced in a sustainable manner.

Contents

Preface VII
List of Figures and Tables IX
Contributors XIII

1 A Sixteenth-Century Ottoman Compendium of Astronomical Instruments: Seydi Ali's *Mir'at-ı Kâinat* 1
Gaye Danışan

2 Eastern and Western Instruments in Osman Efendi's *Hadiyyat al-Muhtadī* (The Gift of the Convert), 1779 16
Mahdi Abdeljaouad and Pierre Ageron

3 Treatises on *Pergar-ı Nisbe* (the Sector) in Manuscript Collections in Turkey 39
Atilla Polat

4 Measuring Altitudes with an *Alla Franca* Instrument: The Ottoman Engineer Feyzi's Treatise on the Portable Sextant 55
Feza Günergun, Gaye Danışan and Atilla Polat

5 How Did the Turketum (or Torquetum) Get Its Name? 80
Richard L. Kremer

6 A Mingling of Traditions: *Aspects of Dialling in Islam* 108
Anthony Turner

7 Kāshānī's Equatorium: Employing Different Plates for Determining Planetary Longitudes 122
Hamid Bohloul

8 The Introduction of the Telescope into Iran before the Nineteenth Century 142
Seyyed Hadi Tabatabaei

9 Hugo Masing's Golitsyn-Vilip Seismographs: From Tartu to Five Continents 154
Janet Laidla

10 Instruments and Laboratories in the Schools of the Greek Community of Istanbul, 1850–1960 168
Panagiotis Lazos, George N. Vlahakis and Constantine Skordoulis

11 From the Ottoman Empire to Canada: George Petrovic's Metrological Instruments in the Canada Science and Technology Museum 187
Hasan Umut and David Pantalony

12 Instruments of Knowledge and Power in a Colonial Context: Scientific Instruments during the French Occupation of Egypt, 1798–1801 206
Patrice Bret

13 The Magic Lantern as an Ambassador between Cultures and Religions: Imrich Emanuel Roth and the First Dissolving View Shows in the Ottoman Empire, 1845–1846 229
Bernd Scholze

14 Scientific Instrument Retailers in Istanbul in the Nineteenth Century, and Verdoux's Optical Shop 240
Meltem Kocaman

Index 257

Preface

The Scientific Instrument Symposiums organised annually by the Scientific Instrument Commission (SIC) of the International Union for the History and Philosophy of Science and Technology provide an opportunity for scholars and others interested in the history of scientific instrumentation to meet and exchange views. Since the first meeting in Bucharest in 1981 symposiums have been held in a number of Western European and North American cities, and also in Moscow and in Rio de Janeiro.

Silke Ackermann, President of the SIC from 2013 to 2017, had long felt there should be a symposium in Istanbul—a city that arguably like few others has been the focal point of knowledge exchange over centuries—to reflect the rich scientific heritage and scholarship of this part of the world and to establish new links in order to facilitate a vivid scholarly debate. Professional contact and personal friendship encouraged Silke Ackermann to suggest to Feza Günergun, Head of the Department of the History of Science at Istanbul University, that her department might host the 2016 symposium. The suggestion was welcomed, and a formal proposal was developed and was submitted to the SIC by Istanbul University. It was accepted and the 35th Scientific Instrument Symposium was scheduled for 26 to 30 September on a theme of "Instruments between East and West".

Despite a number of unexpected challenges that were beyond the control of the organisers, the symposium went ahead with more than 80 participants and became a huge success. More than half of the participants came from various countries in the Islamic world, especially Turkey and Iran, and had not attended any previous SIC Symposium. Istanbul University does not have a permanent exhibition of instruments and other artefacts relating to the history of science, and the symposium provided the opportunity for a unique exhibition of scientific instruments, manuscripts and prints curated by Feza Günergun, who also edited the catalogue of the exhibition, *Pursuing Knowledge/Bilginin İzinde*, to reflect the mobility of knowledge between Central Asia and Western Europe. One of the goals of this exhibition, which continued after the conference ended, was to attract public attention, especially from young people, to both the local and the wider scientific heritage. It was also a call for institutions and individuals to preserve and care for material pertaining to the history of science and technology in Turkey.

During the symposium about 50 papers were presented in two parallel sessions and an engaging poster session covering a wide variety of subjects. There was lively conversation during the coffee and lunch breaks, and during the programme of visits to institutions and instrument collections which included a brief trip to the Asian side of the country and a Bosphorus cruise and Gala Dinner. The deliberate focus on cross-cultural links between East and West has drawn attention to a little-studied aspect of scientific instrument history and it is hoped that this topic will remain firmly on the agenda.

This volume gathers together a number of the papers delivered at the symposium, the unifying theme being that they connect East and West in some way. It should have included a paper by Inga Elmqvist Söderlund of Stockholm University, who organized a special session of the symposium, but who sadly died of cancer before she could contribute a written paper. The papers by Pierre Ageron and Mahdi Abdeljaouad, Hamid Bohloul, Gaye Danişan, Feza Günergun and her co-authors, Atilla Polat, and Seyed Hadi Tabatabaei are erudite studies of manuscripts in Arabic, Ottoman Turkish or Persian ranging from the fifteenth to the nineteenth centuries—sources that have not featured in previous symposiums and which therefore introduce Western scholars to new aspects of often familiar instruments. Some contributions highlight particular instruments or groups of instruments, including Anthony Turner's paper on sundials in Islam. Janet Laidla's paper on the Galitzyn seismograph of the twentieth century linked neatly to the conference visit to the Kandilli Observatory, where several examples were on view. Other papers about single instruments are those by Richard Kremer on a medieval instrument, the turketum, and Bernd Scholze on a nineteenth-century recreational instrument, the magic lantern. Two papers look at very different collections of instruments: Panagiotis Lazos and his co-authors George Vlahakis and Constantine Skordoulis discuss instruments in the Greek schools in Istanbul and Hasan Umut and David Pantalony focus on an eclectic collection of metrological instruments, many from Ottoman sources, now in Canada after the collector emigrated from the former Yugoslavia. The remaining papers take a different approach. Meltem Kocaman uses local sources to examine the retailing of instruments in nineteenth century Istanbul, and Patrice Bret examines mostly French sources to describe the impact of European instruments in Egypt during the French occupation of 1798 to 1801.

It is a highly varied, but fascinating selection of papers, and the editors hope that it will encourage those interested in scientific instruments to look further into the similarities and differences between their roles in the different cultures, and the links between them.

The editors wish to thank Carole Anne Brown, whose help with linguistic problems and proofreading has eased their task considerably.

Neil Brown
Silke Ackermann
Feza Günergun
June 2019

Figures and Tables

Figures

1.1 *Zatü'l kürsi* (Celestial globe) from Seyyid Ali bin Hüseyin, *Mir'at-ı Kâinat* (Istanbul University Rare Books Library) 9
2.1 *Hadiyyat al-Muhtadī*, first page (Istanbul, Military Museum) 17
2.2 The half cubit scale (*miqyās niṣf al-dhirāʿ*) from *Hadiyyat al-Muhtadī* (Istanbul, Military Museum) 22
2.3 The aerial kūnyā (*al-kūniyā al-hawā'iyya*) from *Hadiyyat al-Muhtadī* (Istanbul, Military Museum) 24
2.4 The table (*al-ṭabla*) in the only copy of *Hadiyyat al-Muhtadī* where this instrument is drawn (Cairo, Egyptian National Library) 27
2.5 The half-circle (*niṣf al-dā'iraʿ*) from *Hadiyyat al-Muhtadī* (Istanbul, Military Museum) 28
2.6 A universal instrument for throwing bombs, from Bernard Forest de Bélidor, *Neuer Cursus Mathematicus*, Pentz, Vienna and Leipzig, 1745 33
2.7 The universal instrument with which one can throw a bomb (*al-āla al-kulliya yumkin bihā ramī al-khumbara*) from *Hadiyyat al-Muhtadī* (Istanbul, Military Museum) 34
2.8 The magnetic compass (*al-būṣula*) from *Hadiyyat al-Muhtadī* (Istanbul, Military Museum) 36
3.1 Mehmed Said Efendi's treatise on the sector: how to use the sector with compasses (Istanbul, Topkapı Palace Museum) 44
3.2 *Risale-i Şerh-i Pergarü'n-Nisbe* (Istanbul University Rare Books Library) 46
3.3 Manuscript C2 (Kütahya, Vahidpaşa Public Library) 48
3.4 Manuscript D2 (Istanbul Atatürk Library, Muallim Cevdet Manuscripts) 50
4.1 *a.* The seal on the *Muhadarat-ı Feyzi* (Istanbul University Rare Books Library); *b.* Sultan Selim III's seal (İsmail Hakkı Uzunçarşılı, *Topkapı Sarayı Müzesi Mühürler Seksiyonu Rehberi*, Istanbul, Şehir Matbaası, 1959) 62
4.2 *a.* Portrait of "Feyzi Bey the engineer" with the sextant (*Muhadarat-ı Feyzi*, Istanbul University Rare Books Library); *b.* detail 66
4.3 The note of appreciation signed by Antoine Juchereau, with its Turkish translation (Istanbul University, Rare Books Library) 75
5.1 Nicholas of Cusa's turketum, 1444; from Hartmann, *Die astronomischen Instrumente des Kardinals Nikolaus Cusanus*, Berlin, 1919 82
5.2 The turketum as depicted in Hans Holbein the Younger, *The Ambassadors*, 1533 (London, National Gallery) 83

5.3 The turketum as depicted in P. Apianus, *Astronomicum Caesareum*, Ingolstadt, 1540 85
5.4 Lorch's reconstruction of the Arabic description of Geber's instrument; from Lorch, "The Astronomical Instruments of Jābir ibn Aflaḥ and the Torquetum", *Centaurus* 20 (1976) 94
5.5 Lorch's reconstruction of the Latin description of Geber's instrument; from Lorch, "The Astronomical Instruments of Jābir ibn Aflaḥ and the Torquetum", *Centaurus* 20 (1976) 95
5.6 Reconstruction of the Arabic equatorial circle; from Brice, Imber and Lorch, *The Dāʿire-yī Muʿaddel of Seydī ʿAli Reʾīs*, Manchester, 1976 96
5.7 "Balzo alla turca" (Turkish-style headdress); Parmigianino, *Portrait of a Young Woman*, c. 1530 (Vienna, Kunsthistorisches Museum) 101
7.1 The ruler, the equatorium and the alidade drawn to scale 125
7.2 Ptolemaic planetary theory for superior planets 131
7.3 Mars in the standard plate 134
7.4 The single apogee plate 135
7.5 Mars on the parallel plate 138
7.6 Equality of the parallel plate with the standard plate 140
8.1 Portrait of Reza Abbasi with spectacles, painted in 1673 by Moiin Mosavvar (Princeton University, USA) 144
8.2 The first page of the Persian manuscript by della Valle; from Kamran Amir-Arjomand, "Enteqal-e Elm Dar Ahd-e Safavi …", *Tarikh-e Elm* 9, 1 (2011) 145
8.3 Title page of the printed edition of *The Mejmua Shemsi*, Calcutta, 1826 148
8.4 Drawings of Venus as seen through a telescope; from Merrick, "*Elm ol-Nojum*" (Tehran, Majlis Library of Iran) 151
8.5 Drawings of Mars as seen through a telescope; from Merrick, "*Elm ol-Nojum*" (Tehran, Majlis Library of Iran) 151
8.6 Drawings of Jupiter as seen through a telescope; from Merrick, "*Elm ol-Nojum*" (Tehran, Majlis Library of Iran) 152
8.7 Drawing of Saturn as seen through a telescope; from Merrick, "*Elm ol-Nojum*" (Tehran, Majlis Library of Iran) 152
9.1 Hugo Masing's Golitsyn-Vilip seismograph from the University of Tartu Museum collection 155
9.2 Hugo Masing's Golitsyn-Vilip seismograph from the University of Tartu Museum collection 156
10.1 The physics laboratory of Zografeion School in about 1910 174
10.2 The building of Theological School of Halki; Drawing by the architect Perklis Fotiadis, around 1894 176

10.3 The chemistry laboratory of the Theological School of Halki in 2016 (Photo: Panagiotis Lazos) 178
10.4 A battery of six Leyden jars in a wooden case, in the collection of the Phanar Greek Orthodox College (Photo: Panagiotis Lazos) 179
10.5 The building of Phanar Greek Orthodox College erected in 1882, also known as the "Red Castle" (Photo: Panagiotis Lazos) 181
10.6 The Bonetti electrical machine of the Phanar Greek Orthodox College, made by A. Gaiffe (Photo: Panagiotis Lazos) 182
10.7 The "Cosmographe de Girod" or tellurion of Zografeion School, made by E. Ducretet (Photo: Panagiotis Lazos) 183
10.8 The electric whirl with seven luminous tubes of the Theological School of Halki (Photo: Panagiotis Lazos) 185
11.1 George Petrovic at the opening for his exhibition "A Vision of the Future", 1986 (Ingenium: Canada's Museums of Science and Innovation) 188
11.2 Exhibition poster for "A Vision of the Future", 1986 (Ingenium: Canada's Museums of Science and Innovation) 194
11.3 Four rules (Ingenium: Canada's Museums of Science and Innovation) 196
11.4 Details from a nineteenth-century mercantile rule attributed to Benghazi, Libya (Ingenium: Canada's Museums of Science and Innovation) 199
11.5 Plumb bobs (Ingenium: Canada's Museums of Science and Innovation) 200
11.6 Fifteenth-century level attributed to Istanbul, Turkey (Ingenium: Canada's Museums of Science and Innovation) 201
13.1 Typical professional dissolving view lanterns 230
13.2 Example of a fine hand-painted dissolving view set, *Spectre of the Brocken*, produced by the Royal Polytechnic Institution, London, 1858 234
13.3 Hand-painted slide *The sweet waters of Asia*, showing a scene in Istanbul, Turkey. English, 1850s 236
13.4 Hand-painted slide with a winter scene. English, 1850s 238
14.1 Jean Verdoux's bill for the Yıldız Palace, December 1907 (Istanbul, *Başbakanlık Osmanlı Arşivi* [Ottoman Archives]) 250

Tables

1.1 Operations that could be performed with the instruments described in *Mir'at-ı Kâinat* 12
2.1 Comparison of Christian and Islamic length units 20
3.1 List of the manuscripts consulted 41
3.2 Distribution of the manuscripts consulted in libraries in Turkey 42

3.3 A comparison between the title of *Usage VII* in Bion's *Traité* and its English, Ottoman Turkish and Arabic translations 51
11.1 Number of instruments in the Petrovic collection by date of production 189
12.1 Instruments, tools and other objects purchased for scientific purposes by General Caffarelli 209
14.1 Retailers of scientific instruments in late nineteenth-century Istanbul 245
14.2 Some of Verdoux's contracts and bills 253

Contributors

Mahdi Abdeljaouad
is a retired professor at the University of Tunis. His research interests include the history of mathematics and mathematics education. He is particularly active in the history of mathematics teaching in the Islamic Mediterranean countries.

Silke Ackermann
is Director of the Museum of the History of Science in Oxford. Her research focusses on medieval and early modern astronomy, astrology and calendars in Europe and the Islamic World. She is co-editor of volumes 4 and 6 of this series and served as President of the Scientific Instrument Commission from 2013 to 2017.

Pierre Ageron
is a senior lecturer in mathematics at the University of Caen (France) and an historian of science. He is a member of the Nicolas Oresme Mathematics Laboratory (LMNO) and of the Research Institute for Mathematics Education (IREM) at Caen. His research interests include on the one hand the work of French Arabists from the sixteenth to the nineteenth century, and on the other hand the penetration and the translation of modern European science in Islamic countries during the same period.

Hamid Bohloul
studied physics at Isfahan University of Technology before switching to the history of the exact sciences in the Islamic era. For his MA from the University of Tehran he worked on al-Kāshānī's treatise on planetary distances and sizes, and for his PhD from the Institute for Humanities and Cultural Studies in Tehran he studied al-Kāshānī's *Nuzhat al-Hada'iq*, editing the Arabic texts, translating them into Persian and writing commentaries on them.

Patrice Bret
is a senior researcher on the History of Science and Technology. He has edited eighteenth-century sources and has worked on the scientific instrument maker François Bienvenu (1758–1831). Many of his publications deal with French-Egyptian cultural exchanges. He is preparing an administrative and cultural dictionary of Egypt under the French occupation (*Le Nil, la République et le Croissant*), alongside a biography of Nicolas-Jacques Conté.

Neil Brown
was a curator at the Science Museum in London for almost thirty years, working mainly on physical instruments of the nineteenth and early twentieth centuries. He maintains his interest in scientific instruments in retirement, with a particular focus on the history and development of diffraction gratings and the ruling engines used to make them.

Gaye Danışan
studied astronomy at the Faculty of Science of Istanbul University where she completed her Master's thesis in 2009 on the history of observatories in Turkey. She is currently assistant professor in the Department of the History of Science, Istanbul University, where she received her PhD in 2016 with a thesis on sixteenth-century Ottoman nautical astronomy and astronomical instruments. She pursued a post-doctoral research project on Ottoman portable astronomical instruments at the Museum of the History of Science, Oxford, from 2017 to 2018.

Feza Günergun
originally trained as a chemical engineer and holds a PhD in medical history from Istanbul University. She is Head of the Department of the History of Science at Istanbul University. Her research interests are the introduction and establishment of modern science in Turkey (eighteenth-twentieth centuries), Turkish translations of European science books, the dissemination of modern science in Turkey, and the role of journals in the diffusion and popularization of science.

Meltem Kocaman
is an assistant professor currently working at Istanbul University, Department of the History of Science. Her main field of research is the history of science and engineering in the Ottoman Empire in the nineteenth and early twentieth centuries. The topics of some of her research are the history of physics, physics education, the reception of the theory of relativity, and the first electrical engineers in the Ottoman Empire and the Republic of Turkey.

Richard L. Kremer
teaches history of science at Dartmouth College, USA, and curates this institution's collection of historic scientific instruments. He studies the instruments, tables and codices of medieval and early modern Latin astronomy. His recent publications examine the paper instruments of Johannes Stabius, astrologer for Emperor Maximilian I, and the place of 'observation' in Hellenistic astronomy.

Janet Laidla
has a PhD in history. Her research has concentrated on early modern historiography in Estonia and Livonia. Her interests include the history of culture and science in Estonia during early modern and modern periods. She is currently working at the University of Tartu Museum and at the Institute of History and Archaeology of the University of Tartu.

Panagiotis G. Lazos
is director of the Laboratory Center of Natural Sciences of Ilioupolis in Athens. He is currently a PhD student in the Faculty of Primary Education of the National University of Athens. His subjects of research are the history of scientific instruments, the history of science education and of science experiments in education, the use of Arduino (a project designed to allow students to build and program their own electronic hardware and software) in teaching science, and the history of sciences in south-eastern Europe from the eighteenth to the twentieth centuries.

David Pantalony
is curator of Physical Sciences and Medicine at Ingenium: Canada's Museums of Science and Innovation. In 2017 he was the lead curator for the Science and Medicine Gallery at the new Canada Science and Technology Museum. He teaches a collection-based seminar in history and digital humanities at the University of Ottawa.

Atilla Polat
studied mathematics at the Middle East Technical University in Ankara. He completed his Master's thesis on the nineteenth-century Turkish mathematician Vidinli Hüseyin Tefik Pasha in 2014 in Istanbul University. He is currently a research assistant at the Department of the History of Science, Istanbul University, where he is working on the mathematical treatise compiled by the sixteenth-century Ottoman mathematician Matrakçi Nasuh for his PhD thesis. His research interests focus mainly on the history of mathematics in Ottoman Turkey.

Bernd Scholze
is a collector of magic lanterns and over the past decade he has developed a passionate interest in their history, searching for previously unknown primary sources, both written and instrumental. He is a member of both the British and the American Magic Lantern Societies. He leads the German operation of a pottery company based in Stoke-on-Trent.

Constantine Skordoulis
is a Professor of Epistemology and Didactical Methodology of Physics in the Department of Primary Education, at the National and Kapodistrian University of Athens. He is the academic coordinator of the postgraduate Secondary Science Teachers Education Program at the Hellenic Open University. He has studied Natural Sciences at the University of Kent at Canterbury, UK, and has worked as a visiting researcher at the Universities of Oxford, Jena and Groningen. His research interests include the history of science and science education from a critical perspective.

Seyyed Hadi Tabatabaei
holds an MSc in the history of science. His special interest, which was the topic of his MSc thesis, is the introduction of modern science, especially medicine and astronomy, into Iran. He is fluent in Arabic, Persian, Turkish and English and has an interest in making sundials. He is working as Director of the Documentation Centre for the History of Medicine at Urmia University of Medical Science, Urmia, Iran.

Anthony Turner
works on the social history of ideas during the *Ancien Régime* and on the history of scientific instruments, clocks, watches, sun-dials and precision technology in general. His catalogue of mathematical instruments in the Bibliothèque nationale de France was published in 2018. Current projects include collaboration in an on-line Bio-Bibliographical Dictionary of French Scientific Instrument-Makers, and in a history of French enamelled watches in the seventeenth century. His latest book (as editor) is *Les Sciences à l'âge des Lumières*, which includes his study of the duc de Chaulnes microscope and its derivatives.

Hasan Umut
holds a BSc in Industrial Engineering from Boğaziçi University in Istanbul and an MA in History from Istanbul Bilgi University. He is currently a PhD candidate at McGill University's Institute of Islamic Studies in Montreal, Canada, where his dissertation project is entitled "Theoretical Astronomy in the Early Modern Ottoman Empire: ʿAlī al-Qūshjī's *al-Fatḥiyya fī ʿilm al-hayʾa*". His fields of research include the history of astronomy and the history of science in Islamic societies.

George N. Vlahakis
is an Assistant Professor at the Hellenic Open University in Patras, Greece, and a Research Fellow at the Institute of Historical Research of the National Hellenic

Research Foundation. He is a corresponding member of the International Academy of the History of Sciences and President of the Commission on Science and Literature in the Division of History of Science and Technology of the International Union of History and Philosophy of Science and Technology. Among the subjects of his research are the history of sciences in south-eastern Europe from the eighteenth to the twentieth centuries, the history of scientific instruments, science and religion, science and society, and science and literature.

CHAPTER 1

A Sixteenth-Century Ottoman Compendium of Astronomical Instruments

Seydi Ali's Mir'at-ı Kâinat

Gaye Danışan

1 Introduction

Seydi Ali bin Hüseyin (d. 1562) also known as Katib-i Rumi (the Anatolian author), Galatalı (from Galata), Seydi Ali Reis (the Admiral),[1] Seydi Ali Kapudan (the Captain) or Seydi Ali Çelebi was an Ottoman mariner who wrote extensively on astronomy, geography and navigation.

Seydi Ali came from a family of mariners: his grandfather and father were superintendents (*kethüda*) at the Galata dockyards following the conquest of Istanbul in 1453 by Sultan Mehmet II (r. 1451–1481). He served in the naval arsenal from an early age and became *Azaplar Katibi* (Secretary of the Naval Troops) and then *Tersane Kethüdası* (Superintendent of the Naval Arsenal). He learned navigation in the Mediterranean while sailing with Hayreddin Pasha (also known as 'Barbarossa', d. 1546) who was the chief Admiral of the Ottoman fleet in the early sixteenth century. After being nominated Admiral of the Ottoman Indian Ocean fleet based at Suez, Seydi Ali was asked in 1553 to bring back to Egypt the fifteen galleys that Admiral Piri Reis (d. 1553) had left in Bassorah, but an encounter with the Portuguese and the adverse weather conditions prevented him doing this. He could only take the Ottoman fleet to the port of Surat in the Indian state of Gujarat, arriving in 1554, and he returned to Istanbul overland.[2]

---

1 The Ottomans used the term *Reis* both for a captain and an admiral.

2 Seydi Ali Reis, *Mir'at-ül Memalik* (trans. by Necdet Akyıldız, Tercüman 1001 Temel Eser, Istanbul, 1975, p. 35; Seyyidî 'Alî Re'îs, *Le Miroir des Pays, une anabase ottomane à travers l'Inde et l'Asie centrale* (tr. J.L. Bacqué-Grammont), Sindbad, Actes Sud, Arles, 1999, pp. 9–35; Mahmut Ak, "Seydî Ali Reis", in E. Çakıroğlu (ed.), *Yaşamları ve Yapıtlarıyla Osmanlılar Ansiklopedisi*, Yapı ve Kredi Yayınları, Istanbul, 1999, 2 vols.: v. 2, pp. 525–527; Svat Soucek, "Sīdī 'Ali Re'īs (Seydi Ali Reis)", in C.E. Bosworth, E. Van Donzel, W.P. Heinrichs *et al.* (eds.), *The Encyclopedia of Islam*, Brill, Leiden, 1954–2005, 12 vols. plus indexes: v. 9 (1997), pp. 535–536.

 | DOI:10.1163/9789004412842_002

Few sixteenth-century Ottoman navigators wrote books on general or nautical astronomy. Seydi Ali is notable for three treatises related to astronomy: *Hülasatü'l-Hey'e* (Epitome of Astronomy), *Kitabü'l-Muhit* (Book of the Ocean) and *Mir'at-ı Kâinat* (The Mirror of the Universe).[3] Although *Kitabü'l-Muhit* is well studied, *Mir'at-ı Kâinat* has attracted relatively little attention although it deserves in-depth study.[4] Why and for whom did Seydi Ali compile *Mir'at-ı Kâinat*? Why did he prefer to introduce instruments such as the astrolabe, sine quadrant, astrolabe quadrant, celestial globe, and equatorial circle? What does *Mir'at-ı Kâinat* reveal about the construction of instruments? Is there any clue to their origin? Is there any connection between the *Mir'at-ı Kâinat* and Seydi Ali's other treatises? How were these instruments related to sixteenth-century Ottoman nautical astronomy? The present study which is based on the Ayasofya 2674 manuscript of *Mir'at-ı Kâinat*,[5] aims first to discuss these questions by analysing it and the various other copies of *Mir'at-ı Kâinat,* and then to compare them with other treatises written by Seydi Ali.

2 Seydi Ali's Entire Corpus

Seydi Ali would now and then mention his interest in the science of the configuration of the heavens (*ilm-i hey'e*), mathematical sciences (*ulum-i*

3 Tuba Uymaz, "Seydî Ali Reis'in Hülâsa el-Hey'e (Astronominin Özeti) Adlı Eseri Üzerine Bir İnceleme", unpublished master's thesis, Ankara Üniversitesi Sosyal Bilimler Enstitüsü Felsefe (Bilim Tarihi) Anabilim Dalı, Ankara, 2009; Himmet Büke, "Seydi Ali Reis—Kitabü'l-Muhit", unpublished master's thesis, Pamukkale Üniversitesi Sosyal Bilimler Enstitüsü Türk Edebiyatı Bölümü Türk Dili ve Edebiyatı Anabilim Dalı, Denizli, 2010. See also papers focusing on some chapters of *el-Muhit*: Luigi Bonelli, "Del Muhît o 'Descrizione dei mari delle Indie' dell'ammiraglio turco Sîdî 'Alî detto Kiâtib-i Rûm", *Rendiconti della Reale Accademia dei Lincei, Classe di Scienze Morali, Storiche e Filologiche,* series 5, v. 3 (1894), pp. 751–777; Maximilian Bittner, Wilhelm Tomaschek, *Die topographischen Kapitel des indischen Seespiegels Mohît,* Verlag der geographischen Gesellschaft, Vienna, 1897; Sidi Çelebi, Joseph von Hammer and Baron Purgstall (eds. and trs.), "Extracts from the Mohit, that is the Ocean, a Turkish work on navigation in the Indian seas", *The Journal of the Asiatic Society of Bengal* 3 (1834), pp. 545–553; 5 (1836), pp. 441–468; 7 (1838), pp. 767–780; 8 (1839), pp. 805–812 and 823–830.

4 William Brice, Colin Imber and Richard Lorch, *The Da'ire-yi Mu'addel of Seydi 'Ali Re'is (Seminar on Early Islamic Science, Monograph No. 1)*, University of Manchester, Manchester, 1976.

5 Seyyid Ali bin Hüseyin, "*Mir'at-ı Kâinat*", Istanbul, Süleymaniye Library, Ayasofya MS 2674, 123 fols.

riyaziye) and the science of the stars (*ilm-i nücum*) in his treatises.[6] Moreover, he associated the art of navigation with these sciences, and considered that his knowledge in astronomy was important for his assignment as Admiral of the Ottoman fleet in the Indian Ocean in 1553.[7]

Seydi Ali probably had time to study astronomy in Aleppo where he had spent the winter after the Eastern Campaign of Suleiman I (r. 1520–1566). During this time Mevlana Hamdullah Ibn Şeyh Cemaleddin, who was knowledgeable about Euclid's *Elements*, Ptolemy's *Almagest* and Aristotle's works, encouraged him to translate Ali Qushji's (d. 1473) *al-Fathiyya* from Arabic into Turkish. Seydi Ali translated the text and added sections on geography. He named it *Hülasatü'l-Hey'e* (Epitome of astronomy),[8] and then presented it to Suleiman I, at an unknown date but probably before his nomination in 1553 as the Admiral to the Ottoman fleet in the Indian Ocean. A year later, in 1554, he completed the *Kitabü'l-Muhit fi İlm el-Eflak ve'l-Ebhur* (Book of the Ocean on the Science of the Spheres [Heavens] and the Seas). Known as *Kitabü'l-Muhit* or *el-Muhit* for short, the book aimed to give practical information about seafaring in the Indian Ocean and was written for Ottoman seamen who had little experience on the open sea. There are some stylistic differences between the texts of *Hülasatü'l-Hey'e* and *el-Muhit,* as well as differences between the purposes of their composition. Seydi Ali's experience on the open sea may be the reason for these differences. In *el-Muhit*, written after 1553, he deals with practical methods rather than the theoretical issues of astronomy that he treated in *Hülasatü'l-Hey'e*. Equally important, in *Hülasatü'l-Hey'e*, he mentioned instruments such as *Hind dairesi* (lit. Indian circle) and astrolabes, but not within the context of seafaring.[9] Conversely, in *el-Muhit*, he recommends seamen to use the astrolabe, the sine quadrant and the equatorial circle.[10] *Mir'atü'l-Memalik* (The Mirror of Countries, published in 1557), the text in which he relates his adventures on his way from India to Istanbul, provides clues about his knowledge of instruments. When he and his crew got lost in the Indian Ocean, he applied the seafaring techniques that he had used for sailing

6 In the Ottoman Empire *ilm-i nücum* was the science of stars and included both astronomy and astrology. See Ahmet Tunç Şen, "Reading the Stars at the Ottoman Court: Bayezid II (r. 886–918 H / 1481–1517 CE) and his celestial interests", *Arabica* 64 (2007), pp. 557–608.

7 Seydi Ali Reis, *Mir'at-ül Memalik*, cit. (n. 2), pp. 33–34.

8 Seyyid Ali bin Hüseyin, "*Hülasatü'l-Heye*", Istanbul, Süleymaniye Library, Ayasofya MS 2591, fols. 1v–94v; copied in 955 H (1548–1549 CE).

9 *Ibid.*, fol. 78v.

10 Seyyid Ali bin Hüseyin, "*Kitabü'l-Muhit*", Istanbul, Süleymaniye Library, Nuruosmaniye MS 2948, fols., 14v, 17r, 26r, 45r.

in the Mediterranean. However, he soon realised that Mediterranean nautical techniques were not appropriate when navigating in the Indian Ocean.[11] This might be the reason why he started to write *el-Muhit*, a guide for the Indian Ocean. Also, in *Mir'atü'l-Memalik*, he mentioned the use of the astrolabe and the equatorial circle on land but not at sea.[12] He also said that he had compiled treatises about astronomy, physics (*fenn-i hikmet*), navigation, and the position of stars, but did not give their titles.[13]

Mir'at-ı Kâinat, which includes information about the construction and use of five portable astronomical instruments, the astrolabe (*usturlab*), the sine quadrant (*rub-ı müceyyeb*), the astrolabe quadrant (*rub-ı mukantara*), the celestial globe (*zatü'l-kürsi*) and the equatorial circle (*daire-i muaddil*), is the most intriguing of Seydi Ali's treatises. The original copy presented to Suleiman I is not extant and the date of its compilation is uncertain but it was probably written after 1554, from the dates of the other works in Seydi Ali's corpus. There are numerous copies of *Mir'at-ı Kâinat*. The oldest surviving manuscript dates back to 1574, eleven years after Seydi Ali's death.[14]

Mir'at-ı Kâinat is not the only treatise produced by Seydi Ali on portable astronomical instruments. The following are also attributed to him: *Risale-i Da'ire-i Muaddil* (Treatise on the equatorial circle), *Risale-i Usturlab* (Treatise on the astrolabe), *Risale-i Rub-ı Müceyyeb* (Treatise on the sine quadrant), *Risale-i Zatü'l-Kürsi* (Treatise on the celestial globe). This suggests the question: did Seydi Ali write independent treatises besides *Mir'at-ı Kâinat* or were they only copies made from *Mir'at-ı Kâinat*'s chapters on instruments? Preliminary analysis of the treatises *Risale-i Usturlab, Risale-i Müceyyeb, Risale-i Mukantarat, Risale-i Daire-i Muaddil*, and the chapter *Zatü'l-Kürsi* of *Mir'at-ı Kâinat* provides the following information.

There are three compendiums (*mecmua*) which include treatises on portable astronomical instruments attributed to Seydi Ali. The first includes three treatises on the astrolabe, equatorial circle and celestial globe. These were put together in no particular order in 989 H (1581–1582 CE) by the copyist Mustafa b. Muhammed.[15] A comparison shows that these texts were copied from *Mir'at-ı*

11 Seydi Ali Reis, *Mir'at-ül Memalik*, cit. (n. 2), pp. 42–45.

12 *Ibid.*, p. 69.

13 *Ibid.*, p. 33.

14 Seyyid Ali bin Hüseyin, "*Mir'at-ı Kâinat*", Istanbul, Süleymaniye Library, Nuruosmaniye MS 2950/1, fols. 1v–92r; copied by Ibn al-Amin Ahmed b. Yahya in 981 H (1573–1574 CE). For other copies of *Mir'at-ı Kâinat* see E. Ihsanoğlu, R. Şeşen, C. İzgi, C. Akpınar and İ. Fazlıoğlu (eds.), *Osmanlı Astronomi Literatürü Tarihi*, IRCICA, Istanbul, 1997, p. 143.

15 Seyyid Ali Çelebi, "*Risale-i Daire-i Muaddil*", Istanbul, Süleymaniye Library, Aşir Efendi MS 470/5, fols. 52v–55v; "*Risale-i Usturlab*", Aşir Efendi MS 470/10, fols. 94r–116v; "*Mir'at-ı*

Kâinat. The second compendium containing treatises on the celestial globe and the sine quadrant, and attributed to Seydi Ali, is different. It starts with a treatise named *Risale-i Zatü'l-Kürsi*.[16] The text of this is almost identical to the chapter of *Mir'at-ı Kâinat* dealing with *zatü'l-kürsi*: the copyist has copied the chapter with its preface. On the contrary, the text of the treatise *Risale-i Rub-ı Müceyyeb*,[17] which is in the same compendium, is not the same as the chapter entitled "*Rub-ı Müceyyeb*" of the *Mir'at-ı Kâinat*.[18] The treatise has 25 sections while the chapter in *Mir'at-ı Kâinat* has 20 sections. These extra sections explain how to perform trigonometric multiplication, division, and square root calculations with the sine quadrant. *Risale-i Rub-ı Müceyyeb* also has an introduction which explains the author's purpose, but it is different from the introduction of *Mir'at-ı Kâinat*.[19] The third compendium comprises two treatises, *Risale-i Ceyb*[20] and *Risale-i Mukantarat*,[21] both attributed to Seydi Ali. These were copied probably in 1278 H (1861–1862 CE). The surviving copies begin with an introduction which is different from that of *Mir'at-ı Kâinat* but a comparison suggests that the information in the treatise was taken from the chapters of *Mir'at-ı Kâinat*.[22]

3 Description and Analysis of the *Mir'at-ı Kâinat*

Mir'at-ı Kâinat is important not only because it introduces five portable astronomical instruments, but also because some sources assert that it includes information about nautical astronomy.[23] The introduction states that Seydi Ali had studied the science of the stars (*ilm-i nücum*), and he believed that the

Kâinat" (it includes only the fourth chapter entitled "*Zatü'l-Kürsi*"), Aşir Efendi MS 470/11, fols. 117v–128v.

16 Seyyid Ali bin Hüseyin, "*Mir'at-ı Kâinat*", Istanbul, Kandilli Observatory, MS 50/1, fols. 1v–13r (it only includes the fourth chapter entitled "*Zatü'l-Kürsi*").

17 Seyyid Ali bin Hüseyin, "*Risale-i Rub-ı Müceyyeb*", Istanbul, Kandilli Observatory, MS 50/9 fols. 63v–79r; copied probably in twelfth century H (about eighteenth century CE).

18 Seydi Ali bin Hüseyin, "*Mir'at-ı Kâinat*", cit. (n. 5), fols. 40v–64v.

19 Seyyid Ali bin Hüseyin, "*Risale-i Rub-ı Müceyyeb*", cit. (n. 17), fols. 63v–64r. See also Gaye Danışan Polat, "A treatise by the 16th century Ottoman Admiral Seydi Ali Reis on Rub-i müceyyeb (sine quadrant)", in Dejanirah Couto, Feza Günergun and Maria Pia Pedani (eds.), *Seapower, Technology and Trade—Studies in Turkish Maritime History*, Piri Reis University Publications & Denizler Kitabevi, Istanbul, 2014, pp. 337–341.

20 Seydi Ali Reis, "*Risale-i Ceyb*", Istanbul, Ali Emiri Library, MS 4622/5, fols. 47–53.

21 Seydi Ali Reis, "*Risale-i Mukantarat*", Istanbul, Ali Emiri Library, MS 4622/4, fols. 37–47.

22 Seydi Ali bin Hüseyin, "*Mir'at-ı Kâinat*", cit. (n. 5), fols. 40v–97r.

23 Ihsanoğlu *et al.*, op. cit. (n. 14), p. 142.

five instruments were relevant to it. He realised that there was no work that introduced all these portable astronomical instruments in one place. Moreover, most of the descriptions were written in Persian or Arabic. According to him, this caused problems for those who wanted to study the science of the stars. So he decided to write a treatise in Turkish that contained brief and useful information about these instruments, and he began by examining the various Arabic, Persian, and Turkish sources.[24]

It is worth noting that Seydi Ali did not describe the target audience of the treatise, although he said that he wrote it to explain the science of the stars (*ilm-i nücum*). The oldest surviving copy of *Mir'at-ı Kâinat* (dated 1574) bears the title *Mir'at-ı Kâinat fi Ilmi'l-Mikat* (Mirror of the Universe on the Science of Timekeeping) which points to its use by timekeepers. It is possible that the copyist extended the title by adding '*fi ilmi'l-mikat*'. The *munejjims*, who were interested in both astrology and astronomy, might have been the target readers, but could this treatise be written for them only? There are two counter-examples. One is the *Mediterranean Atlas* by the famous cartographer from the Maghreb Ali ibn Ahmad ibn Muhammad al-Sharfi (d. after 1571),[25] and the second is the *Kitabü'l-Muhit* the afore-mentioned work of Seydi Ali. It is clear from these two works that seamen also needed to know about the science of the stars. The position of the stars is essential for determining the seasons and the directions of the wind which are important data for navigation. Also, in *el-Muhit* Seydi Ali says that the position of the stars is important for determining the most propitious time to declare war or attack the enemy.[26]

4 The Five Portable Instruments Described in *Mir'at-ı Kâinat*

In the sixteenth century, Ottoman *munejjim*s (astronomers/astrologers) and *muwaqqit*s (timekeepers) used the astrolabe, sine quadrant and astrolabe quadrant extensively. The popularity of these instruments led to the compilation of many treatises describing them and their use. Seydi Ali's *Mir'at-ı Kâinat* fits this Ottoman literary tradition: it describes the parts of the instrument and

24 Seydi Ali bin Hüseyin, "*Mir'at-ı Kâinat*", cit. (n. 5), fols. 2v–3r.

25 Al-Sharfi, "Mediterrenian Atlas 1551", Paris, Bibliothèque Nationale de France, MS Arabe 2273, fol. 1v, and Al-Sharfi, "Mediterrenian Atlas, 1571", Oxford, Bodleian Library, MS Marsh 294, fol. 9r.

26 Seyyid Ali bin Hüseyin, "*Kitabü'l-Muhit*", cit. (n. 10), fols. 67r–70v. For detail see also Gaye Danışan Polat, *16. Yüzyılda Osmanlılarda Deniz Astronomisi ve Astronomi Aletleri* (Nautical Astronomy and Astronomical Instruments of the Ottomans in the 16th Century), PhD dissertation, Istanbul University, Istanbul, 2016, pp. 234–239.

then explains how to use it. These treatises are seldom illustrated, but a number of portable astronomical instruments were depicted in the well-known miniature of the Istanbul observatory where Taqi al-Din (d. 1585) and his staff made observations in late sixteenth century.[27]

The manuscript text of *Mir'at-ı Kâinat* is composed of an introduction (*mukaddime*), five chapters (*makale*) and 120 sections (*bab*). The introduction describes Seydi Ali's purpose in compiling the treatise and its content. The five chapters each deal with the structure and use of one of the instruments. The first chapter is about the planispheric astrolabe which is described as an instrument consisting of three parts: plate (*sahife* or *levha*), main body or mater (*ümm*) and the rete (*şebeke*). For each latitude one needs to use a specific plate to be able to calculate the altitude and the azimuth of a celestial object at that latitude. On the back (*zahr/arka*) of the main body or mater, there is an alidade (*idade*) for measuring the altitude of a celestial object, as well as a shadow square for performing trigonometric calculations. This astrolabe is similar to the one described by Mustafa bin Ali el-Muvaqqit (d. 1571) in his treatise on the astrolabe. According to el-Muvaqqit, it was the best-known astrolabe among experts.[28] This statement implies that Seydi Ali chose to describe the most common type of astrolabe in use in the Ottoman Empire.

In the second chapter, Seydi Ali describes the sine quadrant and explains how to use the observations made with it.[29] The third chapter is about the astrolabe quadrant, which bears arcs and lines for performing astronomical calculations. Tiny headless nails are used for marking important points.[30] Generally, Ottoman quadrants have an astrolabe quadrant on one side and a sine quadrant on the other, but Seydi Ali did not mention this arrangement.

The fourth and fifth chapters deal respectively with the celestial globe and the equatorial circle, instruments that appear less frequently in Ottoman astronomical literature than the other three instruments. These chapters include interesting details, the first being confusion about the meaning of the term *zatü'l-kürsi*. In *Cihannüma* by Katip Çelebi (1609–1657), *zatü'l-kürsi* is illustrated as an armillary sphere with the Earth inside,[31] but the instrument

27 "*Şehinşahname*", Istanbul, Istanbul University Rare Books Library, MS F1404, fol. 57r.

28 Mustafa bin Ali el-Muvaqqit, "*Risale-i Usturlab*", Istanbul, Süleymaniye Library, Nuruosmaniye MS 4971/2, fol. 13v.

29 Seyyid Ali bin Hüseyin, "*Mir'at-ı Kâinat*", cit. (n. 5), fols. 40v–64v.

30 *Ibid.*, fols. 64v–97r.

31 Kâtip Çelebi and İbrahim Müteferrika, *Cihannüma (360 Yıllık Bir Öykü)/The Book of Cihannuma (A 360 Year Old Story)* (ed. by Bülent Özükan, Mustafa Kaçar and Orhan Koloğlu), Boyut Yayın Grubu, İstanbul, 2008, p. 44. Reprinted from the original 1732 edition.

described by Seydi Ali is a celestial globe.[32] The term *zatü'l-kürsi* alone means a stand. If Seydi Ali had used the term *el-küre zatü'l-kürsi*, he would mean 'a globe with a stand'. Seydi Ali mentions different traditions in the construction of Islamic celestial globes. He describes two types of meridian ring (*daire-i nısf-ün nehar*), which passes through the north and south poles, and observer's zenith. One of them is divided in 360° bearing *abjad* numerals, and the globe can rotate within ring, pivoting on the celestial poles. The second type of meridian ring is graduated in 180° and is attached to the horizon ring. In this type, the globe rotates within the stand (*kürsi*) and can be set for the observer's latitude.[33] He mentions that the globes have 48 constellations but some of them are engraved with figures while others just have the names written on the globe.[34]

In a seventeenth-century copy of *Mir'at-ı Kâinat*, there is a figure of *zatü'l-kürsi* which has been added by the copyist (*Fig. 1.1*). The globe is mounted with the north pole up for use in northern latitude. The meridian ring is graduated in 360° bearing *abjad* numerals. The quadrant of altitude (*rub-ı irtifa*) is divided into 90° and is perpendicular to the horizon. The horizon is labelled in two segments of 90°. It is numbered from east towards the north and south from 90° to 0° in *abjad* letters. The globe has only the great and lesser circles: ecliptic, equator, and polar circles. Zodiacal names are inscribed along the ecliptic in two lines in Arabic but with some misspellings. The circles are graduated in 2° intervals with every thirtieth labelled. The two poles of the ecliptic are labelled the "North pole of the zodiac" (*kutb-i burc-ı şimali*) and the "South pole of the zodiac" (*kutb-i burc-ı cenubi*). The instrument depicted on this seventeenth-century copy has two additional pieces which are described neither in this copy nor in other copies. One of these pieces is probably a plumb line, and the other is an hour circle attached to the North pole.[35] The plumb line represents the prime vertical, the great circle passing through the zenith and the east and west points of the horizon. The plumb line also intersects the celestial meridian at about 49°. This angle is equal to the angle between the zenith and north celestial pole or the complement of observer's latitude. Thus, this figure represents the celestial globe for the latitude of 41° N. The other piece is an hour circle with a pointer. It is fixed on the meridian circle, at the North pole, and it is graduated into 24 equal hours in

32 Seyyid Ali bin Hüseyin, "*Mir'at-ı Kâinat*", cit. (n. 5), fols. 97r–99r.

33 *Ibid.*, fol. 97r.

34 *Ibid.*, fol. 97v.

35 Seyyid Ali bin Hüseyin, "*Mir'at-ı Kâinat*", Istanbul University Rare Books Library, MS T1804, fol. 101r.

FIGURE 1.1 *Zatü'l kürsi* (Celestial globe) from Seyyid Ali bin Hüseyin, *Mir'at-ı Kâinat*
ISTANBUL UNIVERSITY RARE BOOKS LIBRARY, MS T1804, FOL. 101R

two parts (midnight to noon and noon to midnight) with every second hour labelled in *abjad* numerals in a counter-clockwise direction. It represents the hours of the day. The pointer is moveable to any required position on the hour scale. Thus it serves to point out the time. This system was preferred especially for astronomical calculations such as the determination of the hour angle. For example at 7 o'clock in the morning, apparent time, the hour angle of the Sun is 5h (= 12h–7h). However, the hour circle on this celestial globe probably represents European tradition rather than Islamic tradition, and indicates possible European influence.[36]

Seydi Ali describes the equatorial circle as a semi-circle or a circle made from copper or silver. On its periphery are inscribed the names of famous towns and the *qibla* from each of them. In the middle of the instrument there is a case for a compass needle. A black line within this case shows the north. However, Seydi Ali said that the black line should be offset 7° east of north and that he had verified this with a *zij* (a book of astronomical tables).[37] This allows for magnetic declination. Some copies of the *Mir'at-ı Kâinat,* include notes about the declination which were added later by copyists. Ibn al-Amin Ahmed b. Yahya, who copied *Mir'at-ı Kâinat* in 981 H (1571–1572 CE), noted a declination of about 7°, and pointed out that this value was used in Seydi Ali's time. Ahmed b. Yahya also said that Muhammad Sami, who was contemporary with him and probably a *muwaqqit* or *munejjim,* did not refer to this declination.[38] The same note appears in a copy of the treatise *Risale-i Daire-i Muaddil,* which is possibly borrowed from a chapter of *Mir'at-ı Kâinat.*[39] In the same treatise, and in the same folio but in different handwriting, there is another note giving a different value of the declination. According to this note, the value of the declination

36 For detailed information on Islamic celestial globes see Emilie Savage-Smith, *Islamicate Celestial Globes: Their History, Construction, and Use*, Smithsonian Institution Press, Washington D.C., 1985; Emilie Savage-Smith, "An Arabic Celestial Globe in Tartu", *Tartu Ülikooli ajaloo küsimsi* 41 (2013), pp. 221–230; Emilie Savage-Smith, "Gerard Turner Memorial Lecture: Of Making Celestial Globes There Seems No End," *Bulletin of the Scientific Instrument Society* 132 (2017), pp. 2–10; Moya Catherine Carey, *Painting the Stars in a Century of Change: A Thirteenth-century Copy of al- Sufi's Treatise on the Fixed Stars British Library Or. 5323*, PhD dissertation, School of Oriental and African studies, London, 2001. See also for European celestial globes Edward Luther Stevenson, *Terrestrial and Celestial Globes: Their History and Construction Including a Consideration of Their Value as Aids in the Study of Geography and Astronomy*, Yale University Press, New Haven, 1921, 2 vols.

37 Seydi Ali bin Hüseyin, "*Mir'at-ı Kâinat*", cit. (n. 5), fol. 43v.

38 Seydi Ali bin Hüseyin, "*Mir'at-ı Kâinat*", cit. (n. 14), fol. 86v.

39 Seydi Ali Çelebi, "*Risale-i Daire-i Muaddil*", cit. (n. 15), fols. 52v–55v.

was 12° in 1164 H (1750–1751 CE) in Istanbul.[40] This note implies that this copy was examined in the eighteenth century by someone who found a difference in the value of magnetic declination and noted it down.

These notes added by copyists were probably addressed to authors who had written on the equatorial circle without being aware of the variation in magnetic declination because Seydi Ali did not directly deal with this in *Mir'at-ı Kâinat*. However, he did give detailed information about the 7° declination between true north and magnetic north in his *Kitabü'l-Muhit*, and advised navigators to measure declination with the equatorial circle. Moreover, he noted that this variation was known in Portugal and other regions of Europe (*Firengistan* or Western countries), and that the *qiblanuma* (*qibla* indicators) made in "German lands" (*Diyar-ı Alman*) and imported into Anatolia (*Diyar-ı Rum*) were calibrated to work correctly with a magnetic declination of 7° east. The declination of 7° would not pose a problem when navigating in the Black Sea and the Mediterranean, but could do so when sailing in the Indian Ocean. If navigators were unaware that magnetic declination changed from place to place, they would suppose that navigational errors were caused by currents. Seydi Ali did point out the different requirements of navigators, astronomers/astrologers and timekeepers. He noted that some authors who were engaged in the science of the stars (*ilm-i nücum*) did not need to use a compass, therefore they did not mention it in their treatises.[41] As he had recourse to various Arabic, Persian and Turkish sources he probably learned about the 7° declination from Islamic texts, but he was also aware that some compasses and *qibla* indicators produced in Europe took the 7° declination into account.

---

40 *Ibid.*, fol. 52v. The magnetic declination given by Evliya Çelebi in the seventeenth century, and Ibrahim Hakkı Erzurumlu in the eighteenth century was 11°. Ibrahim Müteferrika stated that the first measurement of declination by the Turks was carried out in Bebek (Istanbul) in 1727. The details are given in a long note (*Teyzil-al-Tabi*) he added to *Cihannuma*, the book he edited and published in 1732 (see n. 31). Later, he mentioned a declination of 11.5° in his *Füyuzat-ı Mıknatısiye* (Benefits of the Magnetism), published in 1732. This was a translation he made either from a book printed in Leipzig or from William Whiston's *The Longitude and Latitude found by the inclinatory or dipping needle* (London, 1721). The declination of 11.5° may be the value cited in the source book. See Niyazi Berkes, *The Development of Secularism in Turkey*, McGill University Press, Montreal, 1964, repr. Hurst and Company, London, 1998, p. 46, n. 37; Şahabeddin Demirel, "İbrahim Müteferrika'nın *Füyuzât-ı Mıknatısiye* (Mıknatısın Yararları) adlı kitabı", in *Dil ve Tarih Coğrafya Fakültesi Atatürk'ün 100. Doğum Yılına Armağan Dergisi*, Ankara, 1992, pp. 265–330; F. Özçep and T. Özçep, "Notes on the History of Geophysics in the Ottoman Empire", *History of Geo- and Space Sciences* 5 (2014), p. 166.

41 Seyyid Ali bin Hüseyin, "*Kitabü'l-Muhit*", cit. (n. 10), fols. 16v–17r.

5 The Problems Solved with the Astronomical Instruments Described in *Mir'at-ı Kâinat*

The five instruments described in *Mir'at-ı Kâinat* have some common features but also some different features which help us to understand why Seydi Ali compiled a book about them. The descriptions also allow us to discuss the place of these instruments in sixteenth-century Ottoman nautical astronomy. *Table 1.1* summarises the various operations that could be performed with the five instruments, based on the information given by Seydi Ali.

TABLE 1.1 Operations that could be performed with the instruments described in *Mir'at-ı Kâinat*

Operations	*Usturlab* (Astrolabe)	*Rub-ı müceyyeb* (Sine quadrant)	*Rub-ı mukantara* (Astrolabe quadrant)	*Zatü'l-kürsi* (Celestial globe)	*Daire-i muaddil* (Equatorial circle)
Finding the height of an accessible or inaccessible tower	X	X	X		
Finding the distance between two towers which are not on the same ground level; finding the distance to a tower whose height is known	X	X	X		
Finding the width of a stream	X	X	X		
Finding the depth of a pit	X	X	X		
Performing trigonometric multiplication and division, and square root calculations		X			
Determining the first day of the lunar months, and the times of lunar and solar eclipses				X	
Determining azimuth, altitude, time and direction	X	X	X	X	X

The analysis of *Mir'at-ı Kâinat* shows that three groups of problems could be solved with the five portable astronomical instruments described in the treatise: firstly, determining the altitude of the Sun and calculating solar parameters such as the Sun's declination and obliquity of the ecliptic; secondly, calculating the time such as noon (*zeval*) and the period of midday (*asr*); thirdly, finding the cardinal points of the compass and the direction of the *qibla*. These operations are solved by means of spherical triangles. Although the angular positions of celestial bodies can be measured with astronomical instruments such as the astrolabe quadrant and the astrolabe, the trigonometric equations necessary for the spherical astronomy cannot be solved with these instruments. The advantage of the sine quadrant is that it can solve trigonometric multiplication and division, and also square root operations, and it allowed the user to cope with these trigonometric equations. The observer could solve many astronomical problems with any of five instruments but the equatorial circle was the easiest to use because it required fewer calculations.[42] The celestial globe can also be used for the other purposes such as timekeeping, and also as a kind of calendar which allows the user to calculate the first day of each lunar month, and the dates of lunar and solar eclipses. Star positions can be determined with a celestial globe and also with an astrolabe, but the celestial globe has an advantage because it can be used at any latitude. There were types of astrolabe that could be used at any latitude, but the astrolabe which Seydi Ali described in *Mir'at-ı Kâinat* is not of this kind.

Some of these instruments were also useful for making terrestrial measurements, as seen from the table. Could these terrestrial measurements be useful for Ottoman navigators as well? There is an evidence for this in Piri Reis's *Kitab-ı Bahriye* (Book on Seafaring) in which he uses the heights of mountains for navigational purposes.[43] However, he does not describe the methods and instrument he used.

42 Iz al-din b. Muhammad al-Wafai also mentioned its convenience in his treatise entitled *Risaletun fi Dairet il-Muaddil.* See Sevim Tekeli, "Îzzüddin b. Muhammed al-Vefaî'nin 'Ekvatör Halkası' adlı makalesi ve Torquetum (The 'Equatorial Armillary' of Iz al-din b. Muhammad al-Wafai and the Torquetum)," *Ankara Üniversitesi Dil ve Tarih-Coğrafya Fakültesi Dergisi* 18 (1960), p. 230.

43 Piri Reis, *Kitab-ı Bahriye* (transliterated by Ahmet Demir), T.C. Başbakanlık Denizcilik Müsteşarlığı, Ankara, 2002, pp. 35–36; Danışan Polat, *16. Yüzyılda Osmanlılarda Deniz Astronomisi ve Astronomi Aletleri*, cit., (n. 26), pp. 265–270.

6 Conclusion

Seydi Ali believed that the astrolabe, the sine quadrant, the astrolabe quadrant, the celestial globe, and the equatorial circle were useful portable astronomical instruments for *ilm-i nücum,* the science of the stars. The fact that treatises accounting for these instruments were rare, scattered and written in Arabic and Persian motivated him to write a single book in Turkish. The chapters in *Mir'at-ı Kâinat* on portable astronomical instruments were written in the Ottoman astronomical literary tradition in which the parts of the instruments are described, and then their functions are explained. Comparison with Seydi Ali's other treatises suggests that *Mir'at-ı Kâinat* was probably written after 1554. This is the period when Seydi Ali arrived the Indian state of Gujarat, returned to Istanbul overland, and also compiled the *Muhit.* It is possible that he decided to include the celestial globe and the equatorial circle, which were less popular than the astrolabe, the astrolabic quadrant, and the sine quadrant among the Ottomans, in *Mir'at-ı Kâinat* after he came across the tradition of these portable instruments during his travels.

On the other hand, the mention of a magnetic declination of 7° in the *Mir'at-ı Kâinat* leads to the suggestion that Seydi Ali both borrowed from Islamic sources and knew about German compasses made with allowance for that declination. Besides, Seydi Ali describes a celestial globe made in the Islamic tradition in his treatise but the instrument depicted in the seventeenth-century copy shows European influence because of the hour circle. This brings up the question of the availability of European portable astronomical instruments on the Ottoman market. We very much hope, that future research will provide new evidence about the use of the celestial globe (*zatü'l-kürsi*) and the equatorial circle (*daire-i muaddil*) by the Ottomans, and the possible interaction between European and Ottoman portable astronomical instruments.

The analysis of *Mir'at-ı Kâinat* allow us to discuss whether the five portable astronomical instruments described by Seydi Ali were suitable for use by mariners in seafaring. Because of their properties only four of them, the astrolabe, the sine quadrant, the celestial globe, and the equatorial circle seem to have been useful for navigators. In *Muhit,* he suggested that sailors should use the astrolabe and sine quadrant for measuring the altitude of the Sun and they should use the equatorial circle for determining the variation of magnetic declination on distant seas. Thus, even if these portable astronomical instruments were used by Ottoman sailors, they would had been used primarily for these measurements. However, Seydi Ali writes about the variation of magnetic declination in the chapter about the equatorial circle in *Mir'at-ı Kâinat,* and he gives detailed information about the variation of magnetic declination for sailors, and points out that this declination was not mentioned by *munejjims* in *Muhit*. This suggests that, Seydi Ali might have considered sailor's needs while he was compiling *Mir'at-ι Kâinat*. Nevertheless, it is difficult to say whether *Mir'at-ι Kâinat* was popular among Ottoman sailors because surviving copies seem to have been reproduced for timekeeping purposes.

Acknowledgments

The study of the celestial globe described in this paper was carried out in the History of Science Museum, University of Oxford, as part of a project supported by the Scientific and Technological Research Council of Turkey postdoctoral research scholarship programme (TUBITAK-2219) entitled *Ottoman portable astronomical instruments and their field of application: a comparative study for the 16th century*.

CHAPTER 2

Eastern and Western Instruments in Osman Efendi's *Hadiyyat al-Muhtadī* (The Gift of the Convert), 1779

Mahdi Abdeljaouad and Pierre Ageron

1 The Gift of the Convert

The aim of the present paper is to discuss the origin of the instruments described in *Hadiyyat al-Muhtadī* (The Gift of the Convert), an Arabic-language mathematical treatise completed in Belgrade on 28 January 1779. This treatise was never printed, but the eleven extant manuscript copies we could locate, very similar in appearance to traditional Arabic manuscripts (*Fig. 2.1*), suggest a widespread circulation throughout the former Ottoman Empire.[1] Although known to historians of Ottoman modern sciences,[2] it had never been thoroughly studied before we decided to investigate its content and tackle its many riddles. In particular, its author, a certain Osman Efendi bin Abdülmennan, remained mysterious and its sources remained likewise unknown. We published a first comprehensive overview of *Hadiyyat al-Muhtadī* in 2016.[3] There, we

1 The eleven extant copies are: Istanbul, Military Museum, MS 3027, undated, ca. 1785; Medina, King Abdelaziz Library, MS 2879, dated 9 April 1785; Istanbul, Kandilli Observatory, MS 65, dated 16 August 1785; Doha, Qatar National Library, no shelfmark, dated 30 March 1791; Cairo, Egyptian National Library, MS DR 695, dated 4 November 1791; Cairo, Egyptian National Library, MS TR 137, dated 31 December 1796; Princeton, University Library, MS Garrett 162 L, formerly in Medina, undated; Kütahya, Vahid Paşa Public Library, MS 786, undated; Istanbul, Istanbul University Library, MS AY 703, undated; Riyadh, Prince Sultan Library undated, lacking its end; Baghdad, Awqaf Library, MS 6292, dated 1804, may be burnt or stolen. We had access to all but the last two.

2 See: Ekmeleddin İhsanoğlu, Ramazan Şeşen, Cevat İzgi, *Osmanlı Matematik Literatürü Tarihi*, IRCICA, Istanbul, 1999, 2 vols.: v. 1, pp. 243–246; Ramazan Şeşen, "The Translator of the Belgrade council Osman b. Abdülmennan and his place in the translation activities", in Ekmeleddin İhsanoğlu (ed.), *Transfer of Modern Science and Technology to the Muslim World*, IRCICA, Istanbul, 1992, pp. 371–383.

3 Mahdi Abdeljaouad, Pierre Ageron & Mahmoud Shahidy, "Émergence d'un savoir mathématique euro-islamique: *L'Offrande du converti pour ranimer la flamme éteinte*", *Philosophia Scientiae* 20, 2 (2016), pp. 7–32.

 | DOI:10.1163/9789004412842_003

بسم الله الرحمن الرحيم
الحمد لله الذي علم الانسان ما لم يعلم والصلوة والسلام على محمد الذي هو
مدينة العلم والحكم وعلى آله واصحابه الذين هم كالنجوم في الظلم وبعد

FIGURE 2.1 *Hadiyyat al-Muhtadī*, first page
ISTANBUL, MILITARY MUSEUM, MS 3027

revealed that Osman Efendi was in fact born in Germany in 1713—unfortunately, we still do not know his birth name and place of origin. He moved to Belgrade, probably during the Austrian occupation (1717–1739), converted to Islam and worked as an interpreter for the Ottoman governor of the city from 1747 onwards. He welcomed diplomats going through Belgrade, exchanged letters with governors of the Habsburg border provinces and asked merchants to bring him books from Vienna, Leipzig or Breslau. As well as authoring *Hadiyyat al-Muhtadī*, he translated three books from German to Turkish, respectively on geography, botany and chemistry.[4] He died in Belgrade on 19 October 1784 and is said to have been mourned by all.[5]

Osman Efendi's mathematical treatise is the only book he wrote in Arabic. Its full title is *Hadiyyat al-Muhtadī li-īqād al-sirāj al-munṭafi'*, which can be translated as "The Gift of the Convert to Rekindle the Extinguished Lamp". Indeed, Osman sets out his ambition to "revive and renew" geometry, "a forgotten and neglected science" in Islamic countries. He reveals at the end of the book that he translated some German and French books for this purpose. However, he does not refer explicitly to any author or any book. After a thorough analysis, we discovered that *Hadiyyat al-Muhtadī* is actually an elaborate personal work going far beyond a simple translation process. It predominantly appears as a careful blending of chapters of several European books, freely translated and commented on. Furthermore, we realized that Osman incorporated in his book a wide range of non-European features. While the latter seem to be taken directly from Islamic knowledge of the living tradition, we identified with certainty the three European books that served as main sources:

- Bernard Forest de Bélidor, *Neuer Cursus Mathematicus zum Gebrauch der Officiers von der Artillerie und der Ingenieurs*. This is the translation into German, published in 1745, of the French book *Nouveau cours de mathématique, à l'usage de l'artillerie et du génie*, published in 1725.[6] The translator Johann Theobald Bion supplemented the original with additional paragraphs which allows us to say with certainty that Osman used the German translation. We will refer to the 1745 edition; however, he may have used the 1759 or the 1775 edition, which do not differ significantly.

4 The three books translated by Osman Efendi are: Johann Hübner, *Kurtze Fragen aus der neuen und alten Geographie*, an edition previous to that of 1718, translated in 1751 (wrongly identified by Adnan Adıvar in 1943 as Varenius's *Geographia Generalis*, an error repeated by all scholars until one of us—Ageron—discovered the actual source in 2016); Pietro Andrea Mattioli, *Commentarii in libros sex Pedacii Dioscoridis*, probably from one of its two German translations entitled *Kräuterbuch*, translated in 1770; Gottfried Heinrich Burghart, *Zum allgemeinen Gebrauch wohleingerichtete Destillierkunst*, 1736, together with *Neue Zusätze*, 1748, translated in 1782. These translations all remained in manuscript.

5 These details are taken from various Austrian newspapers of the eighteenth century. For bibliographic information, see Abdeljaouad, Ageron & Shahidy, op. cit. (n. 3).

6 Bernard Forest de Bélidor, *Neuer Cursus Mathematicus zum Gebrauch der Officiers von der Artillerie und der Ingenieurs* [...] *in die Teutsche Sprache übersetzt und mit nöthigen Zusätzen versehen von Johann Theobald Bion*, Johann Joseph Pentz, Vienna and Leipzig, 1745. The French original is: Bernard Forest de Bélidor, *Nouveau cours de mathématique, à l'usage de l'artillerie et du génie*, Claude Jombert, Paris, 1725.

- Christian Freiherr von Wolff, *Anfangs-Gründe aller Mathematischen Wissenschaften*. We will refer to the seventh edition, published in 1750,[7] but Osman may have used an earlier or a later edition.
- François Blondel, *Die Kunst Bomben zu werffen* (The Art of Throwing Bombs). This is the very faithful and anonymous translation into German, published in 1686, of the French book *L'Art de jetter les bombes*, published in 1683.[8] We will refer to the latter, although it seems likely that Osman used the German translation.

Osman's treatise is arranged in two parts, each divided into three books. The first part contains a book on plane Euclidean geometry with an appendix on elementary arithmetic, a book on surveying, and a book on solid geometry. The second part, oddly titled "conclusion", contains a book on the motion of bodies, whose first chapter deals with conical sections, a book on the throwing of bombs, beginning with some theorems of Galilean dynamics, and a book on the mathematical aspects of digging and blasting of mines.[9] Descriptions of instruments are to be found in five out of these six books. In many cases, but not always, they are accompanied by neat drawings. In what follows, we list all of these instruments, translate Osman's descriptions of the lesser known ones, and discuss their origin: East or West?

2 Instruments in the Book on Plane Euclidean Geometry

Hadiyyat al-Muhtadī begins with a long book on plane Euclidean geometry, mostly borrowed from Wolff's and Bélidor's treatises. However, roughly halfway through this book there is a specific chapter that is not a translation of any of those: it is entirely devoted to geometrical instruments. Inclusion of this particular topic at this place is justified from its outset:

> Exposition of some instruments needed in geometry. This chapter serves as a preamble to the following chapters where search for values

7 Christian Freiherr von Wolff, *Anfangs-Gründe aller Mathematischen Wissenschaften*, Renger, Frankfurt & Leipzig/Halle, 1750, 4 vols.

8 François Blondel, *Die Kunst Bomben zu werffen*, Johann Hofmann, Sulzbach, 1686. The French original is: François Blondel, *L'Art de jetter les bombes*, chez l'auteur et Nicolas Langlois, Paris, 1683.

9 For details on the book on mines, see Pierre Ageron, "Mathématiques de la guerre souterraine: Bélidor dans l'Empire ottoman", in É. Barbin, D. Bénard and G. Moussard (eds.), *Les Mathématiques et le Réel : expériences, instruments, investigations*, PUR, Rennes, 2018, pp. 211–224.

of lengths and widths is explained, which can only be done with specific instruments.[10]

The author starts by comparing Christian and Islamic units of length measurement. He points out that Christian surveyors have chosen to decimalize their units "because of the difficulty of computing with the foot divided into 12 inches". Addressing Muslim mathematicians, he advises them to undertake a similar reform and suggests a hybrid system, which he makes use of throughout all sections of his work. We summarize his findings in *Table 2.1*.

Osman considers the Islamic inch and the Christian inch as equal, so that the *dhirāʿ* (cubit) becomes equivalent to two feet. Occasionally, he also mentions the spear (*rumḥ*), which he considers to be equal to 5 cubits or 10 feet. Furthermore, in Bélidor's and Blondel's books, lengths are always expressed in *toises*, a *toise* being approximately the distance between the fingertips of a man with outstretched arms. *Toise* was appropriately translated by *Klafter* in German versions of these books, but in Osman's treatise, it was quite inconsistently rendered by *qadam* (foot), *dhirāʿ* (cubit) or *dhirāʿān* (two cubits). Recall that the correct ratios are the following: 1 toise is approximately equal to 4 cubits or 6 feet (or 1.8 metres).

TABLE 2.1 Comparison of Christian and Islamic length units, and Osman's suggestion for the use of decimal fractions

Islamic units		**Christian units**		**Christian decimal units**		**Osman's suggestion**	
1 cubit (*dhirāʿ*)	24 inches	1 toise (*dhirāʿ*)	6 feet	1 perch (*qaḍīb*)	10 feet	1 cubit	20 inches
½ cubit	12 inches	1 foot (*qadam*)	12 inches	1 foot	10 inches	½ cubit	10 inches
1 inch (*iṣbaʿ*)	6 barley grains	1 inch (*iṣbaʿ*)	12 lines	1 inch	10 lines	1 inch	10 lines
1 barley grain (*shaʿīra*)		1 line (*khaṭṭ*)		1 line	10 points (*nuqāṭ*)	1 line	10 points

10 We chose to translate the text of the oldest dated and most reliable copy we know of (Istanbul Military Museum, MS 3027). Since its folios are not numbered, we shall not give any other indication of the location of a quote than the book it pertains to. The other manuscripts may contain a few errors, but no important variants.

Eleven mathematical instruments are described in this book:

1. *al-miqyās* (the scale, the gauge, the unit of measurement);
2. *al-mimsaḥa al-mustaʿmila fī misāḥat al-arḍ* (the land surveying instrument; see details below);
3. *miqyās niṣf al-dhirāʿ* (the half cubit scale; see details below);
4. *al-burkār* (the compasses for drawing circles);
5. *al-misṭara* (the straight edge);
6. *misṭara al-muwāzāt* (the parallel rulers);
7. *al-muthallath al-qāʾim al-zāwiya* (the right-angled triangle, the triangular set square);
8. *al-zāwiya al-qāʾima* (the right angle, the L-square ruler);
9. *al-kūniyā* (a triangular levelling instrument; see details below);
10. *al-kūniyā al-hawāʾiya* (the aerial *kūniyā*, see details below);
11. *al-minqala* (the protractor).

al-mimsaḥa al-mustaʿmila fī misāḥat al-arḍ (the instrument used in land surveying)

The term *mimsaḥa* can have various meanings and no drawing of this instrument is provided in the manuscripts, but its description makes it clear that the surveyor's chain is meant:

> A common practice among geometers is to make a measuring instrument [*mimsaḥa*] of twenty, thirty or forty cubits, made of rope or wood, each cubit being hooked onto the next by a mobile ring of iron, or the like, with two large rings of iron at its extremities, through each of which passes a stick. Placed on the ground, the instrument can be moved thanks to the sticks.

miqyās niṣf al-dhirāʿ (the half cubit scale)

Although he does not give it a specific name, Osman is particularly keen on this instrument:

> Every geometer, whether master or apprentice, should make a measuring instrument of half a cubit in length—or one foot, both are equal to each other—made of wood, copper or similar materials as bone, cardboard or others, divide it into ten equal parts expressing inches, also divide one of these parts into ten equal parts indicating lines, and keep it with him until the time he needs it.

He then proceeds to give a more precise description of this *miqyās* and three examples of its use. A drawing is also included (*Fig. 2.2*). All this makes

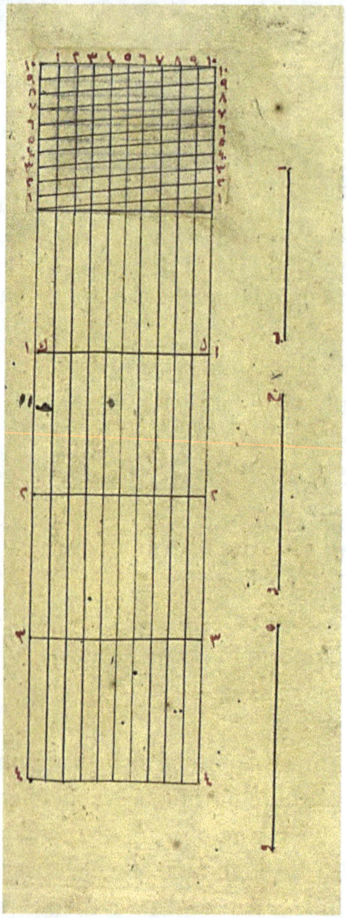

FIGURE 2.2
The half cubit scale (*miqyās niṣf al-dhirāʿ*) from *Hadiyyat al-Muhtadī*
ISTANBUL, MILITARY MUSEUM, MS 3027

quite clear that it is nothing but the ruler called in Europe *échelle de dixme, échelle géométrique, verjüngter Maßstab* or 'plane diagonal scale', invented in the seventeenth century in connection with the use of decimal fractions. It is graduated in two directions, allowing the precise measurement of lengths using the principle of similar triangles. Although the *verjüngter Maßstab* appears in Wolff's *Anfangs-Gründe*,[11] a book to which Osman resorted on many other occasions, his account of this instrument is definitely not a translation of this source. It is also not translated from one famous treatise of the time that at first seemed to us a plausible source, namely Nicolas Bion's comprehensive *Traité de la construction et des principaux usages des instruments de mathématiques*, nor from its German translation and supplements by Johann Gabriel

11 Wolff, op. cit. (n. 7), v. 1, p. 136 (sect. 68), pp. 200–201 (sect. 193) and pl. 14, fig. 115.

Doppelmayr. It remains possible, of course, that Osman consulted these sources, but his inspiration is different.[12]

al-kūniyā

This is a level in the shape of an isosceles triangle with a plumb line attached to the vertex of the two equal legs. Probably of Greek origin (*kuniyā* is derived from *γωνία*, meaning angle), it was a common instrument in the medieval Islamic world. For example, clear descriptions of it can be found in the works by two well-known authors of the thirteenth century, Abū al-Ḥasan al-Marrākushi and Quṭb al-dīn al-Shirāzī. The same instrument was also known in several European traditions. In France it was widely known until recently under various names: *niveau de maçon, niveau à perpendicule, niveau de Salomon, archipendule, libellule.*

al-kūniyā al-hawā'iyya (aerial *kūniyā*)

This is also a level, essentially similar to a *kūniyā* but turned upside down: the plumb line now falls from the middle of the base while the triangle is suspended from the base by a string fixed between two vertical posts (*Fig. 2.3*).

In contrast to the plain *kūniyā*, this version is quite rare in Western literature or material culture and was probably hardly used in practice. Its only occurrences known to us are a drawing in a booklet printed in Poland in 1573,[13] one actual instrument probably made in Austria circa 1590,[14] and another drawing on the frontispiece of an edification book printed in Lower Saxony in 1655.[15] However, it is widely documented in Islamic countries, from Spain

---

12 *Neu-eröffnete mathematische Werck-Schule Nicolai Bion*, translation by Johann Gabriel Doppelmayr of Nicolas Bion's *Traité de la construction et des principaux usages des instruments de mathématiques*, Monath, Leipzig & Nürnberg. The first edition (1713) is a plain translation of Bion's book, published in 1709. The second edition (1717) includes a *Weitere Eröffnung*. The third edition (1726), the fourth edition (1741) and the fifth edition (1765) further add a *Dritte Eröffnung*. On the *scala decimale*, see p. 15 and fig. 4.4 (all editions).

13 Olbrycht Strumieński, *O Sprawie, sypaniu, wymierzaniu i rybieniu stawów*, Łazarz Andrysowic, Cracow, 1573 (no page numbers). The instrument is called *blaszka* (small plaque). See some comments in: Félix Kucharzewski, "Sur quelques niveaux du seizième siècle", *Bibliotheca Mathematica*, ser. 3, v. 1 (1900), pp. 60–63.

14 Florence, Museo Galileo, inv. no. 3538. We thank Giorgio Strano for pointing this level out to us. It is part of a prestige mining instruments set of Austrian or Hungarian type, brought back by Prince Mattias de' Medici. We are tempted to see it as an isolated Ottoman-influenced artefact.

15 Peter Eichholtz, *Geistliches Bergwerck, das ist andächtige, liebliche und bewegliche Betrachtungen, wie an dem löblichen Bergwerck* [...] *das gantze Christenthumb* [...] *vorgestellet seyn* [...], Dunker, Goslar, 1655. We thank Thomas Morel for pointing out to us this drawing. However, Dr. Morel also confirmed that no printed or manuscript German

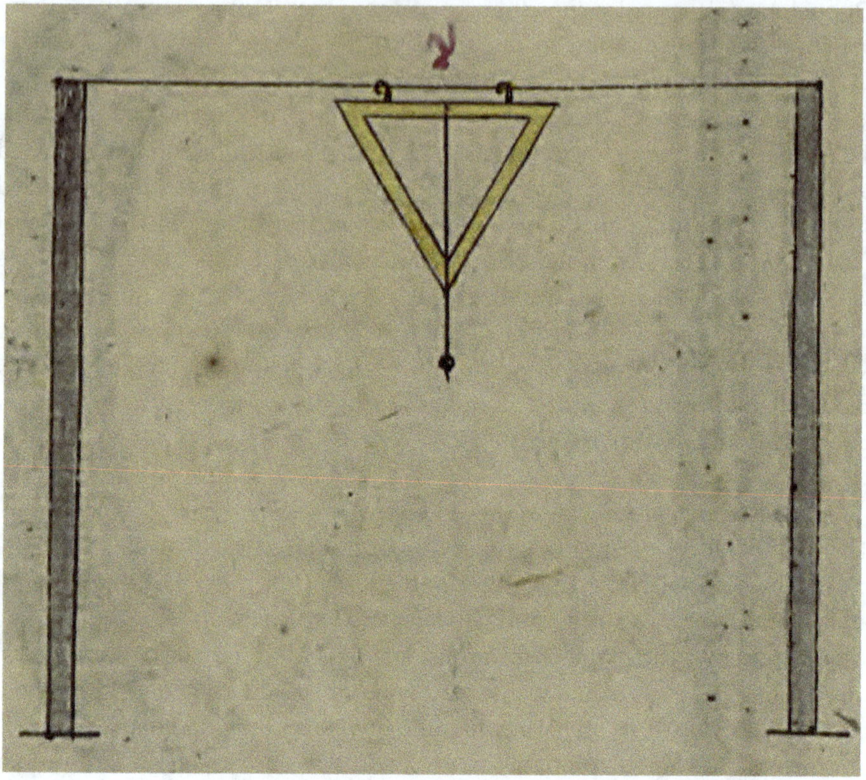

FIGURE 2.3 The aerial kūnyā (*al-kūniyā al-hawāʾiyya*) from *Hadiyyat al-Muhtadī*
ISTANBUL, MILITARY MUSEUM, MS 3027

to Iran, under the following names: *murjiqāl* (from Spanish *murciélago* meaning *bat*), *mizān* (balance in Arabic), *terazi* (balance in Turkish), *havai terazi* (aerial balance). Among the most widely circulated late mathematical sources in which this instrument is discussed, let us mention the textbooks by Bahāʾ al-dīn al-ʿĀmilī (d. 1621) called *Khulāṣat al-ḥisāb* (The Very Essence of Arithmetic) and ʿAbd al-laṭīf b. Aḥmad al-Dimashqī (d. 1749) entitled *Sharḥ Nukhbat al-tuffāḥa fī ʿilm al-misāḥa* (Commentary on the Best of the Apple in the Science of Measurement). It was still commonly used in the nineteenth century, as can be seen from the very precise account by the French traveller Antoine François Andréossy of the methods of Constantinople's fountain-makers.[16]

language treatise on underground surveying of the sixteenth and seventeenth centuries ever mentioned any instrument similar to an aerial *kūniyā*, so that its actual use among practitioners (*Markscheider, Schiner*) is dubious.

16 Antoine François Andréossy, *Voyage à l'embouchure de la mer Noire ou essai sur le Bosphore*, Plancher, Paris, 1818, pp. 237–241 and pl. 6. On this instrument see also the chapter by Umut and Pantalony, *infra*, pp. 201–202 and *Fig. 11.6*.

An intriguing fact is that Osman concludes his description of this instrument by expressing the following personal opinion:

> For me, *i.e.* in my opinion, it is preferable that the *kūniyā* be made in the shape of a half circle, that the two rings hang at its diameter's extremities, that the half circle be divided into two quarters and that each of the two quarters be divided into 90 equal parts. This is more adequate and more perfect for the practice, as cannot be concealed.

This variant, Osman's favourite, does not seem to be attested in the Ottoman Empire. However, it turns out to be identical to the level and inclinometer typically used by German mine surveyors called *Gradbogen* (graduated arc) or *Wasserwaage* (water balance)—though not using water. Could Osman have seen this instrument in some German treatise about underground surveying?[17]

3 Instruments in the Book on Surveying

The book on surveying, the second book of part 1, is rather oddly arranged. It contains three chapters. The first chapter and the first half of the second chapter are translated from Wolff's *Anfangs-Gründe*. They aim at measuring the distance between two places or the height of a tower or mountain, accessible or not. This problem is treated by means of stakes serving as milestones, then with two modern instruments, both relying on the use of similar triangles: *al-ṭabla* (the table), *niṣf al-dāʾira* (the half-circle). The second half of the second chapter describes much more rudimentary ways to evaluate heights, using either stakes (*khashabāt*) or optical reflection through a mirror (*mirʾā*). Finally, the third chapter applies trigonometric methods to surveying problems, with the use of *al-rubʿ al-mujayyab* (the sine quadrant).

al-ṭabla (the table)

Here is our translation of Osman's detailed account of this instrument:

> The table is a composite instrument, well known among geometers, with which one determines distances between objects and their heights. It is a wooden board whose surface is plane—it is necessary that the upper surface be extremely plane—, square in shape, *i.e.* its length and its width

17 There is such a picture in: Nikolaus Voigtel, *Perfectionirte Geometria Subterranea*, Johann Dietzel, Eisleben, 1686, between pp. 17 and 18. We thank Thomas Morel for pointing it out to us.

> are equal, generally of 20 inches, although more or less is possible, and two inches thick, as much as suffices. It is put on another board, circular in shape. Under the latter is a tripod [*uthfiyya*], which they call in Persian *sepeh*.[18] The diameter of the circular board is the half of the side of the plane table. There is a hole in its centre, and the plane table also is pierced underneath in the middle. The table is fixed firmly to the circular board by means of a screw [*lawlab*], in Turkish *burma*, inserted into it from below. It is lying flat if I want to determine distances of things parallel to the horizon, or standing upright, *i.e.* set vertical on the circular board, if I want to determine the vertical and its height, and for this, it is necessary that there be another hole in the middle of one side of the plane table to hold it in place by inserting the screw into it. The way to use the plane table is to put on its surface a sheet of paper to draw the lines and to stick the sheet of paper onto its edges with wax or anything similar in order to draw lines at its two extremities.

Somewhat unusually, Osman does not tell the reader about any picture of the *ṭabla*. Indeed, most manuscripts of *Hadiyyat al-Muhtadī* do not contain one, with the exception of a single relatively late manuscript (*Fig. 2.4*).[19] The written depiction by itself immediately calls to mind the European instrument called in German *Tischlein*, in French *planchette* and in English *small plane table*. However, Osman's precise source for this description is unclear. Although the *Tischlein* does appear in Wolff's *Anfangs-Gründe der Geometrie*,[20] no explicit and detailed description of it is provided there; we also discounted Nicolas Bion's *Traité des instruments de mathématiques*, or its translation and supplements by Doppelmayr. Regarding the use of the *ṭabla*, it is translated from Wolff. The illustration showing how to use two small plane tables to measure the height of an inaccessible tower[21] was adapted to the cultural context, turning a fancy medieval tower with a set-back storey into a slender Ottoman minaret with a balcony.

18 The Persian word is *seh pāyeh* (three bases), but it is pronounced *sepāye*, which explains the Ottoman Turkish form *sepeh*. The *h* however is pronounced in the modern Turkish form, which is *sehpa*. We thank Mahmoud Shahidy for these details.

19 Osman Efendi bin Abdülmennan, *Hadiyyat al-Muhtadī*, Cairo, Egyptian National Library, MS DR 695, fol. 44r.

20 Wolff, op. cit. (n. 7), v. 1, pp. 206–208 (sect. 199) and pl. 17, fig. 120.

21 *Ibid.*, pl. 17, fig. 122.

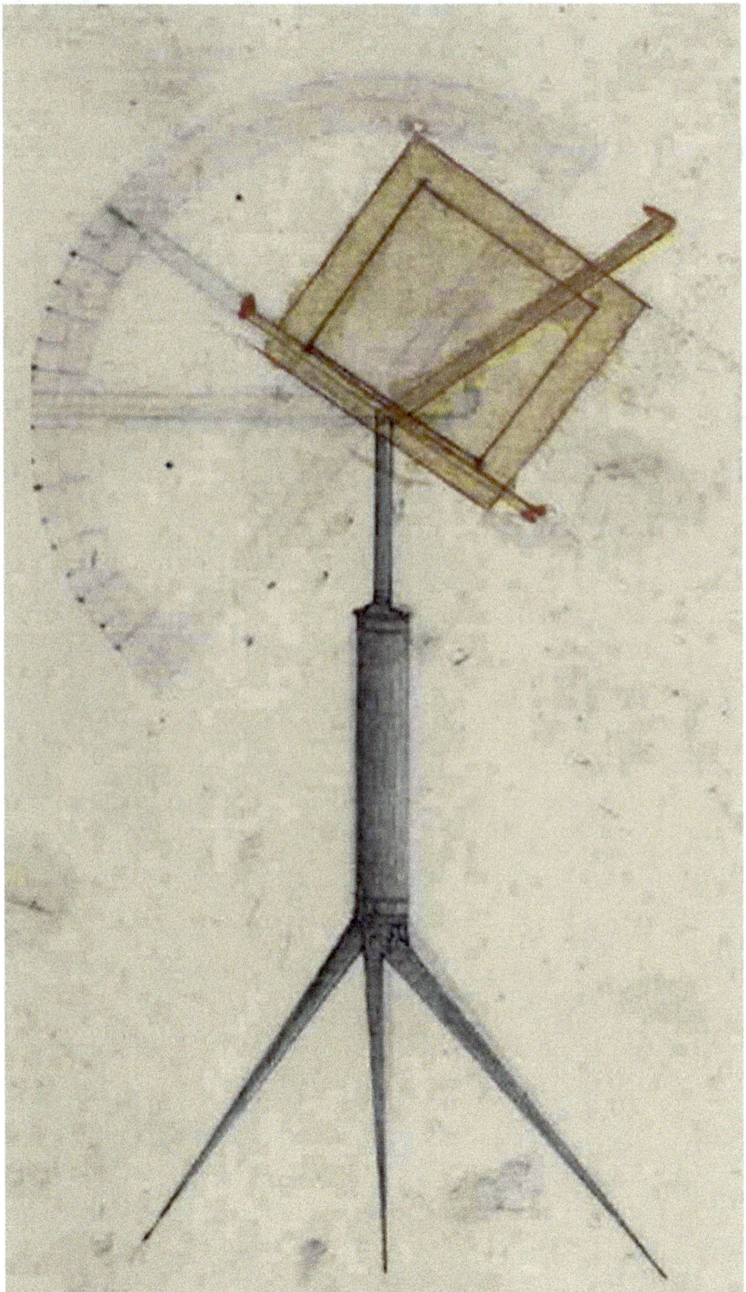

FIGURE 2.4 The table (*al-ṭabla*) in the only copy of *Hadiyyat al-Muhtadī* where this instrument is drawn
CAIRO, EGYPTIAN NATIONAL LIBRARY, MS DR 695, FOL. 44R

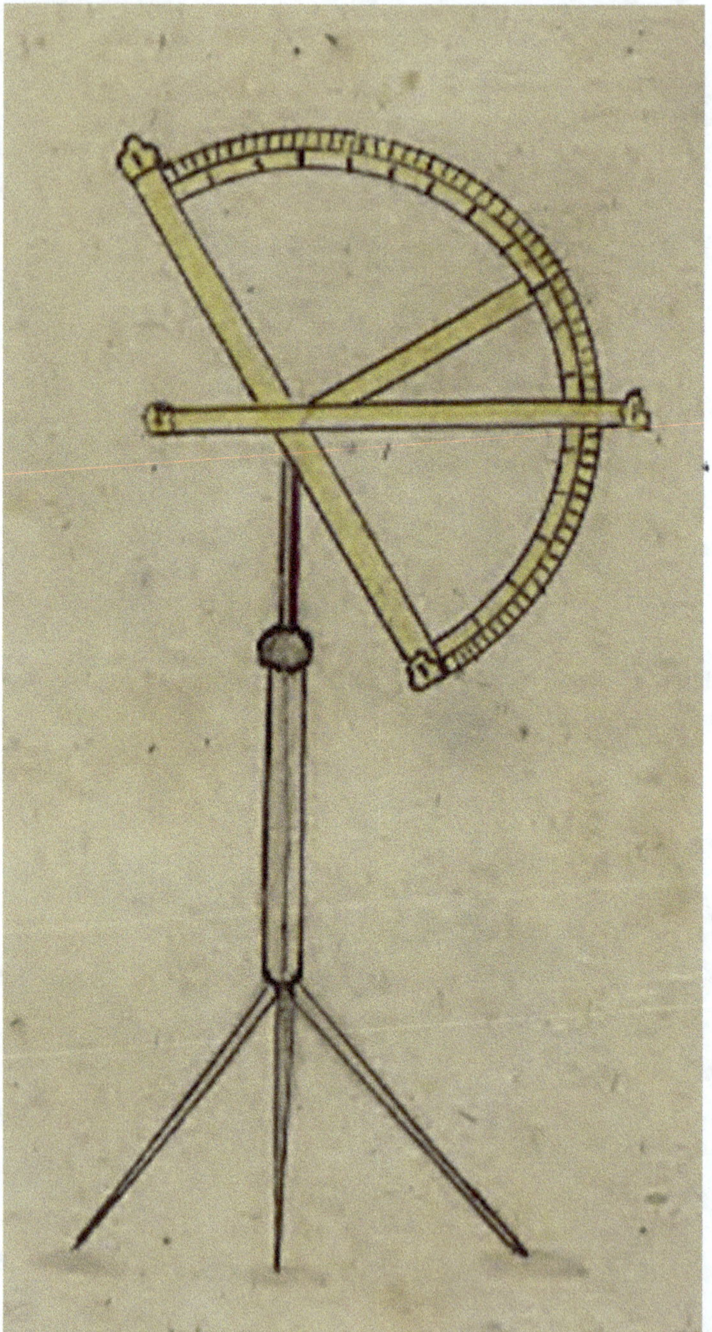

FIGURE 2.5 The half-circle (*niṣf al-dāʾira*) from *Hadiyyat al-Muhtadī*
ISTANBUL, MILITARY MUSEUM, MS 3027

niṣf al-dāʾira (the half-circle)

In contrast to the previous instrument, Osman's description of the half-circle is followed by a neat and precise drawing (*Fig. 2.5*):

> The half-circle, called by French geometers the simple astrolabe [*al-usṭurlāb al-basīṭ*], is an instrument specifically dedicated to determination of angles, when distances or heights are measured. It is a half-circle made of brass subdivided in 180 equal parts in either direction. A mobile ruler, also in brass, is fastened to the centre, bearing two sight vanes. For the sightings, the half-circle is fitted horizontally or vertically on a tripod [*uthfiyya*], *i.e. sepeh*, at the top of a stick. Its picture is like this.

This again is a well-known European surveying instrument. In this case also, Osman's source is unclear. To be sure, the half-circle appears in Wolff's *Anfangs-Gründe*, but there it bears the name *Winckelmesser* (angle measurer) and is built differently—in particular, it is supported by a cross base.[22] Despite some similarities, we also discount Bion's treatise or its translation by Doppelmayr as a direct source. Moreover, we could not find a French source where the half-circle is called *astrolabe simple* as stated by Osman. In France, its most common name was *graphomètre*, while *demi-cercle* was also used by some authors.

khashabāt (stakes), *mirʾā* (mirror)

The traditional methods using stakes or mirrors described by Osman are those that were still in common use in the Ottoman Empire in the eighteenth century. For example, we found a very similar account in ʿAbd al-laṭīf b. Aḥmad al-Dimashqī's *Sharḥ Nukhbat al-tuffāḥa fī ʿilm al-misāḥa*, already mentioned above, and in Muḥammad al-Aqkirmānī's (d. 1760) *Risāla fī maʿrifat al-abʿād* (Treatise on the Knowledge of Distances). These methods have their origin in the Greek tradition.

al-rubʿ al-mujayyab (the sine quadrant)

The sine quadrant is the iconic instrument of Islamic mathematics. It is said by Osman to be "famous" and is only very briefly described, with no drawing, as if the author took for granted that his reader was already familiar with it. Besides the book on surveying, it appears in the book on throwing of bombs and the book on mines. In different situations, Osman explains in detail how it can be used as an alternative to the methods drawn from European sources, *e.g.* the small plane table, sine tables and Pythagoras' theorem.

---

22 *Ibid.*, p. 134 (sect. 64) and pl. 7, fig. *.

4 Instruments in the Book on Throwing of Bombs

The book on throwing of bombs, the second book of part 2, is a subtle blend of the aforementioned books by Bélidor, Blondel and Wolff. The first two chapters are chiefly translated from Blondel. The main result proved is the following: if a body be projected either parallel to the horizon or in any oblique direction it will, by its motion, describe a parabola; furthermore, its horizontal range will be proportional to the sine of double the angle of elevation. An adapted sine table, also borrowed from Blondel, is given.[23] Osman makes the crucial point that, in this table, the sine of twice 45° is taken to be equal to 10000 and not to 60, as was usual in Islamic countries. He works out four numerical examples, similar to, but different from those in Blondel. Afterwards, he explains that the following specific instrument can be used in place of a sine table.

niṣf dā'ira li-ramī al-khumbara (half circle for the throwing of bombs)

Osman provides a drawing of this instrument together with the following description:

> In cases where the sine table is not available, wherever throwing a bomb is required and at any time, we want a specific instrument with which one can extract the elevation necessary to throw a bomb and its range without resorting to the aforementioned sine table. It consists in a half circle, made of copper or wood, whose circumference is divided into 90 parts, *i.e.* degrees, beginning at point *A* and finishing at *B*, and whose diameter *AB* sticks out and is extended until *H* so that it goes in the instrument. A half-diameter *DC* is placed [perpendicular to *AB*] at its middle *D* and divided into 200 equal parts, replacing the sines. Lines parallel to the diameter *AB* are drawn from the subdivisions of the circumference, every 5 degrees. These parallels cut the half-diameter *DC*; they are similar to the sines and cosines of the sine quadrant.

This description is faithfully translated from a chapter of Blondel's book.[24] Blondel attributes this instrument to Torricelli, but Osman does not mention names. The three numerical examples worked out by Osman are exactly those

23 Blondel, op. cit. (n. 8), pp. 80–82 (pt. II, bk. I, ch. III: *Table des sinus servant au jet des bombes*).

24 *Ibid.*, pp. 107–112 (pt. II, bk. II, ch. III: *Par un autre instrument sans le besoin des sinus*).

given by Blondel. For instance, in the last one, he asks for the elevation angle needed to hit a target located on the horizon line at a distance of 1200 feet, knowing that an elevation of 30° is needed for a target at a distance of 1500 feet. He obtains two solutions, namely 21° 56' and 68° 4'. Then comes the more complicated case when the target is not on the horizon line but makes an angle *i* with it.

Osman chose to publicize two only out of the many methods described by Blondel. One of them involves rather complicated trigonometric calculations.[25] Although no symbolic notation at all is used in Osman's treatise, it is convenient to sum up his results in the following formula, giving the elevation angle e needed for a target at a distance d, knowing that the maximal reach of the throw is d_{max}:

$$e = \tfrac{1}{2}(i + a) \text{ or } \tfrac{1}{2}(i + 180° - a) \text{ where } a = \arcsin[\sin(i) + (d/d_{max})\cos(i)].$$

Osman works out the examples where $i = \pm 15°$, $d = 620$ cubits and $d_{max} = 1200$ cubits. They are essentially those considered by Blondel, although he took $d = 310$ toises and $d_{max} = 600$ toises. But instead of using a sine table to six decimal places as Blondel does, he decides to work with the sine quadrant. The alternative method avoids or simulates these calculations by using a modified version of the previous half circle:[26]

> Concerning extraction of such an elevation if the target is located above or below the horizon line without using the sine table, one divides in two a circle made exactly as the previous half circle [...] except that an arm tangent to the half circle, like the arm *BW*, is fixed at point *B*, equal to the half-diameter *DC*, divided in 200 equal parts as *DC* is, on which is fixed a thread movable from *B* to *W*, and another thread with a plumb-bob at point *A*.

al-āla al-kullīya yumkin bihā ramī al-khumbara (the universal instrument with which one can throw a bomb)

In the third chapter of the book, Osman depicts an instrument meant to solve all problems in the art of throwing bombs, whether the target be on a plane

25 *Ibid.*, pp. 127–128 (pt. II, bk. III, ch. VI: *Troisième pratique par les sinus*).

26 *Ibid.*, pp. 133–137 (pt. II, bk. III, ch. VIII: *Cinquième pratique par le cercle de Torricelli rectifié*).

above the horizon, below it or parallel to it. The chapter begins with the following description of this universal instrument:

> It is a complete circle, made of brass, wood or the like, as circle *ABCD*, whose circumference is divided into 360 equal parts, its numbers being written from *A* to *D*. A ruler *AE* is placed on the circle, perpendicularly to the diameter *AC* and tangent to the circle at the point *A*, such that the length of the ruler has the same value as the diameter of the circle. A movable thread with a plump-bob is attached to the ruler so that it can be moved towards its two extremities, according to what the practice requires.

This is the instrument invented by Blondel.[27] The above lines are very close to Blondel's words, but also to Bélidor's who himself heavily relied on Blondel on this matter.[28] Immediately after that, Osman suggests two alleged improvements which seem to be fully his own:

> If the rule were extended from its extremity and its length set to twice the diameter, then it would be preferable and easier to take the elevation when it is introduced into the cannon. And it is more convenient to place against the diameter *BD*, perpendicular to the diameter *AC*, a half circle *HKJ*, to divide each of the cords *KH, KJ* into 90 equal parts or degrees and to hang another thread with a plump-bob at the centre *T*.

For this reason, Osman's drawing of the universal instrument (*Fig. 2.7*) differs from both Blondel's and Bélidor's drawings (*Fig. 2.6*). Then Osman proceeds to explain in detail how to use this instrument. His account seems to be a combination of Blondel[29] and Bélidor[30] together with many personal

27 *Ibid.*, pp. 164–165 (pt. II, bk. VI, ch. I: *Construction d'un instrument universel pour le jet des bombes*).

28 Bélidor, *Neuer Cursus Mathematicus*, cit. (n. 6), p. 432 (sect. 944: *Ein Universal-Instrument zu machen, welches dienet, die Bomben auf unterschiedene Arten Flächen zu werfen*) and fig. 374. In the original French edition: Bélidor, *Nouveau cours de mathématique*, cit. (n. 6), p. 431 (sect. 742: *Construction d'un instrument universel pour jeter les bombes sur toutes sortes de plans*) and pl. 25, fig. 346.

29 Blondel, op. cit. (n. 8), pp. 166–169 (pt. II, bk. VI, ch. II: *Usage d'un instrument universel pour le jet des bombes*).

30 Bélidor, *Neuer Cursus Mathematicus*, cit. (n. 6), pp. 433–434 and 438–439 (sects. 945, 946 and 950: *Von dem Gebrauch dieses Universal-Instruments*), and pl. 25, fig. 375, pl. 26, figs. 379, 380 and 381. In the original French edition: Bélidor, *Nouveau cours de mathématique*,

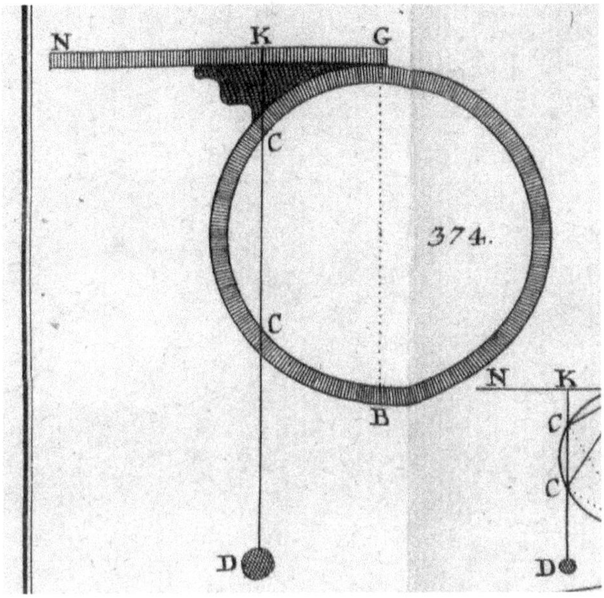

FIGURE 2.6 A universal instrument for throwing bombs from Bernard Forest de Bélidor, *Neuer Cursus Mathematicus*, Pentz, Vienna and Leipzig, 1745, pl. 25, fig. 374

additions. Finally, he works out an example of his own with the universal instrument. By taking the values $i = 10\frac{1}{4}°$, $d = 610\frac{1}{6}$ cubits and $d_{max} = 800$ cubits, he obtains $e = 37\frac{3}{4}°$ or $62\frac{1}{4}°$. If we use the above formula, we find instead $e = 39\frac{1}{5}°$ or $61°$.

5 Miscellaneous Instruments

al-āla ʿalā ṣūrat jāmāt al-ḥammām (the instrument resembling the cups of glass in the dome of the public bath)

In the first book of part 2, Osman digresses on various non-mathematical subjects. Among other things, he describes a rather unexpected device in the

cit. (n. 6), pp. 432–436 and 439–440 (sects. 743, 744, 745 and 747, prop. 20: *Usage de l'instrument universel pour le jet des bombes*), and pl. 25, fig. 347, and pl. 26, figs. 351, 352 and 353.

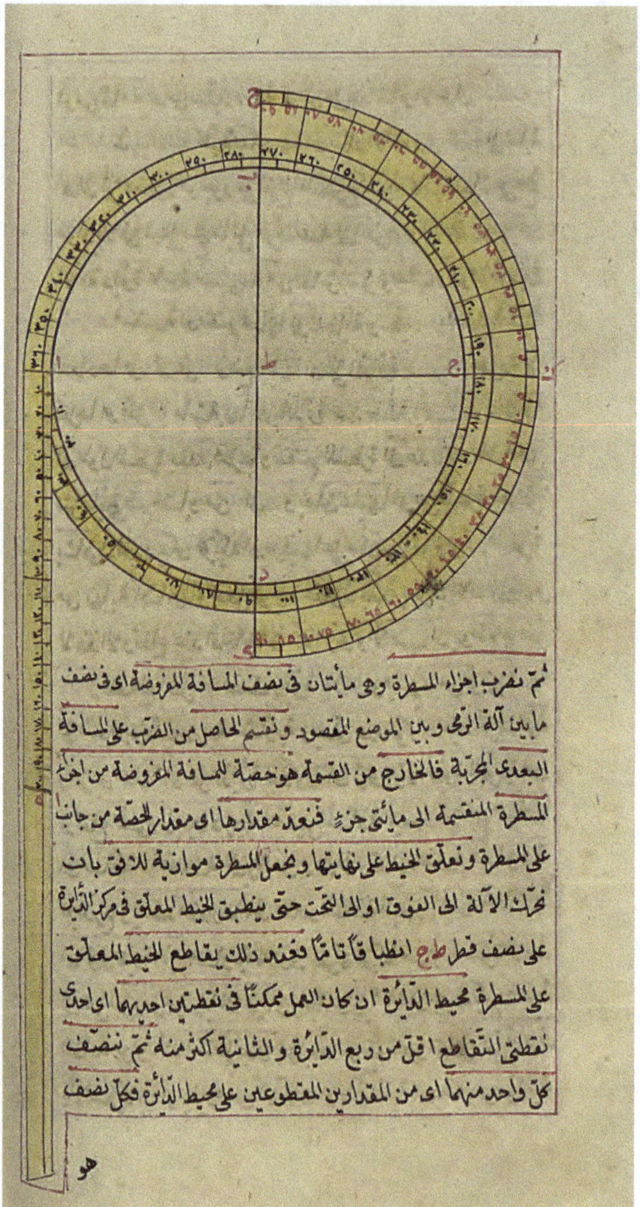

FIGURE 2.7 The universal instrument with which one can throw a bomb (*al-āla al-kulliya yumkin bihā ramī al-khumbara*) from *Hadiyyat al-Muhtadī* (Istanbul, Military Museum, MS 3027). The artist omitted the plumb line, shown only in the later copy (Cairo, Egyptian National Library, MS DR 695)

context of his book. For this purpose, he resorts to a surprising comparison, but obviously one which was familiar to his readers:

> When the late philosophers experimented with these three [heat, air, and magnetic force] with a large instrument resembling the cups of glass in the dome of the public bath (*ḥammām*), they placed underneath a living thing and removed the air by means of another machine. They noticed that the air did not pass through the glass and did not affect what was below, for they saw that without air the living being was dead. They found that the heat passes through the glass and influences through it, because they placed snow and something approaching, without any barrier other than this machine: they judged that the heat is more penetrating and subtle than the air. They observed that the magnetic force passes through the glass and through the machine, and sets in motion the [magnetized] needle which is underneath, even from a far distance and deeply buried, in contrast to heat which has no influence below the earth, as is seen with snow and ice put underground. They established that the subtlety of the magnetic force is higher than that of heat, and developed a long research on it, which cannot be contained here.

Clearly, the instruments described above are a glass bell jar together with a vacuum pump. It seems to us that Osman is reporting on an experiment in which Robert Boyle tested in 1659 the effects of rarefied air in different situations, especially on living beings and magnetism, but his precise source is unknown to us.

al-būṣula (the magnetic compass)

The third book of part 2, dedicated to mines, contains only one new instrument description. To determine the right direction while digging a gallery, Osman says, one needs a magnetic compass. This instrument is drawn (*Fig. 2.8*) and described as follows:

> It is a specialised instrument used among engineers and navigators when a desired direction is required, whether onshore, offshore or underground. The method for manufacturing it is that you make a board of wood, brass, bone or something similar, its surface being square or rectangular, that you let a circle turn on it, in the middle of it, and that you draw its diameter across the board, dividing the circle in two equal halves. Then, you divide each of the two halves of the circle into 180 parts, *i.e.* equal degrees. Then, you set up at its centre a pointed pivot [*mayl*] made

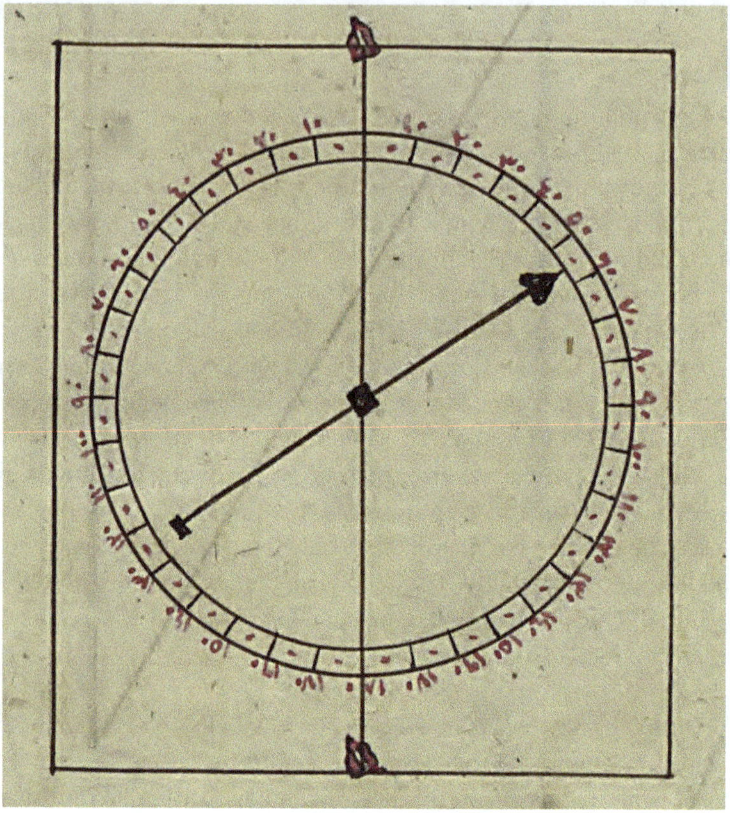

FIGURE 2.8 The magnetic compass (*al-būṣula*) from *Hadiyyat al-Muhtadī*
ISTANBUL, MILITARY MUSEUM, MS 3027

> of brass, on which you put a needle, one end of which has been treated with magnetic stone [...; Osman reports here about the magnetisation process]. Then, you put the board with the needle in a box [*qaʿba*],[31] in Turkish *qūṭu*, and cover the surface of the needle with glass to preserve it from wind, dust and the like. Its picture is like this.

We recall that the magnetic compass was known to medieval Arab navigators, who called it *bayt al-ibra* (house of the needle) or *ḥuqqa* (box). However, the *būṣula* used by Ottomans is of Western origin, as evidenced by its name

31 The Arabic word *qaʿb* or *qaʿba* refers to a large cylindrical drinking vessel, like a mug. In this context, it probably denotes a square box. The word *qūṭu*, in Persian or Ottoman Turkish, unambiguously means box.

derived from Italian *bussola* or French *boussole*.[32] A description of it can be found in Piri Reis's *Kitāb-ı bahriye* (Book of Navigation) written in 1521 and perspective views of two models of *būsula* are displayed in Kātib Çelebī's *Cihānnümā* (Cosmorama) printed in Istanbul in 1732. At an unspecified time, the Ottomans used a magnetic needle in a semi-circle known as *lağım tapası* (mine plug) while digging the gallery of a mine.[33] The exact source of Osman's account of the *būsula* is unknown to us, but it should be noted that it is not necessarily European.

6 Conclusion

Twenty-one scientific instruments are described in *Hadiyyat al-Muhtadī*. They can be divided into four categories:

1. Very basic instruments (compasses, parallel rulers, triangular level, etc.), belonging to the common heritage of East and West;
2. Typically Western instruments, unknown in the Ottoman Empire at the time of writing: the decimal diagonal scale, the small plane table, the surveyor's half circle, the gunner's quadrant;
3. Typically Eastern/Islamic instruments: the aerial *kūniyā*, the sine quadrant;
4. The modern Western magnetic compass, present in the Ottoman Empire since the sixteenth century.

This striking diversity gives new and particularly clear evidence of what we have previously claimed: Osman's treatise is not a plain compilation of European books, but a fascinating witness to an attempt at creating a Euro-Islamic hybrid knowledge.

Very little is known about Osman Efendi himself. Our previous studies of *Hadiyyat al-Muhtadī* lead us to see him as a master in the art of compilation and intertextuality. To some extent, he was a brilliant player and manipulator. Any detailed analysis of any part of *Hadiyyat al-Muhtadī* brings to light enigmas and mysteries concerning its construction. The case of instruments is especially puzzling, since in almost all cases the precise written source used by Osman remains unknown. Is it possible that Osman had a direct basic practical knowledge of them and did not need to resort to any written source? The

32 The modern Arabic spelling is *bawṣala*.

33 See Ekmeleddin İhsanoğlu, *Science, Technology and Learning in the Ottoman Empire*, Aldershot, Ashgate, 2004, pp. xxxii–xxxiii.

only exceptions are the gunner's quadrants and the universal instrument for throwing of bombs: they are clearly borrowed from the books by Blondel and, incidentally, Bélidor. However, even in those cases, Osman's account goes far beyond what can be found in these books. Generally he does not hesitate to suggest modifications of instruments or to make changes to numerical examples. Did he really own or build the instruments he describes? Did he experiment with them himself? Our previous awareness of some of his tricks would make us suspect that he did not.

CHAPTER 3

Treatises on *Pergar-ı Nisbe* (the Sector) in Manuscript Collections in Turkey

Atilla Polat

1 Introduction

The sector, called *pergar-ı nisbe* in Ottoman Turkish, was one of the major mathematical instruments in use in Europe from the second half of the sixteenth century to the nineteenth century. It is an instrument based on one of the properties of similar triangles, that corresponding sides are in the same proportion. The sector had many beneficial uses. The principal use was to find the proportion between quantities of the same kind. The lines on its two legs serve to divide a given line into any number of equal parts or at any given proportion; to augment and diminish any plane figure or solid in a given ratio; to measure angles; or to inscribe a regular polygon in a given circle.[1] Thus, it had many practical applications in mathematics, geometry and surveying, and it also had military uses. As an example, given the diameter of a ball made of a given metal, it helped to find the diameter of another ball made of another metal, but having the same weight.

There are many texts describing the sector and its uses, especially in French, English and German, but until now there has been no study of the introduction of this instrument to Ottoman Turkey. Two sources of information about this have been studied for this paper. Firstly, there are inventories from the late eighteenth and early nineteenth centuries of the Library of the Imperial School of Engineering (*Muhendishane-i Berri-i Hümayun*) in Istanbul, but these only give the name of the instrument. Secondly, there are treatises on the description and the use of the sector held in various manuscript collections in Turkey. This chapter examines the nine manuscripts of the four treatises discovered in various libraries of Turkey in order to show how this instrument was introduced to and used by the Ottomans.

These nine manuscripts are dated from mid-eighteenth to early nineteenth centuries (see *Tab. 3.1* for a list of the manuscripts). Not all of them bear a title,

1 Jacques Ozanam, *Du compas de proportion expliqué et démontré …, Nouvelle edition corrigée et augmentée*, chez Jean Jombert, Paris, 1700, pp. 3–4.

 | DOI:10.1163/9789004412842_004

but were generally named *Risale-i Pergarü'n-Nisbet* (Treatise on the Sector). Only the author of the earliest treatise is known: it was composed in about 1750 in Istanbul by Mehmed Said Efendi (d. 1767), a teacher of mathematics and a timekeeper. The names of the translators or the copyists are only noted on a few manuscripts. Also, none of the manuscripts refer to the original work consulted. However, similarities and connections found between the manuscripts showed that they were all translated from the same source, Nicolas Bion's (1652–1733) well-known work on the mathematical instruments, *Traité de la construction et des principaux usages des instrumens de mathématique*.[2] Thus, Bion was one of the earliest French authors whose works were consulted or translated into Turkish in the eighteenth century, when Ottoman administrators undertook the modernisation of the army by importing technical knowledge through translations or by hiring technicians from Europe.

2 Classification of the Manuscripts

Nine manuscripts were found in catalogues of libraries in Turkey and digitised copies of them were obtained. Two nineteenth-century copies of a treatise on the sector, written by the Ottoman judge and mathematician Kuyucaklızade Muhammed Atıf (d. 1847), were discovered in libraries in Iran and Kuwait, but copies could not be obtained in the time available for this study.[3]

The libraries' own catalogues and the two-volume bio-bibliographical compendium of the mathematical works written during the Ottoman period

2 Nicolas Bion, *Traité de la construction et des principaux usages des instrumens de mathématique*, chez la veuve de Jean Boudot, Paris, 1709. Other editions of this work were printed in 1716, 1723, 1725 and 1752. The English translation by Edmund Stone was also consulted: Nicolas Bion, *The Construction and Principal Uses of Mathematical Instruments*, John Senex and William Taylor, London, 1723. See this work for an explanation of many of the terms used in this paper.

3 This treatise *al-Durrat al-Nafsiyya fi A'mal Bargar al-Nisba* (الدرة النفسية في اعمال بركار النسبة) (Wonderful Pearls on the Use of the Sector) is not mentioned in the Turkish catalogues consulted so far. It was written by Kuyucaklızade Muhammed Atıf in 1231 H (1815–1816 CE). One copy is at the Jaber al-Ahmad Central Library Manuscript Collection in Kuwait and the other is at the Shah Cheragh Library in Shiraz in Iran. For a brief biography and list of works of Kuyucaklızade see İhsan Fazlıoğlu, "Mehmed Atıf Efendi, Kuyucaklızade" in *Diyanet İslam Ansiklopedisi*, Türkiye Diyanet Vakfı, Istanbul, 1988–2013, 44 vols.: v. 28 (2003), p. 444; Ekmeleddin İhsanoğlu, Ramazan Şeşen *et al.*, *Osmanlı Astronomi Literatürü Tarihi*, IRCICA, Istanbul, 1997, 2 vols.: v. 2, pp. 588–590; and Ekmeleddin İhsanoğlu, Ramazan Şeşen and Cevat İzgi, *Osmanlı Matematik Literatürü Tarihi*, IRCICA, Istanbul, 2 vols.: v. 2 (1999), pp. 290–293.

TABLE 3.1 List of the manuscripts consulted

	Persons mentioned in the mss.	MS Title	Date	Language
A1	Mehmed Said (author/translator)	*Pergar-ı Nisbetin Aʿmali ve Bazı Bürhanı Beyanındadır*	about 1750 (1740s?)	Turkish
B1	Sadi (translator)	No title	second half of 18th *c.*	Turkish
B2		*Risale-i Şerh-i Pergarü'n-Nisbe*	second half of 18th *c.*	Turkish
C1	Muğlalı Mehmed (copyist)	*Risale-i Pergarü'n-Nisbet*	1775	Turkish
C2	Mehmed Sadık (copyist)	*Hendeseden Pergar-ı Nisbet*	1770–1780	Turkish
C3		*Risale-i Pergarü'n-Nisbet*	1793	Turkish
C4		No title	early 19th *c.*	Turkish
D1		*Risale-i Pergarü'n-Nisbet*	early 19th *c.*	Arabic and Turkish
D2		*Risale-i İstiʿmal-i Pergarü'n-Nisbet*	early 19th *c.*	Arabic and Turkish

(fourteenth to twentieth centuries)[4] did not provide enough information to establish the existing connections between the manuscripts. After examining all the manuscripts it was possible to put them into four groups. The manuscript kept in the Topkapı Palace Museum Library is a unique copy: it will be designated A1. The two manuscripts kept in the Millet Library and Istanbul University Rare Books Library are copies of the same work and will be designated B1 and B2 respectively. The two manuscripts in the Vahidpaşa Public Library in Kütahya, and the manuscripts in the Archaeology Museum Library in Istanbul and at Kandilli Observatory Library are copies of the same work and will be designated C1, C2, C3 and C4 respectively. Lastly, the manuscripts in the Süleymaniye Library and the Atatürk Library in Istanbul are based on the same work and will be designated D1 and D2 respectively (see *Tab. 3.2* for the distribution of the manuscripts in libraries in Turkey).

4 Ekmeleddin İhsanoğlu, Ramazan Şesen and Cevat İzgi, *Osmanlı Matematik Literatürü Tarihi*, IRCICA, Istanbul, 1999, 2 vols.

TABLE 3.2 Distribution of the manuscripts consulted in libraries in Turkey

	City	Library	Collection	MS number
A1	Istanbul	Topkapı Palace Museum Library	Hazine	1753/11
B1	Istanbul	Millet Library	Ali Emiri Riyaziye	82
B2	Istanbul	Istanbul University Rare Books Library	Turkish Manuscripts	T6846
C1	Kütahya	Vahidpaşa Public Library		740
C2	Kütahya	Vahidpaşa Public Library		728
C3	Istanbul	Archaeology Museum Library		602
C4	Istanbul	Kandilli Observatory Library	Turkish Manuscripts	82/2
D1	Istanbul	Süleymaniye Library	Yazma Bağışlar	741
D2	Istanbul	Atatürk Library	Muallim Cevdet	K.153/1

3 The Manuscripts

Manuscript A1 is bound with several other works by the same author, Mehmed Said Efendi (d. 1181 H/1767 CE). It forms folios 68r–71v of the whole volume and has 25 lines to the page with a page size of 12 × 21 cm. The name of the whole work is given in folio 2r as *Mecmuʿa-i Resail-i Saʿidiyye* (مجموعة رسائل سعيديه). There is no date on the manuscript, but examination of the other works in the volume shows that it was written in about 1750 by the same author.[5] This treatise is probably the earliest in the list. The author titled the treatise as *Pergar-ı Nisbetin Aʿmali ve Bazı Bürhanı Beyanındadır* (On the Uses and Some Proofs of the Sector). The term used for the sector in the text is *pergar-ı nisbet* (پركارنسبة) which is a literal translation of *Compas de proportion*.[6] There is neither an introduction listing the sources consulted and the advantages of the instrument nor any mention of the reasons for writing about it. The author starts by giving the number of the 'lines' on each leg of the instrument: three on one face and

5 The latest date given in the whole volume is 1163 H (1749–1750 CE) on fol. 57r, but Mehmed Said Efendi might have prepared the work at an earlier date and written this unfinished copy in about 1750.

6 The terms used for the sector in the various manuscripts vary slightly. As this was a newly introduced instrument, the authors coined terms having same meaning but linguistically they used different compound expressions.

three on the other. Then he gives a short description of the first line, the "Line of Equal Parts" (*aksam-ı mütesaviye hattı*), and explains its seven uses. After that he describes the second line, the "Line of Planes" (*hattü's-sütuh*) which has six uses, but the text ends unexpectedly in the middle of the third use without any explanation.

Comparison of Mehmed Said Efendi's text and the section devoted to the Sector in Nicolas Bion's *Traité* shows that Mehmed Said Efendi started to translate the chapter on the use of the sector from Bion with some comments and additions, possibly in order to clarify the subject for the reader (*Fig. 3.1*).

Mehmed Said Efendi had a traditional *madrasa* education, and was a scholar well-versed in the mathematical sciences. He taught geometry at the Corps of Salaried Bombardiers (*Ulufeli Humbaracı Ocağı*), newly established within the Imperial School of Engineering under the supervision of the French general Claude Alexandre de Bonneval (1675–1747), alias Ahmed Pasha the Bombardier. The Corps was a short-lived institution and fell into disuse in 1738, three years after its launching. Mehmed Said also served as the timekeeper (*muwaqqit*) at the Laleli Mosque until his death. He wrote on astronomy, geometry, surveying and scientific instruments in both Arabic and Turkish. In 1737, he invented a geodetic instrument which he named *müsellesiye* (derived from *müsellesat,* trigonometry in Arabic) and which was a combination of a sine quadrant (*rubʿ-ı müceyyeb*) and two arms with two sights each. He presented his work to Sultan Mahmud II (r. 1730–1754). The quadrant was widely used by the Ottomans, especially for astronomical measurements, and it seems that he tried to modify the traditional instrument for calculating remote distances. The instrument was used mostly by Ottoman gunners to calculate the distance to the target.[7] Although his translation from Bion is dated about 1750, it is possible that he came across the *Traité* earlier, at the time he taught geometry at the Corps of Salaried Bombardiers, from 1735 to 1738. This suggests that the introduction of the sector may be related with the teaching of geometry to the gunners at the Imperial School of Engineering.

Manuscript B1 has 13 folios with 19 lines to the page, and a page size of 11 × 16.8 cm; written area 6 × 12.7 cm. It has no title and no date. The terms used for the sector are *pergar-ı nisbiyy* (پرکار نسبیّ) and *mikyas* (مقیاس), an Arabic term used for measure in general. There is no sketch or illustration of the instrument,

7 For a detailed biography of Mehmed Said Efendi see Mustafa Kaçar and Atilla Bir, "Ottoman Engineer Mehmed Said Efendi and His Works on a Geodesical Instrument (Müsellesiye)", in E. İhsanoğlu, K. Chatzis and E. Nicolaidis (eds.), *Multicultural Science in the Ottoman Empire*, Brepols Publishers, Turnhout (Belgium), 2003, pp. 71–87. For the list of works of Mehmed Said Efendi see İhsanoğlu, Şeşen *et al.*, *Osmanlı Astronomi*, cit. (n. 3), v. 2, pp. 468–461, and İhsanoğlu, Şeşen and Cevat, *Osmanlı Matematik*, cit. (n. 3), v. 1, pp. 212–214.

FIGURE 3.1 Mehmed Said Efendi's treatise on the sector. The diagram on the left shows how to use the sector with compasses
MANUSCRIPT A1; ISTANBUL, TOPKAPI PALACE MUSEUM, MS HAZINE 1753/11, FOL. 68R

but folio 2r is left empty, possibly for an illustration. In folio 13v there are two names, Sa'di for the translator and Veliyyüddin for the copyist, about whom we could not find any information. The treatise begins with a twelve-line preface which includes a short prayer. In the preface it is mentioned that the sector was invented in *eyadi-yi efrence*, a phrase which was used to describe European countries. The translator says he translated a book on the sector because it had many wonderful uses.[8] The treatise comprises an introduction in which the lines on the sector are described briefly, and six sections on the uses of the sector.[9]

Manuscript B2 has 11 folios with 21 lines per page and a page size of 13.2 × 22.8 cm; written area 8.5 × 17.5 cm. It has been copied around mid-eighteenth century. The title on folio 1r reads *Risale-i Şerh-i Pergarü'n-Nisbe* (رسالهٔ شرح پرکار النسبه; A Commentary on the Sector). There is no date and no mention of the translator or copyist. The layout of this copy is more elaborate when compared with the others and suggests that it might have been prepared for an eminent person. There is an illustration of a sector on folio 2r, but there are some diagrams missing from the manuscript (*Fig. 3.2*).

Manuscript C1 has 24 folios[10] with 17 lines per page and a page size of 17.5 × 11.5 cm; written area 12.5 × 5.5 cm. The title is given as *Risale-i Pergarü'n-Nisbet* (رساله پرکار النسبة; Treatise on the Sector) on folio 1r. Next to the title there is a transliteration of the French *compas de proportion* which reads *kumpas dö puruporsyum* (قُومپآس دُو پُرُوپُرسْيُوم). The term used for the sector in the text is *pergar-ı nisbet* (پرکار نسبت). In folio 24v the copyist gives his name as Muğlalı Mehmed Efendi. He writes that he finished the treatise in the *madrasa* of Rüstem Paşa in 1189 H (1775–1776 CE).[11] The treatise is bound with a work

---

8 "Eyadi-yi efrencede zuhur iden pergar-ı nisbiyy, san'at-ı 'acibeyi müştemil oldıgına bina'en terceme olınup", Istanbul, Millet Library, Ali Emiri Riyaziye MS 82, fol. 1v.

9 The chapter entitled "Des usages de compas de proportion" (Of the Use of the Sector) in Bion, *Traité*, cit. (n. 2), p. 29 ff., is divided into seven sections but the translator of B1 did not translate the section on the construction and use of other kinds of gauging-rods.

10 The folios are unnumbered except the last one. Here there are two numberings: ٢٤ (24) in arabic numerals and 23 in Latin script. We relied on ٢٤ (24) and the numbering of the treatise will be considered as fols. 1r–24v.

11 An author named Müftizade Palabıyık Mehmed Efendi (d. 1804) is mentioned in İhsanoğlu, Şeşen and Cevat, *Osmanlı Matematik*, cit. (n. 3), v. 1, pp. 261–263. He was from the city of Muğla, as was the copyist of C1. He was educated in a *madrasa* and was a member of the *ulema* (scholars') class. Müftizade Palabıyık Mehmed Efendi was known for his deep knowledge of the mathematical sciences. He taught at the newly-established naval engineering school. There is thus a possibility that the copyist of C1 and Müftizade Palabıyık Mehmed Efendi are the same persons but there is not enough evidence to be certain.

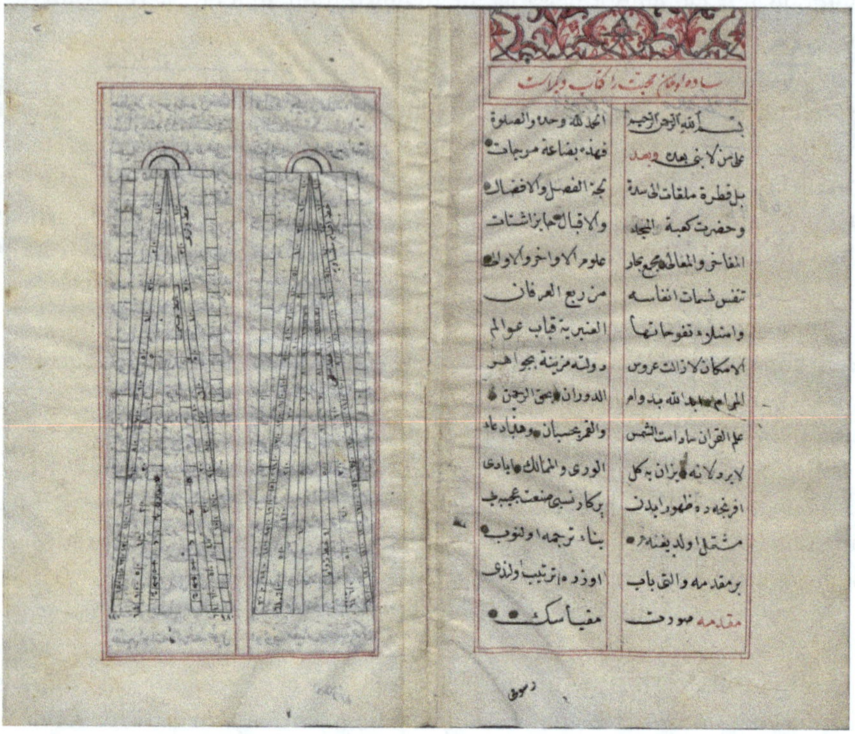

FIGURE 3.2 *Risale-i Şerh-i Pergarü'n-Nisbe*
MANUSCRIPT B2; ISTANBUL UNIVERSITY RARE BOOKS LIBRARY, MS T6846, FOLS. 1V–2R

on geometry which was copied by Mehmed Emin Vahid Efendi (d. 1828), the founder of the Vahidpaşa Public Library.

The treatise does not begin with prayer. Instead, the anonymous author gives brief information on the need for a treatise on the sector. He describes the sector as a curious invention (*acibü'l-ihtira*ʿ) and he adds that the construction and the use of the instrument depend on well-known geometric principles. He explains that he had written the treatise because he had never seen any treatise in Arabic, Persian or Turkish describing the use of this instrument, which had many beneficial uses, and he wanted to extend its use to Islamic cities.[12]

12 Kütahya, Vahidpaşa Public Library, MS 740, fol. 1v. The author does not say that he is relying on a European source or that he is making a translation.

After this preface the text begins with an introduction describing the lines on the sector followed by six sections on the uses of the sector.[13]

Manuscript C2 comprises folios 87r–106v in a volume of works on astronomy, timekeeping, geometry and surveying. It has 21 lines per page and a page size of 21.5 × 15.5 cm; written area 15.5 × 9.5 cm. There is no date and neither the author nor the copyist are named in the manuscript. On the first folio of the volume, the title of manuscript C2 is given as *Hendeseden Pergar-ı Nisbet* (هندسه دن پرکار نسبة) which means roughly "the sector, a geometrical tool". To determine the name of the copyist the first manuscript in the volume was examined.[14] Comparison of the handwriting shows that the two works were written by the same copyist. He gives his name in the colophon of this first treatise as Mehmed Sadık bin Mehmed Emin Cihangiri. The estimated range of the dates of the other works in the volume suggest that manuscript C2 was probably copied in about 1770–1780. Marginal notes and errors in the texts suggest that C1 is somewhat earlier than C2, and that C2 was copied from C1 with some corrections. Within the text of C2 there are various geometric diagrams. An illustration of the two sides of a sector is on folio 101v, followed by plates of elementary geometric figures (folios 102r–105r, folio 106v), some of which have no direct relation to the text about the sector, but they confirm the connection of the manuscript to Bion's *Traité*. These figures are identical to those given in the first seven plates of the *Traité*. There is a two-page text on folios 105v–106r about the use of the compass for nautical purposes (*Fig. 3.3*).

Manuscript C3 has 21 folios with 21 lines per page and a page size of 15 × 21.7 cm; written area 7 × 15.5 cm. The date copying ended was 27 Cemaziyelevvel 1208 H (31 December 1793 CE). The copyist is unknown and the title is given as *Risale-i Pergarü'n-Nisbet* (Treatise on the Sector) on folio 1v. Some of the scribal misreadings and errors in the text are similar to the ones in C2, which suggests that C3 was probably copied from C2, but C3 does not include the figures unrelated to the sector which are included in C2. Manuscripts C1 and C2 include only an untitled table listing seventeen metals and their weights per cubic

13 Like manuscript B1, C1, C2, C3 and C4 all lack the section on the construction and use of other kinds of gauging-rods, which was included by Bion.

14 The first manuscript, Kütahya, Vahidpaşa Public Library, MS 726, is a text on sundials written by the astronomer Taqi al-Din (1526–1585). The colophon is on fol. 72r. This treatise is dated as 14 Safer 1191 H (24 March 1777 CE). There is no information about the writer but the treatises in Kütahya Vahidpaşa Public Library which he compiled or copied show that he had an interest in astronomy and scientific instruments. Mehmed Sadık bin Mehmed Emin Cihangiri must not be confused with Mehmed/Muhammed Sadık Cihangiri (d. 1812), who was the chief astrologer of his time. For the latter see İhsanoğlu, Şeşen *et al., Osmanlı Astronomi*, cit. (n. 3), v. 2, pp. 561–562.

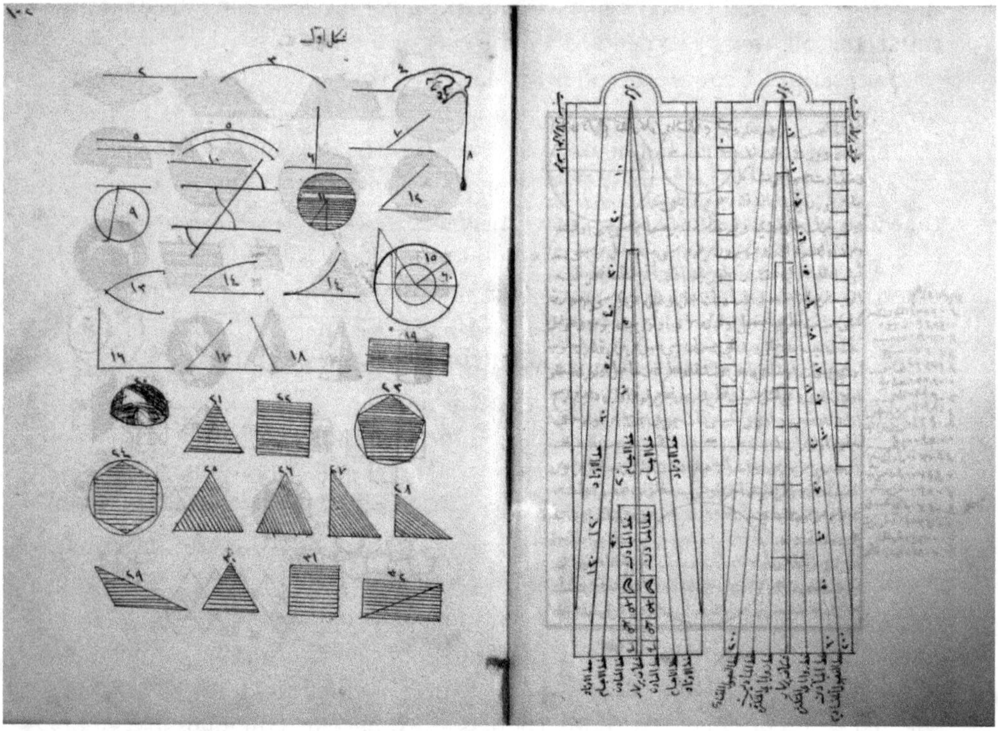

FIGURE 3.3 Manuscript C2
KÜTAHYA, VAHIDPAŞA PUBLIC LIBRARY, MS 728, FOLS. 101V–102R

foot, copied from the *Traité,* but C3 includes additionally tables for the "Line of Planes" (*Hattü'l-meyadin*), "Line of Polygons" (*Hatt-ı zevaya-yı mütekessire*), "Line of Chords" (*Hattü'l-evtar*), "Line of Solids" (*Hattü'l-ecsam*) and "Line of Metals" (*Hattü'l-maadin*), all from Bion.[15]

Manuscript C4 comprises folios numbered 62v–73v in a volume of works on logarithms, navigation and mathematics. It has a page size of 14 × 21.5 cm; written area 9.2 × 18 cm. There is no title or date and the copyist is unknown.[16] The layout of the treatise and the tables included at the end resemble manuscript C3. It includes all the tables in C3. The copyist of C4 must have been more

15 See Bion, *Traité*, cit. (n. 2), pp. 31–41 for an explanation of these terms.

16 Istanbul, Kandilli Observatory Library, Turkish Manuscripts, MS 82/2, fol. 1r includes a short note which reads "*merhum Hoca Tahsin Efendi'nin mecmuʿasıdır bi-hattihi*". This tells us that the volume belonged to the late Hoca Tahsin Efendi (1811–1881) and was written by him. Hoca Tahsin Efendi was one of the men who contributed to the introduction of modern astronomy into Turkey. There is no certainty about the accuracy of this note, and C4 might have been written by another scribe.

careful than the copyist of C3, because he has corrected misreadings and errors made in C3.

Manuscript D1 has 32 folios with 23 lines per page and a page size of 20.4 × 21.7 cm; written area 14.5 × 7.5 cm. The date and the author are unknown.[17] The title is given in folio 1r as *Risale-i Pergarü'n-Nisbet* (Treatise on the Sector). The main difference between this treatise and the others is that it is the only translation of Bion which includes the two chapters devoted to the sector. The other treatises investigated so far only deal with the uses of the sector. Another feature of D1 is the language used. The text is written in both Arabic and Turkish. From the beginning of the manuscript to the end of the construction of the "Line of Planes" (folios 1r–4v) the text is in Turkish but the titles of chapters and their first sentences are given in Arabic. From folio 4v up to the end, the text is written in Arabic. Although the tables given by Bion are included (though some are incomplete), all the figures are missing.

Manuscript D2 has 26 folios with 27 lines per page and a page size of 16 × 22 cm; written area 8.5 × 15.8 cm. The date and the author are unknown. The title is given in folio 1b as *Risale-i İsti'mal-i Pergarü'n-Nisbet* (Treatise on the Use of the Sector). It is bound with two other treatises related to surveying and geometry. This manuscript is a copy of the same treatise as D1 but contains only the chapter on the use of the sector. At the start of the manuscript the short descriptions of the "Lines" are given in Turkish (folios 1v–2r), the rest of the work is in Arabic, like D1. Unlike D1, D2 includes an illustration of the sector and the figures related to the sector. However, it has only the table listing the seventeen metals and their weights per cubic foot which is also given in C1, C2, C3 and C4 (*Fig. 3.4*).

17 There is a small possibility that D1 might have a connection with Mustafa Sıdkı Efendi (d. 1769), who copied and compiled a large number of scientific treatises. It is known that he had translated a chapter from Bion's *Traité* on the construction and use of an instrument showing the eclipses (see *Section 5* below). This instrument was invented by Phillippe de La Hire (1670–1718). Bion wrote a chapter on this instrument with the title *De la construction et usage d'une machine qui montre les éclipses, tant du soleil que de la lune, les mois et les années lunaires avec les épactes*; Bion, *Traité*, cit., (n. 2), pp. 229–236 (bk. 6, ch. 4; different pagination in later editions), and Bion, *The Construction*, cit. (n. 2), pp. 168–171. A copy of Mustafa Sıdki Efendi's translation with the title *Devair-i İctima' ve İstikbalin Resm u İsti'mali* is held in Edirne, Selimiye Library, MS 560/4, fols. 64r–72v. It is noteworthy that the handwriting of D1 resembles that of this manuscript. For Mustafa Sıdkı Efendi's translation see Feza Günergun, "The Ottoman Ambassador's Curiosity Coffer: Eclipse Prediction with de La Hire's 'Machine' Crafted by Bion of Paris", in F. Günergun and D. Raina (eds.), *Science Between Europe and Asia*, Springer, Dordrecht, 2011, pp. 103–123.

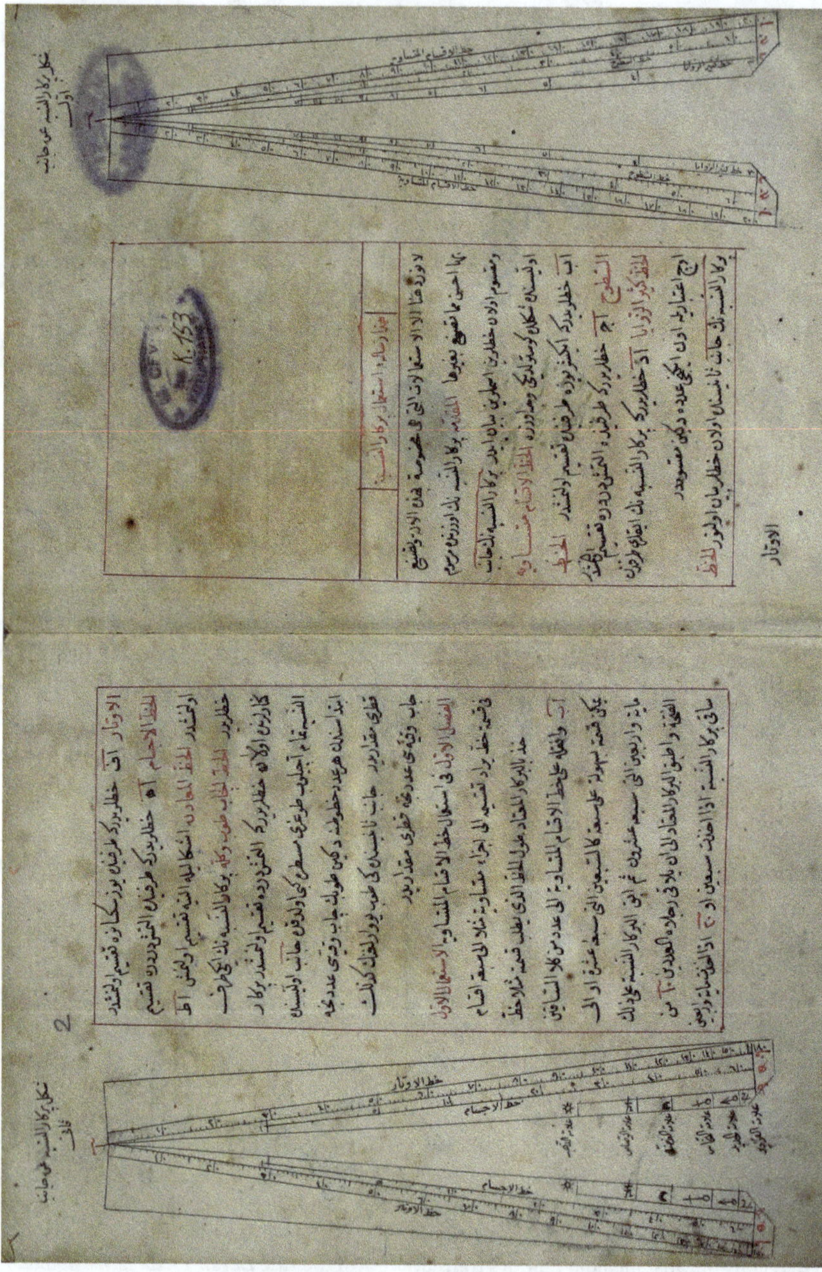

FIGURE 3.4 Manuscript D2
ISTANBUL ATATÜRK LIBRARY, MUALLIM CEVDET MANUSCRIPTS, MS K.153/1, FOLS. IV–2R

4 The Language Used in the Manuscripts

The manuscripts investigated so far show that the knowledge of the sector was transferred to Ottoman Turkey mainly through Nicolas Bion's *Traité*. There are four distinct translations. The author of A1 seems to have been preparing a commentary of Bion's two chapters, and added extra information for readers unfamiliar with the instrument, but as mentioned above, it is an incomplete treatise. The translator of D1 and D2 mostly followed Bion's text. The translator of C1, C2, C3 and C4 mostly followed the order and headings given by Bion but made minor changes during the translation. The translator of B1 and B2 shortened the text given by Bion, and this makes it the least intelligible of the four translations. A detailed comparison of these manuscripts would provide information about Turkish scientific terminology. *Table 3.3* compares the subtitles of the seventh use in the section devoted to the use of the line of equal parts.

All except B1 and B2 translate "the circumference of a circle". In B1 and B2, only *da'ire* (circle) is used, "*circonférence*" is omitted, and this does not mean the same thing. In A1, "*circonférence d'un cercle donné*" is translated correctly as "*muhit-i da'ire-i ma'lume*", whereas in the C and D manuscripts only "*circonférence d'un cercle*" is translated.

TABLE 3.3 A comparison between the title of *Usage VII* in Bion's *Traité* and its English, Ottoman Turkish and Arabic translations

Bion's *Traité*	Usage VII—Trouver une ligne droite égale à la circonférence d'un cercle donné.
Edmund Stone's English translation	Use VII—To find a right line equal to the circumference of a given circle.
A1	Amel-i sabi'—muhit-i daire-i ma'lumeye müsavi bir hatt-ı müstakim tahsil.
B1, B2	Amel-i sabi'—bir da'ireye müsavi hatt-ı müstakim ihdası beyanındadır.
C1, C2, C3, C4	Yedinci amel—bir da'irenin muhitine müsavi takriben bir hatt-ı müstakim bulmak.
D1, D2	al-isti'māl al-sābi'—li-kay naḥṣul khaṭṭan mustaqīman musāwiyan li-muḥīṭ dāira.

5 How Did Bion's *Traité* Reach Turkey?

Bearing in mind the flux of individuals (diplomats, traders, travellers, artisans, technicians, officers) from France to the Ottoman Empire in the eighteenth century, Bion's *Traité* might have reached Istanbul in various ways. As none of the nine manuscripts consulted for this paper mentioned how the source was obtained, it is hard to determine which route the *Traité* took. The fact that the earliest translation was made by a teacher of mathematics at the Corps of Bombardiers created under the supervision of a French army officer suggests that *Traité* was already available to the teachers of the Corps in late 1730s. The *Traité* may have been taken to Istanbul by Mehmed Said Efendi (d. 1761),[18] who was appointed Ottoman ambassador to France in 1741–1742. He was interested in scientific instruments and visited the shop of Nicolas Bion (1652–1733), which was run at that time by his son Jean-Baptiste Nicolas Bion. Mehmed Said Efendi ordered and purchased some instruments from Bion's workshop. Although we do not know which instruments he bought, it is likely that he owned an eclipsarium of the de La Hire type, and the "*petit livre*" published by Nicolas Bion to describe its use.[19]

Some years after he returned to Istanbul, the ambassador asked Mustafa Sıdkı Efendi (d. 1769), a bureaucrat competent in mathematics, for a translation of a text on the construction and use of an eclipsarium. Mustafa Sıdkı finished the translation in 1161 H (1748 CE). He states that the text he translated was from Nicolas Bion, and it was about an instrument invented by the Frenchman Monsieur de La Hire.[20] The Turkish translation is an abridged version of the chapter V of the *Traité*, introducing the eclipse calculator. Mustafa Sıdkı either used the manual brought by Mehmed Said Efendi or the related chapter in Bion's *Traité*.[21] In either case, this shows that a work or works by the French King's instrument maker Nicolas Bion reached Ottoman Turkey by about the middle of the eighteenth century, because of the interest of an Ottoman statesman. There could be other routes used for the translations which are still to be identified.

18 The ambassador Mehmed Said Efendi (d. 1761) and Mehmed Said Efendi (d. 1767), the author of the manuscript A1 introducing the sector are not the same person.

19 Günergun, op. cit. (n. 17), p. 113.

20 *Ibid.*, pp. 110–111.

21 Mustafa Sıdkı, op. cit. (n. 17); and Günergun, op. cit. (n. 17), p. 115.

6 Was the Construction or Use of the Sector of More Interest?

Of the nine manuscripts only one (D1) includes both the Chapter 1 (on the construction) and Chapter 2 (on the use) of the *Traité*. The other eight manuscripts describe only the uses of the sector taken from Chapter 2. However, C3, C4 and D1 contain most of the tables given in the 'construction' chapter of the *Traité*.

Because D1 is a translation of both chapters it is not possible to draw a definitive conclusion about the objectives of the authors and translators. This analysis of the manuscripts supports the thesis that there was more interest in the chapter on the uses of the sector because it met the practical needs.

7 Other Traces of the Sector in Ottoman Turkey

As well as the manuscripts investigated for this paper there are other sources of information about the sector. The earliest of these is a manuscript entitled *Fünunü'l-Harb* (Warfare Techniques).[22] This is translated from *Commentarii Bellici* (Vienna, 1718), part of Raimondo Montecuccoli's (1609–1680) *Memorie della Guerra*, but with four additional chapters borrowed from other European authors. The last of these has the title *Eşkalü'l-Alati'l-Hendesiyye* (On Geometrical Instruments). This chapter is translated from the first volume of Allain Manesson Mallet's (1630–1706) *Les travaux de Mars ou l'art de la guerre*.[23] One of the instruments described here is the sector, and it is translated as *pergal-i tenasüb*, which resembles *pergar-ı nisbe*.[24] The date of the translation is possibly around the middle of the eighteenth century.

Another late example of the sector in Ottoman Turkey can be found in the first volume of Hoca İshak Efendi's (1774–1835) *Mecmua*ʿ*-i Ulum-ı Riyaziye* (Compendium of Mathematical Sciences). İshak Efendi was a professor of mathematics and engineering in the Imperial School of Engineering (*Mühendishane*). He translates the sector as *pergar-ı nisbiyy* and gives a very short description and an illustration of it.[25]

22 "*Fünunü'l-Harb*", Istanbul, Istanbul University Rare Books Library, MS T6854.

23 Allain Manesson Mallet, *Les travaux de Mars ou l'art de la guerre divisé en trois parties*, Denys Thierry, Paris, 1684–1685, 3 vols.: v. 1 (1685), pp. 28–29 and 39–42.

24 "*Fünunü'l-Harb*", cit. (n. 22), fol. 235r.

25 Hoca İshak Efendi, *Mecmua*ʿ*-i Ulum-ı Riyaziye*, Bulak Matbaası, Cairo, 1841–1845, 4 vols.: v. 1, p. 450 and pl. 6.

So far no actual sector from Ottoman times has been found. A 1784 inventory of the Library of the Imperial School of Engineering includes two sectors.[26] There is another sector in the list of 1801 which includes the large collection of scientific books and instruments of the late Ebubekir Râtib Efendi (d. 1799), whose collection was bought for the *Mühendishane,* and added to its inventory.[27] The question how the sector was used in the engineering school remains unanswered for now.

There is one last clue in an Ottoman dictionary. A short entry with the title *Pergarü'n-nisbe* is included in *Lugat-ı Tarihiyye ve Coğrafiyye* (Dictionary of History and Geography).[28] In it the usefulness of the instrument is emphasized, and it is remarked that there was a sector in every case of drawing instruments (*ustunc takımı*). As the sector was not a popular instrument in the nineteenth century in Turkey, there is a possibility that this statement has been translated from a European source.

8 Conclusion

Most of the Ottoman treatises on scientific instruments compiled prior to the nineteenth century are in manuscript form. Based on the nine manuscripts about the sector found so far, the only firm conclusion is that the source of the translations is Nicolas Bion's *Traité*. To assess the dissemination of knowledge about the sector through Ottoman Turkey and the use of the instrument is not yet possible. There are two possibilities, either more evidence will be found to bridge the gaps in this study or we will have to accept that the Ottomans were not generally interested in this instrument. The reason for this indifference, if any, could be that Ottoman mathematicians or engineers knew mathematical methods that allowed them to solve the problems without having recourse to the sector. Despite this, the manuscripts investigated here show how the efforts of a small circle of learned men contributed to the introduction of scientific instruments from Europe to the Ottoman Empire.

26 Mustafa Kaçar, *Osmanlı Devleti'nde Mühendishanelerin Kuruluşu*, unpublished PhD dissertation, Istanbul University, 1996, p. 65.

27 Kemal Beydilli, *Türk Bilim ve Matbaacılık Tarihinde Mühendishane: Mühendishane Matbaası ve Kütüphanesi (1776–1826)*, Eren Yayıncılık, Istanbul, 1995, pp. 376 and 409.

28 Ahmet Rıfat, *Lugat-i Tarihiyye ve Coğrafiyye*, Mahmud Bey Matbaası, Istanbul, 1299 H (1881–1883 CE), 7 vols.

CHAPTER 4

Measuring Altitudes with an *Alla Franca* Instrument

The Ottoman Engineer Feyzi's Treatise on the Portable Sextant

Feza Günergun, Gaye Danışan and Atilla Polat

1 Introduction

Measuring the Sun's altitude was of primary importance for the Muslim populace of the Ottoman lands, eager to establish the prayer hours in a precise way. The most common instrument for this measurement was the portable wooden quadrant, known as *rubʿu tahtası* or *rubʿu* in brief. The quadrant also helped to orient the user towards the *qibla*, the direction of Mecca, a condition of the daily prayers. One side of the quadrant, called the *muqantarat*, bears the projections of *almucantars*, the imaginary circles parallel to the horizon (parallels of altitude), and serves to measure the altitude of celestial bodies, especially the solar altitude from which time can be determined. The other side, the *mucayyab*, known as the sine quadrant, bears a graph that serves to solve various trigonometrical problems and to perform basic arithmetical operations. The *mucayyab* side also bears other markings which allow it to be used to determine the time. An astrolabe would also serve these purposes, but the large number of users' manuals for quadrants and the number of instruments both in museums and private collections prove that Ottomans in Turkey favoured the quadrant rather than the astrolabe. Light to carry and convenient to use, wooden quadrants were cheaper and easier to produce than quadrants or astrolabes made of metal. In short, the quadrant, together with the astronomical tables (*zijes*), was the main instrument of the timekeeping office (*muwaqqitkhane*). Timekeepers were also involved in the design and construction of the instrument. Quadrants were probably used by Ottoman seamen to measure the angle between a heavenly body and the horizon.[1] The fact that portable

1 The Ottoman mariner Seydi Ali refers to the sine quadrant and the astrolabe when writing about altitude measurements; Seydi Ali Reis, "*Kitabü'l-Muhit*" (Book of the Ocean), Istanbul, Süleymaniye Library, MS Nuruosmaniye 2948, fol. 45r. The seventeenth-century Ottoman historian Hezarfen Hüseyin Efendi's complaint about the decrease in number of mariners able to take altitudes with quadrants and astrolabes implies that these instruments were used at sea by the Ottomans prior to Hezarfen's time; Hezarfen Hüseyin Efendi, *Telhisü'l-beyan fî*

 | DOI:10.1163/9789004412842_005

quadrants figure in a list of mathematical instruments made in the workshops established around the Barracks of the Bombardiers[2] and the Imperial School of Engineering (*Mühendishane-i Berri-i Hümayun*) in Istanbul during the late eighteenth century implies that this instrument was also used by engineers and gunners to measure heights and distances.

The introduction of mechanical clocks into Turkey from late sixteenth century onwards did not overturn the supremacy of the quadrant, which was probably regarded as a more reliable instrument than a clock that needed frequent adjustment, if not repair. Although mechanical clocks were present in the principal timekeeping offices in the second half of the nineteenth century, and pocket watches flooded the market, quadrants were still produced and pamphlets on their use were still written. Professors at the Military Academy (established in 1834) and military high schools compiled Turkish texts describing the measurements and calculations to be made with quadrants, and also taught these operations in the classroom.[3] The chief timekeeper/astrologer (*müneccimbaşı*) Hüseyin Hilmi Efendi (1858–1917) designed (*tersim*) 41 portable quadrants during his lifetime.[4] The last detailed study of their design and construction techniques was written in 1921, by Major Ahmed Ziya Akbulut (1869–1938), a professor of mathematics and astronomy at military high schools, and the last timekeeper of the Empire.[5]

It is not clear when and how the portable sextant was introduced into Ottoman Turkey.[6] Bearing in mind that sextants are mostly used in celestial

---

kavanin-i Al-i Osman (trs. Sevim İlgürel), Türk Tarih Kurumu Basımevi, Ankara, 1998, p. 160. See also Gaye Danışan Polat, "A treatise by the sixteenth-century Ottoman Admiral Seydi Ali Reis on Rub-i müceyyeb (sine quadrant)", in Dejanirah Couto, Feza Günergun and Maria Pia Pedani (eds.), *Seapower, Technology and Trade—Studies in Turkish Maritime History*, Piri Reis University Publications/Denizler Kitabevi, Istanbul, 2014, pp. 337–341; Gaye Danışan Polat, *16. yüzyılda Osmanlılarda Deniz Astronomisi ve Astronomi Aletleri* (*Nautical Astronomy and Astronomical Instruments of the Ottomans in the 16th Century*), unpublished PhD thesis, Istanbul University, Istanbul, 2016, pp. 270–279.

2 Kemal Beydilli, *Türk Bilim ve Matbaacılık Tarihinde Mühendishane, Mühendishane Matbaası ve Kütüphanesi (1776–1826)*, Eren Yayıncılık, Istanbul, 1995, pp. 353–355.

3 Ekmeleddin İhsanoğlu, Ramazan Şeşen *et al., Osmanlı Astronomi Literatürü Tarihi*, IRCICA, Istanbul, 1997, 2 vols.: v. 2, p. 633 (Yunus Efendi), p. 687 (Vidinli Tevfik Efendi), p. 699 (Ali Sulhi Bey).

4 Hüseyin Hilmi, Ahmed Ziya [Akbulut], *Takvim-i Sal, 25 Şaban 1343 [1925]–3 Ramazan 1344 [1926]*, Matbaa-i Kırımi Yusuf Ziya, Istanbul, 1343 H (1925 CE), p. 3.

5 Ahmed Ziya bin Rıza [Akbulut], *Rubu Tahtası Kullanım Kılavuzu* (ed. and tr. by M. Şinasi Acar, Atilla Bir, Mustafa Kaçar), Biryıl Kültür Sanat Ltd., Istanbul, 2010.

6 In order to observe Venus, and relying on the 10th book of the *Almagest*, the astronomer Taqi al-Din (1526–1585) designed a kind of three-dimensional sextant that he named *mushabbaha bi'l-manatık*. See Mustafa Kaçar, M. Şinasi Acar and Atilla Bir, *Takiyüddin'in Gözlem*

navigation, one expects that its introduction was through the Ottoman navy. However, there is no evidence for this. An order made by Baron de Tott (1733–1793) in 1775 for the purchase of measuring instruments for the School of Geometry (*Hendesehane*) at the Maritime Arsenal on the Golden Horn includes twelve sea octants and other instruments, but no sextants.[7] The absence of these instruments in the early nineteenth-century inventory lists of the school implies that they were not purchased. Baron de Tott left Istanbul and returned to France in 1776. He relates in his *Memoirs* that before his departure his students embarked on the ship to say farewell, and asked him to give them a parting lesson. They sailed with him for about two leagues (about 10 kilometres) conversing on various topics before finally leaving the ship. One of his students had a sextant with him that he used to measure the altitude of the Sun.[8] It is possible that de Tott taught the use of a sextant in the school, as well as geometry. Thus portable sextants, *alla franca* (European style) instruments, were available at the arsenal in the 1770s. Abbé Giambattista Toderini (1747–1814), who lived in Istanbul between 1781 and 1786, does not mention any sextants among the many instruments he noticed at the Engineering School of the Maritime Arsenal, but he does mention "*un très-bel octant anglois de Jean Hadley*".[9] It is possible that individuals (Ottomans or foreigners) who had close contacts with Europe were involved with its introduction.[10] Following

Araçları—Âlât-ı Rasadiyye li Zîc-i Şehinşâhiyye, Türkiye İş Bankası Yayınları, Istanbul, 2011, pp. 44–46, 78, 109 and 139. Taqi al-Din did not record the dimensions of this instrument. If the proportions of the instrument and the observer have been shown correctly by the illustrator on the miniatures, the frames of the instrument would have measured about 4 meters each.

7 For the list of books and instruments ordered from Paris via Comte de Saint-Priest (1735–1821), the French ambassador to the Ottoman Empire between 1768–1785, see Mustafa Kaçar, "Osmanlı İmparatorluğu'nda askeri teknik eğitimde modernleşme çalışmaları ve mühendishanelerin kuruluşu (1808'e kadar)", *Osmanlı Bilimi Araştırmaları* 2 (1998), pp. 69–137: 84.

8 Baron de Tott, *Memoirs of Baron de Tott*, G.G.J. and J. Robinson, London 1785, 2 vols.: v. 2, pp. 203–204.

9 [Giambattista] Toderini, *De La Littérature des Turcs* (French tr. by Antoine de Cournand), Poinçot, Paris, 1789, 3 vols.: v. 1, p. 164. The Italian original edition was printed in Venice in 1787. Among other instruments Toderini named "*un globe céleste qui marquoit les constellations & les étoiles de premiere grandeur, par des signes & des caracteres turcs en or sans figures, ouvrage du professeur; une sphere armillaire de métal faite à Paris, quelques astrolabes arabes, des cadrans solaires turcs & francs de plusieurs especes* [...], *différentes boussoles turques avec la correction & à Azimut, avec la chaise d'Irvin, le compas de Galilée, avec d'autres instrumens de marine*".

10 It is known that telescopes were presented to the Ottoman court as early as 1613, see Claudia Swan, "Birds of paradise for the sultan. Early seventeenth-century Dutch-Turkish encounters and the uses of wonder", *De Zeventiende Eeuw. Cultuur in de Nederlanden in*

the establishment of the Imperial School of Engineering by Sultan Selim III (r. 1789–1808) in 1795 to train engineers for the Ottoman Army, books and instruments were sent to its library by the Imperial Palace's treasury. Among the items sent were two portable sextants, a terrestrial globe, a celestial globe and an alarm clock.[11] How these instruments found their way to the Imperial Palace is unknown. Were they gifts from European diplomatic envoys or purchased by the Treasury for the Imperial School of Engineering when it was created in 1795? Did the Sultan have them procured for his personal interest? Whatever the reason is, it appears that two sextants were first available in the Palace and then in the engineering school at the turn of the nineteenth century. One of these sextants captured the attention of Feyzi, an engineer at the court of Selim III, and stimulated him to compose in 1220 H (1805–1806 CE) a treatise, the *Muhadarat-ı Feyzi* (Feyzi's discourse) which is probably the earliest Turkish account of the instrument.[12] Feyzi named it *alet-i irtifaiye* (the altitude instrument), *alet-i rasadiye* (the observational instrument) and also *alet-i sekstant* (the sextant instrument).

2 Feyzi Bey, the Engineer

The caption of the author's portrait (*Fig. 4.2a*) at the start of the treatise reads that Feyzi Bey was an engineer (*mühendis*) at the *Hane-i Hassa* (*Has Oda*, literally Imperial Chamber) of the *Enderun*[13] section of the Imperial Palace in Istanbul.[14] While religious scholars and judges were educated in the *madrasas*,

interdisciplinair perspectief 29, 1 (2013), pp. 49–63; DOI: http://doi.org/10.18352/dze.8464 (accessed 01 February 2019).

11 These instruments are mentioned in an archival note of 1804, see Beydilli, op. cit. (n. 2), pp. 279 and 281.

12 Bursalı Mehmet Tahir, *Osmanlı Müellifleri* (tr. by İsmail Özen), Meral Yayınevi, Istanbul, 1975 (first ed. 1915), 3 vols.: v. 3 pp. 260–261. A short introduction to the treatise has been published in Feza Günergun, *Pursuing Knowledge—Scientific Instruments, Manuscripts and Prints from Istanbul University Collections*, Istanbul Universitesi, Istanbul, 2016, p. 54. See also notes 19 and 20.

13 *Enderun* (Persian *Andarun*), was a term used to designate the "Inside Service" of the Imperial Household of the Ottoman Sultan. Thus it was used to denote the complex of officials engaged in the personal and private service of the Sultan. Palace Schools were also included therein. See V.J. Parry, "Enderun", in J. Bearman, T. Bianquis *et al.* (eds.), *Encyclopaedia of Islam*, Brill, Leiden and New York, 1954–2005^2, 12 vols. plus indexes: v. 2 (1991), pp. 697–698.

14 Feyzi is depicted wearing the costume and the cap (with earlocks) reserved for those in charge in the *Has Oda*. See İzzet Kumbaracılar, *Serpuşlar*, Türkiye Turing ve Otomobil Kurumu, Istanbul, no date, p. 12 and pl. 5, fig. 19.

personnel trained in the *Enderun* were appointed to administrative, military and technical posts. As those employed in the Sultan's private service were selected from among the members of *Has Oda*, one can assume that Feyzi was in the Sultan's inner circle. According to the date (1805–1806) of the treatise, he was in the inner palace during Sultan Selim III's reign (1789–1808). However, little is known about Feyzi's life in the palace.[15] A biography of him relates that he was a mathematician who authored a text on the use of the sextant and died in Kilitbahir on the Dardanelles while in charge of strengthening the fortifications. The inscription on a gravestone ascribed to him bears the date 1230 H (1814–1815 CE).[16] Although not officially documented, Feyzi may have had some connection either with the engineers at the Istanbul maritime arsenal or with the Imperial School of Engineering created by the Sultan in 1795 for his new armed forces, or with both. As will be mentioned later, Antoine de Juchereau, who wrote a note praising Feyzi's skill in mathematics and his treatise *Muhadarat,* was assigned by Selim III as an engineer to the Imperial School of Engineering.

3 Feyzi's Treatise on the Sextant: *Muhadarat-ı Feyzi*

As mentioned above, a European portable sextant kept at the treasury of the Imperial Palace caught Feyzi's attention. More than that, he developed an "affection for this observational instrument named sextant kept in a box" (*sekstant nam alat-ı rasadiyye dahil-i sanduka taaşşukum olub*).[17] This fascination and his mathematical knowledge prompted him to compile his *Muhadarat* describing its working principles and its use for measuring altitudes, and to provide numerous mathematical and astronomical tables needed to make the necessary calculations.

Two copies of *Muhadarat* are available in Istanbul libraries.[18] The copy at Istanbul University Rare Books Library[19] was acquired along with all the other

15 A search in the archival catalogues of the Topkapı Palace Museum and the Ottoman Archives of the Turkish Prime Ministry has not provided any information on Feyzi as yet.

16 Bursalı Mehmet Tahir, op. cit. (n. 12), v. 3, p. 261.

17 Feyzi, "*Muhadarat-ı Feyzi*", fol. 3r; see nn. 19 and 20.

18 The presence of two copies had been already mentioned in 1915; Bursalı Mehmet Tahir, op. cit. (n. 12), v. 3, p. 261. However, only the copy kept at Istanbul University Rare Books Library is referred to in E. İhsanoğlu, R. Şeşen *et al., Osmanlı Tabii ve Tatbiki Bilimler Literatürü Tarihi*, IRCICA, Istanbul, 2006, 2 vols.: v. 1, p. 147.

19 Feyzi, "*Muhadarat-ı Feyzi*", Istanbul, Istanbul University Rare Books Library, MS T6833, 70 folios, dated 1220 H (1805–1806 CE). Two plates with 5 and 9 figures respectively are appended to the text.

material in Abdulhamid II's (r. 1879–1909) private library at Yıldız Palace in 1924, a year after the proclamation of the Turkish Republic. In all likelihood the manuscript found its way into Yıldız Palace Library in the late nineteenth century when Abdulhamid II ordered the transfer of many manuscripts and books from the library at Topkapı to the Yıldız Palace, his new residence. This copy, which bears Feyzi's portrait and a note of appreciation written by the British engineer Antoine de Juchereau, seems to be the author's personal copy.

A second copy of the *Muhadarat* is in the Hüsrev Pasha Collection of the Süleymaniye Library.[20] Hüsrev Pasha (d. 1855),[21] a modernist Ottoman military officer, was trained in the *Enderun* like Feyzi, and they might have known each other. How this copy of *Muhadarat* found its way into Hüsrev Pasha's collection is not clear.[22] This copy does not bear the seal of Hüsrev Pasha, but the name of a certain Eyüb Pasha is written on the first blank folio. It is possible that it was made for Eyüb Pasha,[23] and then came into the possession of Hüsrev Pasha. On the same blank folio is written in pencil *Risale-i irtifa*, literally "treatise for altitude [measurement]". This second copy also includes the portrait of Feyzi with the sextant, but not the note written by Juchereau. The text and its pagination are identical to the copy kept in Istanbul University.

Did Feyzi keep the manuscript now in the Istanbul University Rare Books Library as his personal copy or present it to Selim III, the reigning Sultan, or

20 Feyzi, "*Muhadarat-ı Feyzi*", Istanbul, Süleymaniye Library, MS Hüsrev Pasha 233, 70 folios, dated 1220 H (1805–1806 CE). Two plates with five and nine figures respectively are appended to the text.

21 Mehmet Hüsrev Pasha, known as Koca Hüsrev Pasha, had taken part in the creation of Selim III's modern army *Nizam-ı Cedid* in 1793 and had fought in Egypt with the British against the French in 1801 when he was nominated Governor of Egypt. A keen supporter of military reforms, he was also appointed to high posts under Sultan Mahmud II (r. 1808–1839). He was twice appointed admiral-in-chief of the Ottoman Navy (1811 and 1822) and commander-in-chief to *Asakir-i Mansure-i Muhammediyye*, the new army that Mahmud II created after the abolition of Janissaries in 1826; see Halil İnalcık, "Hüsrev Paşa, Koca," in *TDV İslam Ansiklopedisi*, Türkiye Diyanet Vakfı, Istanbul, 1988–2013, 44 vols.: v. 19 (1999), pp. 41–45. During his term of office as admiral-in-chief, in 1830, Hüsrev Pasha sent his four young protégés to Europe to be trained in sciences and military arts, thus pioneering the sending of Ottoman students to Europe for education; Feza Günergun, "Science in the Ottoman world", in George N. Vlahakis *et al.*, *Imperialism and Science*, ABC Clio, Santa Barbara—Denver—Oxford, 2006, p. 103.

22 Hüsrev Pasha constructed a library to house his collection in 1255 H (1839–1840 CE) and endowed it for public use. The library was located in the Eyüp district, on the Golden Horn, Istanbul. An inventory of the collection which included about a thousand manuscripts and books has been published; see *Defter-i Kütüphane-i Hüsrev Paşa*, Mahmud Bey Matbaası, Istanbul, no date, also available online at https://archive.org/stream/defteriktbhooistauoft#page/62/mode/2up (accessed 4 February 2019).

23 Further research is needed to identify Eyüb Pasha.

to an Ottoman dignitary who later left it to the Palace Library? Presenting a work with a portrait of himself to dignitaries would have been considered impertinent at the time, unless the author was a member of the Sultan's family or a noble person living in the Palace. To our present knowledge no other Ottoman manuscript bearing the author's portrait has ever been found. It is not clear how the idea of letting himself be portrayed with the sextant occurred to Feyzi. Was Selim III's interest in portraits his motivation? Konstantin Kapıdağlı, the court painter, had portrayed Selim III in 1803–1804 with scientific instruments displayed in a showcase.[24] Thereafter the painter was commissioned by the Sultan to produce portraits of former Ottoman sultans, a project Kapıdağlı worked on between 1804 and 1806.[25] Bearing in mind that Feyzi completed his treatise in 1804–1805, this might have encouraged him to have himself portrayed with his instrument. Unfortunately, his portrait is not signed and might have been made by any artist in service in the court.

The seal (*Fig. 4.1a*) on the first blank page of the manuscript is enigmatic. It does not look like a seal that an Ottoman individual would use. Such a seal would usually bear the name of the individual and a date, all engraved in Arabic characters. As for institutional seals, the name of the institution is often written in full.[26] The seal on the *Muhadarat* bears only symbols: a crescent with an eight-pointed star and a crown-shaped cap. It is known that Selim III ordered the addition of such a star to the flags of the Ottoman army and navy.[27] This links the seal to the reign of Selim III, even though it does not bear a date. The crown-shaped cap, differs drastically from the 176 known examples of Ottoman headgear,[28] and resembles a crown similar to those worn by European rulers. Why were these figures coupled together and from where do they come? The crown and the crescent and star on Selim III's imperial seal (*Fig. 4.1b*)[29] with the Sultan's *tughra* and the Ottoman coat of arms seem to be

24 As far as we know Selim III is the only Sultan whose portrait includes scientific instruments, although only a few. Oil on canvas, Topkapı Palace Museum, İstanbul, TSM 17/30.

25 Wendy M.K. Shaw, *Ottoman Painting: Reflections of Western Art from the Ottoman Empire to the Turkish Republic*, I.B. Tauris, London, 2011, pp. 15–17.

26 The seal of the Engineering School's library bears the *tughra* of Selim III and reads "*Min Kütüb-i Mühendishane-i Hümayun der Hasköy 1216*" (From the Books of the Imperial School of Engineering at Hasköy, 1801–1802).

27 Feyzi Kurtoğlu, *Türk Bayrağı ve Ay Yıldız*, Türk Tarih Kurumu, Ankara, 1992, p. 111; Orhan F. Köprülü, "Bayrak", in *TDV İslam Ansiklopedisi*, cit. (n. 21): v. 5 (1992), p. 253.

28 İzzet Kumbaracılar, *Serpuşlar*, Türkiye Turing ve Otomobil Kurumu, Istanbul, [1979].

29 Selim III's imperial seal (*mühr-i hümayun*) bears the date 1213 H (1798–1799 CE) and is said to have been engraved in London. This seal and its copy are in Topkapı Palace Museum TSM 48/28. See İsmail Hakkı Uzunçarşılı, *Topkapı Sarayı Müzesi Mühürler Seksiyonu Rehberi*, Topkapı Sarayı Müzesi, Istanbul, 1959, p. 16.

FIGURE 4.1 *a.* The seal on the *Muhadarat-ı Feyzi* (1.8 cm × 1.6 cm; Istanbul University Rare Books Library, MS T6833, fol. 2a); *b.* Sultan Selim III's seal (4.9 cm × 5.7 cm; İsmail Hakkı Uzunçarşılı, *Topkapı Sarayı Müzesi Mühürler Seksiyonu Rehberi*, Istanbul, Şehir Matbaası, 1959, p. 16)

the source of inspiration for the seal on the *Muhadarat*. However, while the star of the imperial seal is six-pointed, that on Feyzi's treatise is eight-pointed. Also, the ornate wreath on Selim III's seal (4.9 cm × 5.7 cm) contrasts with the thinner wreath of the seal on the *Muhadarat*, probably because the latter seal (1.6 cm × 1.8 cm) is smaller. The question is to whom did the seal belong? Could Feyzi, eager to show that he was in the imperial service, have had such a personal seal made? This is not impossible bearing in mind that his treatise opens with his splendid portrait. Or, to speculate further, did it belong to Selim III who wished to have a smaller personal seal (a seal ring?) similar to those of previous sultans? On the other hand, it is known that when a member of the *Enderun* died, his or her belongings were included in the Sultan's possessions. Thus, if Feyzi died before Sultan Selim III, who died in 1808, it is possible that the treatise was given to the Sultan's collection and stamped with this small seal.[30] Further research is needed to find the seal's owner.[31] As for the simultaneous use of European and Ottoman symbols of power, one can

30 This depends on the assumption that Feyzi the engineer is the same person as the Feyzi the intendant who died in 1807. Both were sent to Gallipoli. See n. 68.

31 There is no seal similar to that on the *Muhadarat* in the Topkapı Palace Museum collections. It is also absent from the endowment seals on manuscripts. See Günay Kut, Nimet Bayraktar, *Yazma Eserlerde Vakıf Mühürleri*, Kültür ve Turizm Bakanlığı Yayınları, Ankara, 1964.

argue that Selim III, by adding a crown above the crescent and star, intended to upgrade it so that it was a full equivalent of a European symbol such as the monogram of King George III (GR with a crown over it), his contemporary and ally at the time.[32]

4 Sources and Layout of the *Muhadarat-ı Feyzi*

Feyzi does not mention the sources he consulted in compiling his text. Sextant users' manuals—if available—would have been among his sources. As the treatise includes numerical tables copied from Jacques Cassini (1677–1756), James Bradley (1693–1762), Nicolas Louis de Lacaille (1713–1762), and Jérôme Lalande (1732–1807), one wonders if Feyzi consulted their original works or other astronomical books that included their tables. Archival documents attest that a large number of European scientific and technical books were in the *Enderun* during the years Feyzi was in the Palace.[33] These books were later sent to the Imperial Engineering School. The lot of thirty titles (84 volumes) dispatched in 1804 included, among others, works by the French authors Bélidor, Bouguer, Cassini, Clairaut, Lalande, and Le Monnier, as well as the *Encyclopédie* of Diderot and d'Alembert.[34] A second lot dispatched in 1806, the year the school was given new regulations, consisted of twenty titles (350 volumes), among them the French dictionaries of the Enlightenment, the mathematical works of

32 Sultan Selim III, eager to establish a permanent Ottoman presence in Egypt, had rewarded Admiral Horatio Nelson with an order-like jewel for his victory on the Nile in 1798 against the Navy of Napoleon Bonaparte.

33 Beydilli, op. cit. (n. 2), pp. 280–283.

34 Bernard F. de Bélidor, *La science des ingénieurs dans la conduite des travaux de fortification*, Henri Scheurleer, Paris, 1734; Pierre Bouguer, *Nouveau traité de navigation contenant la théorie et la pratique du pilotage*, Hippolyte-Louis Guerin & Louis François Delatour, Paris, 1753; Jacques Cassini, *Eléments d'astronomie*, L'Imprimerie Royale, Paris, 1740; J. Cassini, *Tables astronomiques du soleil, de la lune, des planètes …*, L'Imprimerie Royale, Paris, 1740; Alexis C. Clairaut, *Tables de la lune calculées suivant la théorie de la gravitation universelle*, Durand, Paris, 1754; Joseph Jérôme le François de Lalande, *Exposition du calcul astronomique …*, L'Imprimerie Royale, Paris, 1762; Pierre Charles Le Monnier, *Histoire céleste ou recueil de toutes les observations astronomiques*, Briasson, Paris, 1741; Denis Diderot, Jean Le Rond d'Alembert (eds.), *Encyclopédie, ou Dictionnaire raisonné des sciences, des arts et des métiers*, Paris, 1751–1572, 28 vols. See Feza Günergun, "Les ingénieurs et la modernisation: La transmission des savoirs techniques entre la France et la Turquie au tournant du XIXe siècle", in Claudine Fontanon, Irina Gouzévitch (eds.), *La construction d'un espace transnational: Les ingénieurs civils en Europe, Fin XVIIIe–XXe siècles*, « Histoire des Techniques », Editions Garnier, Paris, in press.

Bézout, Bossut, Lalande and Cagnoli and an anonymous book on logarithms.[35] The title of the latter was recorded as *Mufassal logaritma* (Complete logarithms). The knowledge of astronomy, trigonometry and logarithms embodied in these works was undoubtedly sufficient for Feyzi to compile his *Muhadarat*. On the other hand, he probably consulted the French astronomical ephemerides, *Connoissance des temps* from which he borrowed the "Table de réfraction pour les zones tempérées [...] suivant Messieurs Cassini, Bradley, la Caille" as will be mentioned in the coming pages. Did Feyzi read French himself, or did a European knowledgeable in scientific issues assist him in his project? Could this person have been the engineer Antoine de Juchereau whose appreciative note was inserted between the pages of Feyzi's treatise?

Muhadarat-ı Feyzi starts with a foreword (fols. 2v–3r) referring to the sciences, especially mathematics and astronomy. Feyzi expresses his gratitude to God, the creator of the heavens and the stars, for having given wisdom to men so that they can learn about the heavens and celestial bodies with observational instruments (*alat-ı rasadiyye*). He also acknowledges the Prophet Muhammad in guiding men to learn astronomy for calculating the prayer times. The reigning Sultan Selim III, as illuminative as the Sun, is described as the "re-inventor" of geometry, a science that has fallen into oblivion in the Ottoman Empire. The Sultan, Feyzi says, by spending heavily from the treasury, reinvigorated the study and the practice of geometry (i.e. engineering) and favoured those knowledgeable in this science. Selim III's generous patronage of mathematics had been eulogized previously by Seyyid Mustafa,[36] one of the first students of the Imperial School of Engineering created by the Sultan. Selim III had a keen interest in instruments. The military reforms generated by the modernisation movement of the end of the eighteenth century and the esteem given to mathematics seem to have encouraged Feyzi to write about an instrument which provided observational data that needed to be worked mathematically in order to be used in engineering and timekeeping.

The foreword is followed by an introduction (*mukaddime*), two chapters (*makale-i evveli* and *makale-i saniye*) and a conclusion (*hatime*). Feyzi did

35 J. d'Alembert, D. Diderot *et al.* (eds.), *L'Encyclopédie méthodique, ou par ordre de matières ...*, Panckoucke, Paris, 1782–1832, 197 vols.; Étienne Bézout, *Cours de mathématiques à l'usage du corps royal de l'artillerie*, Imprimerie Royale, Paris, 1770–1772, 4 vols.; Charles Bossut, *Cours complet de mathématiques*, Didot, Paris, 1795–1801, 7 vols.; Antonio Cagnoli, *Traité de trigonométrie rectiligne et sphérique*, Didot L'Aîné, Paris, 1786; J. de Lalande, *Abrégé d'astronomie*, Desaint, Paris, 1774; see also Günergun, op. cit. (n. 34), in press.

36 Seyyid Mustafa, *Diatribe de l'ingénieur Séid Moustapha sur l'état actuel de l'art militaire du génie et des sciences à Constantinople*, [Imprimerie Impériale], Paris, 1807^2, pp. 4–5 and 8–9.

not confine himself to describing the sextant, but aimed also to provide the mathematical knowledge the observer or the reader needed to make use of the measurements made with the sextant to calculate astronomical or navigational parameters. To give an example, he begins by explaining the optical principle of the instrument, then he explains in several paragraphs how to use logarithms and gives a list of mathematical symbols before describing the structure of the instrument (*tarifat-ı alat-ı sekstant*). Because of the insertion of mathematical issues between paragraphs related to the sextant, the layout of the treatise appears to be erratic. This arrangement of information was probably due to Feyzi's intention to gather together all the mathematical knowledge necessary to correct observed data and for the calculations that would follow. Such unusual presentation may explain the title Feyzi chose for his work. *Muhadara* originally meant "transmitting knowledge in the presence of an audience and participating in scientific discussions". The plural of this term, *Muhadarat*, is generally used for literary works in which the author transmits to the reader, by appropriate arrangements and quotations, knowledge he has borrowed from others.[37] Thus, the title reflects well Feyzi's intention of transmitting knowledge and the layout of the text.

5 Feyzi's Sextant

The manuscript includes only two drawings related to the sextant: one is on the portrait of Feyzi (*Figs. 4.2a* and *4.2b*), the other (fol. 4v) is a geometric figure describing the sextant's working principle. The fact that the celestial object indicated in the portrait is the Sun suggests that Feyzi's main concern was to measure the Sun's altitude. A text description of the sextant (*tarifat-ı alet-i sekstant*) is also provided (fols. 8v–10r). By looking at the textual information and the drawing in Feyzi's portrait together, one can partially reconstruct the sextant that Feyzi had in his hands.

Feyzi does not specify the material the sextant was made of but the yellowish colour used by the artist on the portrait suggests that it was made of brass. The most enigmatic part of Feyzi's instrument is its frame. It has a bar across only half of the frame. Has the artist omitted to draw a second horizontal bar? The strangely-drawn tangent screw and magnifier and the odd placement of the index glass suggests that the lack of the second bar may result from the artist's unfamiliarity with the instrument.

37 Hüseyin Yazıcı, "Muhadarat," in *TDV İslam Ansiklopedisi*, cit. (n. 21): v. 30 (2005), p. 391.

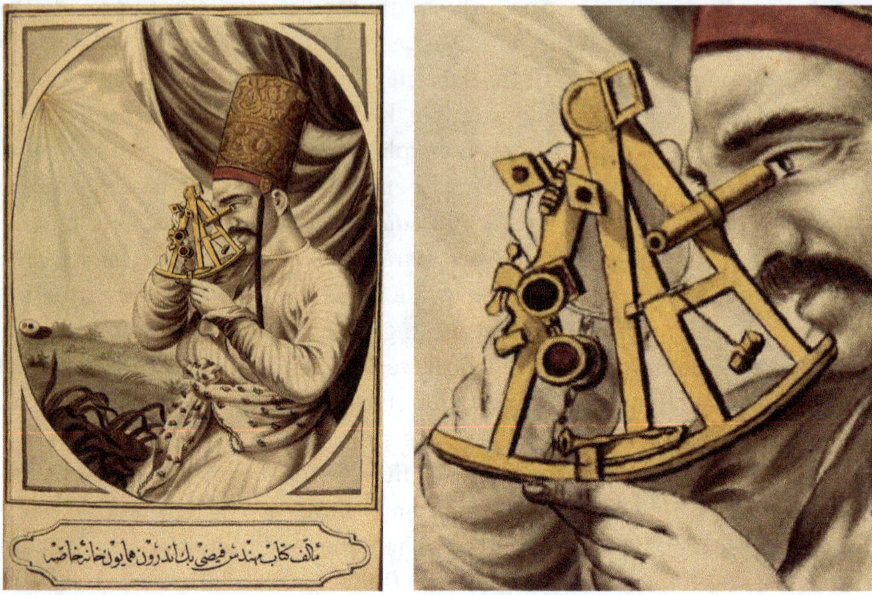

FIGURE 4.2 *a.* Portrait of "Feyzi Bey the engineer" with the sextant (*Muhadarat-ı Feyzi*, Istanbul University Rare Books Library, MS T6833, fol. 1v); *b.* detail from the same portrait

According to the text description, there was a wooden handle to hold the sextant. Feyzi named it "wood with a screw" (*vidalı ağaç*). He indicated that this handle was fixed behind the frame of the sextant that Sultan Selim III sent to the Imperial School of Engineering, but was detachable in the sextant he examined (fol. 8v). This handle was not depicted in the portrait, probably because it was hidden in Feyzi's palm. The text reads that the sextant has three index shades and three horizon shades. Feyzi noted that some of them were red (fol. 8v), but does not specify the other colour. The colours of the shades on the portrait are uncertain but they seem to be red and green. Feyzi mentioned that the sextant had a sight-tube (*camsız dürbün*) and a telescope with lenses (*camlı dürbün*). The sight tube is recommended for nearby observations, the telescope for distant observations. There was a magnifier (*pertavsız*) attached to the middle of the index arm (*idade*) to facilitate the reading of the scale on the arc (*kavs*). There is also a tangent screw (*vida*) to connect the index arm to the arc and another one to drive the horizon glass (fol. 9r). The arc of the sextant in Feyzi's portrait is not graduated, but the text states that each degree of the scale on the arc was divided into three parts of 20 arcminutes each, and that the arc was fitted with a vernier scale which allows to read the number of arcseconds of the angle observed. Feyzi did not explain in detail how to read the exact value of the observed altitude on the scale but said that it was easier to learn this by practising with the instrument (fol. 9r). The information he gives about the graduation of the vernier scale is confusing. Bearing in mind the precision of the best sextants of the time, it is likely that Feyzi's sextant could measure the altitudes to 20 arcseconds. This is probably the reason why Feyzi was fascinated by the sextant: the highest precision a standard size (12 cm × 12 cm) Ottoman wooden quadrant could provide was only 30 arcminutes, or 15 arcminutes at the best.

There is an artificial horizon at the bottom left of Feyzi's portrait. It is not clear if this is a glass horizon or a container filled with mercury. The manuscript reads that one would need mercury to provide a reflective surface: this suggests that a mercury artificial horizon was depicted. On the other hand, the second sextant sent to the Imperial School of Engineering had a level (*tesviye*) used to level the artificial glass horizon. The presence of the green and red spots on the surface of the artificial horizon seems odd, but these can be linked to the green and red shades of the sextant. Feyzi might have asked the artist to paint these spots intentionally, to show the observer or reader that the artificial horizon will look red or green when looking through the telescope and the shades.

Unfortunately, Feyzi does not mention the name of the sextant's maker. Either he did it on purpose because he was not interested in the master who crafted the instrument, or his sextant was not signed, like many others. Feyzi, however, mentioned that Sultan Selim III dispatched a sextant with fixed handle to the Imperial School of Engineering (fol. 8v). An archival document dated 1804 reads that two sextants were dispatched in that year.[38] It is possible that there were more than two sextants at the Palace and that two of them were given to the Imperial School of Engineering and Feyzi examined the third one left. Thus Feyzi's sextant was still in the Palace in 1805–1806 when he wrote the treatise. As the scientific instruments of the Imperial School of Engineering are currently untraceable, and the Topkapı Palace Museum does not hold a sextant, we were bound to rely on the information given in archival documents.

The archival document dated 1804 mentioned above describes the first sextant as follows: "An altitude instrument with a telescope made in England by a craftsman named Ramzdn [sic]. In its mahogany box". This would be an instrument made by Jesse Ramsden (d. 1800), London's leading scientific instrument

38 Ottoman Archives of the Turkish Prime Ministry (*Başbakanlık Osmanlı Arşivi*, henceforth Ottoman Archives), Istanbul, C. MF 69/3419, 19 Zilkade 1218 (1 March 1804).

maker, who by 1789 had produced about a thousand sextants.[39] The second sextant is described as a made-in-England (*İngilterekari*) altitude instrument bearing a patent sign on it (*patent işaretli*). It has a level and a telescope in its mahogany box. The mention of the patent sign may point to a Gilbert & Wright sextant which were signed "Kings Patent. Gilbert & Wright London" on the limb.[40] The reason why the full inscription is not given was probably the unfamiliarity of the clerk with Latin characters and English names. Thus sextants from both companies had been imported to Istanbul at the turn of the nineteenth century. The thick brass-framed sextant on Feyzi's portrait, suggests that Feyzi's sextant could have been a product of Jesse Ramsden. A sextant signed "Berge London late Ramsden" and crafted by Matthew Berge (d. 1819),[41] Ramsden's head workman, who had taken over the business after his master's death, shows similarities with Feyzi's sextant. However, it is not possible to identify it without any further evidence.

6 Operating the Sextant

Feyzi considered both the use of the sextant on land—i.e. for surveying work—and also its use in measuring the altitude of celestial objects, especially the altitude of the Sun much needed by timekeepers and navigators. He allocated a larger space for the latter than for the former. Feyzi explained how to find the altitude at sea. First the horizon is viewed through the telescope of the sextant and the position of the index arm on the scale engraved on the arc of the sextant is noted. Then the index is moved to bring the image of the Sun down to the level of the horizon and the position of the index arm noted again. The difference between the two readings (after various adjustments) gives the altitude of the Sun. Finding the horizon on land is not always possible. Feyzi proposed two methods. The first is to link six trees with a rope to create a horizon. The telescope of the sextant is directed at the rope. By moving the index,

39 Roderick S. Webster, "Ramsden, Jesse", in Charles C. Gillispie (ed.), *Dictionary of Scientific Biography*, Charles Scribner's Sons, New York, 1970–1980, 16 vols.: v. 11 (1980), pp. 284–285; Anita McConnell, *Jesse Ramsden (1735–1800)*, Ashgate, Aldershot, 2007.

40 There is a Gilbert & Wright sextant in the collections of Mystic Seaport, Connecticut, inv. no. 1951.394, see http://mobius.mysticseaport.org/detail.php?module=objects&type=advanced&category1=instruments+-+sextants&kv=132140&record=25&module=objects (accessed 5 February 2019).

41 Royal Museums Greenwich, inv. no. ID NAV 1144; see http://collections.rmg.co.uk/collections/objects/43356.html (accessed 26 March 2018).

the image of the Sun is brought onto the rope (the artificial horizon) and the altitude is found as before. The second method is to create an artificial horizon by using mercury. A vessel filled with 100 *drachmes* (about 300 g) of mercury is placed on the ground. Its position has to be adjusted until the Sun is reflected from the mercury towards the sextant. Then the reflection of the Sun is observed through the telescope and the altitude is measured by moving the index as before (fol. 9v).

Before dealing with the four examples of the use of the sextant on land Feyzi explains (fols. 10r–10v) how to trisect an angle by using dividers.[42] In the first example the distance between the observer and a landmark is measured (fol. 11v); in the second the height of a terrestrial object—either within reach or not—is determined (fol. 12v); the third example is about measuring the distance between two landmarks, both far away from the observer (fol. 13v); and the fourth example explains how to measure the height of a position when the observer stands at its top (fol. 17r). Geometric figures are used in every case.

7 Mathematical and Astronomical Tables

Muhadarat-ı Feyzi includes about twenty tables providing astronomical and mathematical data. As is well known, in order to calculate the true altitude of the Sun the observed altitude has to be corrected for several small errors by using parameters such as the dip of the horizon, atmospheric refraction, the solar radius, and solar parallax. A number of tables list the corrections to be made: tables for the solar parallax values for each month (*Cedvel-i ihtilaf-ı manzar-ı şems*; fol. 30v); tables for the solar radius values for each zodiac sign (*Cedvel-i esnaf-ı aktar-ı şems*; fol. 31r); tables for dip values (*Cedvel-i zaviye-yi inhirat-ı şems*; fol. 31r), and tables for the atmospheric refraction (*Cedvel-i inkisarat-ı şua'a'at-ı şems*; fol. 30r). The latter tables comprise the refraction values given by Cassini, Bradley, and Lacaille for the Sun's altitude between

42 Trisecting an angle with only an unmarked ruler and dividers was an intriguing problem that mathematicians strived to solve for centuries. The construction and the drawing Feyzi gave in fols. 10r, 10v and 71r show that he obtained an approximate result. Although he claimed to have solved the problem, he stated that he could not prove it. He may have thought that solving this 'unsolved problem' would bring him prestige. His solution was erroneous like all those given by other mathematicians. The impossibility of solving the problem was proved in 1837 by Pierre Laurent Wantzel (1814–1848), but the problem continued to draw interest.

0° and 90°, and are identical to those published in the *Connoissance des temps* between the years 1779 and 1784.[43]

There follow tables listing the changing values of the solar declination (*meyl*) during the course of the year for each degree of the 12 signs of the zodiac, which Feyzi says to have borrowed from Lalande (fol. 31v). Other tables give log sin *x* (*nisbet-i ceybiye*) values of the solar declination (fol. 32r); a third table gives log (1-cos *x*) (*nisbet-i sehmiyye*) values of the solar declination (fol. 33v). Similar tables are given for the complement of the solar declination (*tamam-ı meyl*; fols. 32v, 33r and 34r).

Sample problems were given in order to clarify how to use the tables in the calculations after measurements were taken with the sextant. A sample problem asks for the calculation of astronomical refraction (fols. 41r–41v). The data given is the Sun's declination to the north (*meyl-i şimaliye*) and the observed altitude of the Sun at noon (*hissi irtifa*) for the latitude of Istanbul. The solar declination indicates the date when the observation is made. Where this value is not given, the observer would read it in the tables. Thus knowing the date, the observer could read in the tables the true altitude for that date and time. The difference between two altitudes will give him the astronomical refraction. The sample problem also asks for the conversion into degrees of the value of *fazl-ı dair* (the difference between the Sun's hour angle at a given time and its hour angle at noon time) given in hours. To perform the conversion, one has to refer to the tables of the values *nısf el-tadil el-arz* (the complement of hour angle for a given latitude; fol. 20r). Other tables needed are those giving the logarithms of log(1-cos *x*) (*nisbet-i sehmiyye*) for hour 1 to hour 9, the hours being divided into minutes and seconds (fols. 20v–29r). Two tables, one for the values of the complement of the hour angle in degrees (fol. 34v), and the other giving log sin *x* (*nisbet-i ceybiye*) of these values (fol. 35r), are taken from Lalande.

The treatise ends with a note on the conversion of the Hijri dates to Gregorian dates. This is needed because the Hijri calendar used by the Ottomans is a lunar calendar. The observer who wishes to know the Sun's ecliptic longitude (*derece-i şems*, literally "Sun's degree") at a given date is obliged to make this conversion. First, he has to convert the Hijri date (e.g. 6 Safar 1220) on which he made the observation, to the Gregorian date (6 May 1805). Then he will refer

43 *Connoissance des temps, Pour l'année commune 1779. Publiée par l'ordre de l'Académie Royale des Sciences, et calculée par M. Jeaurat, de la même Académie*, Imprimerie Royale, Paris, 1777, pp. 250–251. Tables borrowed by Feyzi bear the title "Table de réfraction pour les zones tempérées [...] suivant Messieurs Cassini, Bradley, la Caille". Comparative investigation is needed to see whether other tables are borrowed from the *Connoissance des temps*.

to the tables (drawn up according to the solar calendar) giving the values of ecliptic longitude where he will read the ecliptic longitude corresponding to the Gregorian.

To facilitate astronomical calculations, Feyzi created 47 practical formulae of four terms each (fols. 37r–37v). When the values of three terms are known, the fourth term can be calculated by the rule of three. These terms include the solar radius; the trigonometric functions (sine, cosine, tangent and cotangent) of the astronomical parameters including the altitudes (*irtifa*), the azimuth (*semt*), the obliquity of the ecliptic (*meyl-i azam*), the latitude (*arz*), etc. To give an example, the formula to calculate the true altitude is the following: *ceybü'l-arz : ceybü'l-hissetü'l-irtifa = ceybü'l-meyl : ceybü'l-tadilü'l-irtifa* (sine of the latitude : sine of the observed altitude = sine of the solar declination : sine of the true altitude).

The tables include data from Bradley, Lalande and Cassini's observations. This implies that the treatise did not aim only to introduce the sextant as a more precise observational instrument but reflected a concern to calculate the astronomical parameters more accurately by benefiting from the improved data of European astronomers. Feyzi aimed to determine the time as precisely as possible. Was he looking for more accuracy in prayer times? We do not yet have any clues. The correction tables he introduced are mostly used by mariners in Europe. Interestingly he did not refer to the use of the sextant among mariners—except for a reference to the horizon at sea.

8 Logarithms and Mathematical Symbols in Feyzi's *Muhadarat*

In the early years of the nineteenth century when Feyzi wrote his *Muhadarat*, information about logarithms (*nisbet*) was already included in a number of Ottoman astronomical and mathematical texts.[44] Şekerzade Feyzullah (d. 1787), a calligrapher interested in timekeeping and mathematics, is acknowledged for being the first author to have composed in 1165 H (1751–1752 CE) a Turkish text on logarithms.[45] About twenty years later, in 1772, Halifezade Çınari Ismail

44 Salih Zeki (1864–1921) and Salih Murad Özdilek (1891–1967) were among the earliest Turkish historians of science to study the introduction of logarithms in Turkey. See Şeref Etker, "Salih Murat Uzdilek ve 'Logaritmanın Türkiye'ye Girişi'", *Osmanlı Bilimi Araştırmaları* 8, 2 (2011), pp. 55–76.

45 The text was given the title *Maksadayn fi halli'n-nisbatayn* (Two chapters on solving logarithms). Copies available in libraries date from 1751 to 1781. See G. Kut, H. Aynur, C. Üçer, F. Büyükkarcı Yılmaz (eds.), *Kandilli Rasathanesi Elyazmaları, I—Türkçe Yazmalar*, Boğaziçi Üniversitesi Yayınları, Istanbul, 2007, p. 525, no. 209.

Efendi (d. 1790), the timekeeper of the Laleli Mosque, Istanbul, completed the Turkish translation of Jacques Cassini's *Les tables astronomiques*, published in Paris in 1740.[46] Halifezade gave some elementary information on logarithms in the introduction to his translation, and appended (probably by borrowing from French texts) a five-place logarithmic table (*cedvel-i ensab*) of the numbers between 1 and 1000 as well as a table of the logarithms of sines and tangents from 1° to 45°.[47]

More detailed texts in Turkish introducing logarithmic operations were compiled in the late eighteenth century by Gelenbevi Ismail Efendi (d. 1790) and Hüseyin Rıfkı Tamani (d. 1836) both professors of mathematics at the Imperial School of Engineering.[48] Among the works Gelenbevi consulted was Jean François Callet's *Les tables portatives de logarithmes*.[49] All these texts remained as manuscripts and less than ten copies of each are presently available in Turkey's libraries. The fact that most copies of these works were devoid of tables suggests that these were texts only aimed to explain how to operate with logarithms, to introduce this new tool and its properties. The lack of logarithmic tables in these manuscript copies may also issue from difficulties encountered in copying tables. It was only in 1789 that the first printed booklet exclusively composed of tables with eastern Arabic numerals became available in Turkey. The booklet was printed on the press of the Imperial School of Engineering.[50] The regular publication of logarithmic tables from 1850 on

46 The Turkish translation was titled *Tuhfe-i Behic-i Rasini Tercüme-i Zic-i Kasini.* For copies see Kut, Aynur, Üçer, Yılmaz, op. cit. (n. 45), pp. 386–388, nos. 200, 227, 228 and 252. Halifezade had also designed the sundials of the Laleli Mosque and translated the *Théorie de la Lune* of Alexis Clairaut. For Halifezade's biography see Meltem Akbaş, "Halifezade, Çınari İsmail Efendi b. Mustafa", in D. Hoffmann, H. Laitko, S. Müller-Wille (eds.), *Lexikon der bedeutenden Naturwissenschaftler*, Spektrum Akademischer Verlag, Heidelberg, 2004, v. 2, pp. 150–151.

47 Cevat İzgi, *Osmanlı Medreselerinde İlim*, İz Yayıncılık, Istanbul, 1997, 2 vols.: v. 1, p. 253.

48 Gelenbevi İsmail Efendi, *Şerh-i Cedavil el-Ensab ve Nisbet-i Ceybiyye ve Zilliye* (Commentary on the table of logarithms and on the logarithms of sines and tangents), 1201 H (1787 CE). Not all copies of this work include tables for the logarithms of sines and tangents of the angles between 1° and 90° that Gelenbevi is said to have composed. Hüseyin Rıfkı Tamani's *Logaritma Risalesi* (Treatise on logarithms), 1206 H (1792 CE) is also devoid of tables.

49 İzgi, op. cit. (n. 47), v. 1, p. 255. The book that Gelenbevi had in his hands in 1787 was Jean François Callet, *Tables portatives de logarithmes publiées à Londres par Gardiner. Augmentées et perfectionnées dans leur disposition* ..., Didot l'Aîné, Paris, 1783. In the text preceding the tables, Callet explained the logarithms and their usage as well as their application to geometry, astronomy and other sciences. It is highly possible that Gelenbevi made use of this text when compiling his treatise. Comparative studies are needed to clarify the knowledge transmitted.

50 Anonymous, *Logaritma Cedavili*, Hasköy Mühendishane Matbaası, [Istanbul], 1789. The booklet consists exclusively of tables giving the eight-place logarithms of numbers

implies that their frequent use started in Turkey only from mid-nineteenth century. Thus in 1804–1805, when *Muhadarat* was composed, logarithms were not commonly used in Turkey. Therefore, Feyzi may have felt obliged to explain to the reader how to use logarithms in mathematical and astronomical operations. Besides, his adherence to logarithms in the calculations may be regarded as a reflection of his interest in novel techniques.

The introduction of the *Muhadarat* includes a list of mathematical symbols used in the treatise (*Derun-ı mecmuada müstamel bazı işarat beyanıdır*). These include symbols in Arabic letters such as ر for radius, but also symbols such as °, ′, ″, ‴ for degrees, minutes, seconds and thirds of arc, ⊥ for perpendiculars, ∟ for right angles, √ for square roots, signs for the trigonometric functions, etc. As Turkish at the time was written with Arabic characters from right to left, symbols borrowed from European texts changed direction. Seemingly, these symbols were not coined by Feyzi, but borrowed from European texts or Ottoman texts compiled or translated from European texts. Symbols were used by Şekerzade Feyzullah in his work introducing logarithms, mentioned above, and also in the *Usul-i Hendese* (Elements of Geometry) of 1797 that Hüseyin Rıfkı Tamani and İngiliz Selim (Selim the Englishman)[51] jointly translated from John Bonnycastle's *Elements of Geometry*.[52] A detailed comparative study is however needed on the mathematical symbols used by Ottoman engineers in the eighteenth century to evaluate those listed by Feyzi.

9 Geocentrism versus Heliocentrism

Two plates appended to the *Muhadarat* include fourteen figures in total representing the factors for which an altitude measured by a sextant needs correcting to obtain the true altitude (dip of the horizon, atmospheric refraction, the

between 1 and 10,000 (pp. 2–85); tables of the logarithms of sines and tangents of arcs from 0° to 45° (pp. 90–179), and tables of sines and tangents of arcs between 0° and 45° (pp. 182–271).

51 Selim who assisted Tamani in the translation was the Englishman Richard Baily, the brother of the English astronomer Francis Baily (1774–1844). See Augustus de Morgan, *A Budget of Paradoxes* (ed. by David Eugene Smith), The Open Court Publishing Co., Chicago and London, 1915², 2 vols.: v. 2, pp. 15–16. Ottoman archival documents refer to him as "Selim the Engineer". For his activities in the service of the Ottomans between 1792–1807, see Beydilli, op. cit. (n. 2), p. 53.

52 John Bonnycastle, *Elements of Geometry containing the principal propositions in the first six, and the eleventh and twelfth books of Euclid*, J. Johnson, London, 1789. For the Turkish translation see J. Bonnycastle, *Usul-i Hendese* (trans. by Hüseyin Rıfkı and Selim), Hasköy Mühendishane Matbaası, [Istanbul], [1797]. The book was reprinted in Cairo at the Bulaq printing press in 1830, and in Istanbul at the printing press of the Military School in 1852.

solar radius, parallax) and various astronomical parameters (the solar declination, obliquity of the ecliptic, precession, etc.). One of the figures represents the model of the universe. Although the planets are not indicated on the figure, it is a geocentric model, with the Earth at the centre. Feyzi, although he referred to the observations of European astronomers and praised a European instrument that could replace the traditional wooden quadrant, still adhered to the traditional world view when calculating astronomical refraction. Whether he believed in the heliocentric model of the universe cannot be deduced from the text.

About twenty years later a reference to the geocentric model was still being made in the *Mirat-ı Alem* (Mirror of the Universe), published in Istanbul in 1824 by Seyyid Ali Pasha (d. 1846), the Principal of the Imperial School of Engineering.[53] After introducing the Ptolemaic, Copernican and Tychonic systems, Seyyid Ali expressed the view that the geocentric model was favoured in the Islamic world because the *zijes* (the most popular being Ulugh Beg's *zij*),[54] used in making calendars were all based on the geocentric model. Personally he also preferred this traditional system because he was familiar with it. Both examples show that Ottomans drew astronomical data both from traditional tables (i.e. Ulugh Beg's *zij*) and from European ephemerides such as those of Cassini, Lalande and others.[55] Thus, the geocentric universe was not a problem for the Ottomans in performing basic astronomical calculations such as keeping time.

10 A Note of Appreciation by the Engineer Antoine de Juchereau

A note of appreciation written in French is inserted between the folios dealing with logarithmic operations (*Fig. 4.3*). It is signed "Antoine Juchereau, Ingénieur anglais au service de la Porte Ottomane." The note reads that the book was written after much consideration and contained information essential for both the army and the navy, the values given in the tables were exact and the author deserved much appreciation for his work, zeal and insight.

---

53 *Mirat-ı Alem* was the abridged Turkish translation of *al-Fathiyya,* the astronomical work composed in Arabic by Ali Qushji (d. 1474) in 1473. Ali Kuşçu, *Mirāt-ı Alem* (trans. by Seyyid Ali), Darü't-tıbaatü'l-Amire, Istanbul, 1239 H (1824 CE), p. 3.

54 Uluğ Bey, *Zic-i Uluğ Bey/Uluğ Bey'in Astronomi Cetvelleri* (trans. by Mustafa Kaçar, Atilla Bir), T.C. Kültür ve Turizm Bakanlığı, Ankara, 2012, 2 vols.: v. 1, Turkish translation, v. 2, facsimile edition of the MS Revan 1714, Istanbul, Topkapi Palace Museum Library.

55 J. Cassini, *Tables astronomiques*, cit. (n. 34), officially replaced Ulugh Beg's *Zij* during the Selim III's reign.

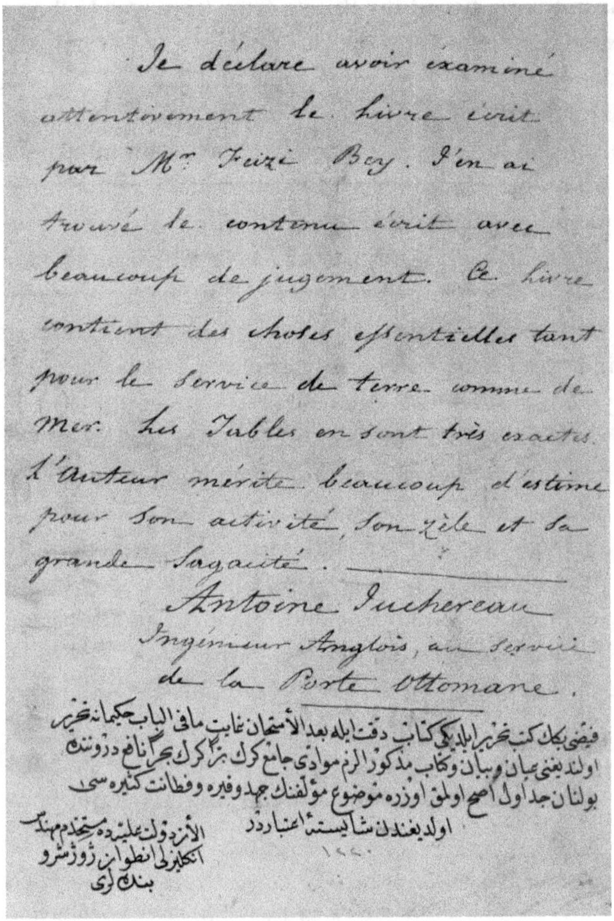
Je déclare avoir examiné attentivement le livre écrit par Mr. Feizi Bey. J'en ai trouvé le contenu écrit avec beaucoup de jugement. Ce livre contient des choses essentielles tant pour le Service de terre comme de mer. Les Tables en sont très exactes. L'Auteur mérite beaucoup d'estime pour son activité, son zèle et sa grande sagacité.

Antoine Juchereau
Ingénieur Anglois, au service de la Porte Ottomane.

فیضی بکك کتب تحریر ایلدیکی کتاب دقت ایله بعد الامتحان غایت مافی الباب حکیمانه تحریر
اولندیغنی عیان و بیان وکتاب مذکور الزم موادی جامع کرک برّاً کرک بحراً نام دروننده
بولنان جداول اصح اولملق اوزره موضوع مؤلفنك جد و فیره و فطانت کثیره سی
اولدیغندن شایسته اعتبار ددر

الآن دولت علیه ده مستخدم مهندس
انکلیزلی انطوان ژوژشرو
بندکلری

FIGURE 4.3 The note of appreciation signed by Antoine Juchereau, with its Turkish translation
ISTANBUL UNIVERSITY, RARE BOOKS LIBRARY, MS T6833, BETWEEN FOLS. 6V AND 7R

The author of the note was Antoine Baron de Juchereau de Saint Denys (1778–1842).[56] He was given the title "*directeur et instructeur en chef du génie militaire*" and he was entrusted with the general inspection of fortifications by the Ottoman administration. He described himself as "*ingénieur*

56 See "Juchereau de Saint-Denis", in Louis Gabriel Michaud (ed.), *Biographie universelle (Michaud) ancienne et moderne* ..., Madame C. Desplaces, Paris, 1843–1865, 45 vols.: v. 21 (undated), pp. 283–285.

anglais", because he was in the service of the British army from 1793 to 1802.[57] Subsequently, he requested employment in the Ottoman army. After his skills and knowledge in engineering and artillery were tested and approved, he was assigned to the Imperial School of Engineering following Sultan Selim III's decree in 1805. He was given a monthly salary of 400 *kuruş* while his interpreter would receive 50 *kuruş* per month.[58] A room was allocated in the school building, a room that was repaired and furnished.[59] Because of his good service in the school and the army, his salary was raised to 650 *kuruş* in 1807,[60] and he was nominated "*inspecteur-en-chef du génie militaire Ottoman*".[61] The same year he took a position in the French Army (*Fransa Devleti maiyetine tayin*),[62] and was called thereafter "Antoine, the French" (*Fransalı Antuvan*). The last archival document about the payment of his dues is dated 1808.[63] Juchereau left Turkey in 1809, after Sultan Selim III's assassination in 1808.[64] Thus he was an eyewitness to many events of the difficult years of Ottoman modernisation that he narrated in the second volume of his memoirs *Révolutions de Constantinople*: the Anglo-Turkish war of 1807–1809, the Kabakçı Mustafa insurrection, the disbanding of the modern *Nizam-ı Cedid* army corps, the dethronement and assassination of Sultan Selim III followed by the dethronement and assassination of Mustafa IV, and the death of Alemdar Mustafa Pasha.[65]

The note praising the author's efforts was probably commissioned by Feyzi himself. It also attests to the connection between an engineer in the *Has Oda* of the Imperial Palace and a foreign engineer serving in the Imperial School of Engineering and the Ottoman army. How did Feyzi and Antoine get to know

57 Antoine de Juchereau de Saint-Denys, *Révolutions de Constantinople en 1807 et 1808*, Librairie de Brissot-Thivars, Paris, 1819, 2 vols.: v. 2, pp. 53–54. For the enlarged edition see Le Baron de Juchereau de St-Denys, *Histoire de l'empire ottoman depuis 1792 jusqu'en 1844*, Au Comptoir des Imprimeurs-unis, Paris 1844, 4 vols.

58 Sultan Selim III's decree of 29 Zilhicce 1219 (31 March 1805); Ottoman Archives, HAT 167/7079.

59 Ottoman Archives, C.AS. 260/10844, 18 Ramazan 1220 (10 December 1805).

60 Ottoman Archives, C.AS. 946/41043, 11 Receb 1221 (14 September 1807).

61 "Juchereau de Saint-Denis", cit. (n. 57), v. 2, p. 53.

62 Ottoman Archives, C.HR. 108/5370, 29 Cemaziye'l-evvel 1222 (3 September 1807).

63 Ottoman Archives, C.HR. 112/5576, 13 Cemaziye'l-ahir 1223 (7 July 1808).

64 Following his return to France he acted as *colonel du génie* in the French army and fought against the Spanish in the Peninsular War of 1808–1814. "Juchereau de Saint-Denis", cit. (n. 57), v. 2, p. 53.

65 He claimed to have had direct and frequent contacts with Ottoman high officials which probably helped him in writing his memoirs, *Révolutions de Constantinople*; A. de Juchereau de Saint-Denys, *Révolutions*, cit. (n. 57), v. 1, p. v. The first volume includes his observations and reflections on Ottoman administration, ulemas, engineers, military institutions, topography of Istanbul, etc.

each other? Where did they meet? In the Imperial Palace, in the Engineering School or elsewhere? As engineers they would have had common interests. What kind of information did they exchange? Unfortunately archival documents do not help to answer these questions. When relating the events of 1807 Juchereau made reference to a Feyzi Efendi, who was sent to the Dardanelles by the Sultan to expedite the fortification works.[66] Could Feyzi Efendi, intendant of the *nizam-i cedid* troops, and Feyzi, the author of the *Muhadarat* be the same person? The biography of Feyzi the engineer reads that he died in 1230 H (1815 CE) when on a mission in the Dardanelles.[67] Juchereau also mentions that Feyzi Efendi, the intendant, was put to death in 1807 because of his negligence and gullibility after the British fleet's passage through the Dardanelles.[68] The dates, however, do not match. Resolving the matter needs further research and exceeds the scope of the present article.

11 Conclusion

Interestingly, the *Muhadarat-ı Feyzi,* one of the earliest texts introducing the sextant into Ottoman Turkey was written not by a mariner or by a timekeeper as one would expect, but out of curiosity by an engineer in charge at the Imperial Palace in Istanbul at the turn of the nineteenth century. The text reveals that the author regarded the sextant as an instrument mainly for observing the Sun's altitude. He did not mention any stellar observations. Measuring distances and heights on land remotely seems to have been a subsidiary issue for him. Moreover, he did not mention the sextant's use at sea. The nature of about twenty tables mostly borrowed from French authors and inserted within the pages of the manuscript points to Feyzi's interest in performing calculations relating to the Sun's altitude, most of them being necessary for timekeeping. To facilitate these calculations, he suggested making use of logarithms. Accordingly, the treatise includes the solutions to many problems related to the calculation of astronomical parameters and logarithmic operations. However, the information Feyzi gave on logarithms was already available in the late eighteenth century in the Ottoman books on mathematics produced in the Imperial School of Engineering.

Feyzi's focus on observing the Sun's altitude and the related astronomical calculations implies that he regarded the sextant as an instrument for

66 *Ibid.*, v. 2, pp. 61, 63, 65 and 67.

67 Bursalı Mehmet Tahir, op. cit. (n. 12), v. 3, pp. 260–261.

68 "Juchereau de Saint-Denis", *Révolutions*, cit. (n. 57), v. 2, pp. 70–71.

timekeeping. However, Ottoman timekeepers, to whom the sextant could have been of great use, did not abandon the traditional quadrant and switch to this modern, more precise instrument from Europe. Although the image of a sextant illustrates the cover of Ottoman almanacs (*takvim-i sal*) of the nineteenth and early twentieth centuries, there is no evidence that Ottoman timekeepers used sextants to measure the Sun's altitude.[69] They seem to have adhered to quadrants and to the almanacs published in Europe, and kept determining prayer times, and time in general, with their traditional instruments. The reason for the reluctance to use new precision instruments may have been economic. Sextants were much more expensive then the wooden quadrants produced in the country. Some timekeepers could make their own quadrants and thus minimise their expenses. Land engineers and military cartographers used plane tables and theodolites to measure distances for their topographical work. Logbooks of Ottoman ships in the mid-nineteenth century show that mariners used sextants to determine latitude and calculate the ship's position.[70] Mariners possibly did not deem it necessary to compose texts in Turkish about the sextant because they learned about its use through practice and from manuals in foreign languages.[71] Therefore it is not surprising to note that no substantial Turkish text was written or book published about sextants after *Muhadarat-ı Feyzi* throughout the nineteenth century. Almost a century after Feyzi, the Ottoman naval officer Süleyman Nutki Bey (1866–1924) would describe the sextant as the most essential of all nautical observational instruments (*alat-i rasadiye-i bahriye*) in his nautical lexicon of 1917.[72]

69 See the almanac prepared by the chief astrologer Hüseyin Hilmi (d. 1924): *Takvim-i Sal, 27 Cemaziyelevvel 1335* [21 March 1917] – *9 Cemaziyelahir 1336* [22 March 1918], Matbaa-i Osmaniye, Istanbul, 1333 H (1917 CE). For a copy see Istanbul, Atatürk Kitaplığı (Atatürk Library), MS K.6489. The late Ottoman almanacs usually included an introduction by the author, sections displaying various calendars, tables for the prayer times, meteorological, agricultural and astrological data, and other useful information. See Avner Wishitzer, *Reading Clocks,* Alla Turca*: Time and Society in the Late Ottoman Empire*, University of Chicago Press, Chicago & London, 2015, p. 165.

70 In 1870s, the sextant was considered to be one of the most essential instruments (*edevat-ı mühimme*) for a ship and was on the list of instruments to be purchased. It is not clear, however, when the sextant was first used in the Ottoman Navy.

71 There are only two sextants at the Maritime Museum, in Istanbul, both from the twentieth century. Sextants would have been used by Ottoman mariners, however they have not found their way into Turkish museums.

72 Süleyman Nutki, *Kamus-i Bahri* (ed. by Mustafa Pultar), Türkiye İş Bankası Yayınları, Istanbul, 2011, p. 254.

Acknowledgments

The authors wish to thank the following for their contributions throughout the research for this paper: Richard Dunn and Willem Mörzer Bruyns for kindly discussing about instruments which might match Feyzi's sextant; the staff of the Topkapı Palace Museum for their collaboration in searching for the sextant described in the *Muhadarat*; Özkan Değer for bringing "the crown and the crescent and star" engraved on Sultan Selim III's seal of 1798–1799 to our attention; Tim Stanley for pointing to a possible connection between the crown on *Muhadarat*'s seal and King George III's monogram; Nurcan Bal for providing information related to the use of the sextant in the Ottoman navy; Tahsin Ömer Tahaoğlu, Berrin Scott, John Scott, Angus O'Neill, Nalan Dönmez Yakarçelik and Solmaz Ceren Özdemir for supplying various pieces of information the authors were looking for. Last but not least the authors would like to thank Şeref Etker and Neil Brown for providing clarity to the text by bringing various issues to the fore.

CHAPTER 5

How Did the Turketum (or Torquetum) Get Its Name?

Richard L. Kremer

1 Introduction

> On sait que le terme même de turquet est resté inexpliqué, les étymologies qui ont été proposées relèvent de la fantaisie.[1]

This is a story about fantasies, about efforts to create an 'origins story' for a medieval astronomical instrument and its name. Our primary sources will be people writing about the instrument, not surviving exemplars of apparatus constructed of wood and brass. We shall analyse, that is, historiography rather than material culture.

The earliest known texts on our instrument were composed in Latin in the 1280s, probably in or near Paris. These texts do not name an inventor and do not explain the Latin term, "turketum", by which they designate the instrument. In 1469, Regiomontanus attributed the device to "Geber the Spaniard", an Arabic astronomer who had lived in Seville in the first half of the eleventh century. In 1598, Tycho Brahe guessed that "Arabians or Chaldeans" in the twelfth century had invented the instrument. By the early twentieth century, German historians designated "Turks" as the inventors. In these stories, the location of our instrument's birth moves between West and East. Was the early turketum a Latin instrument, an Arabic instrument, a Turkish instrument?

Our survey of these stories will suggest that attempts to designate particular inventors of early instruments often tell us more about the designators than about the historical origins of instruments. "What's in a name?" Juliet asks Romeo, in Shakespeare's play. Rather much, we shall see, just as Juliet saw.

2 The Turketum in Brass, Ink and Paint

Before turning to the texts, we shall briefly introduce the instrument. The oldest surviving example of the apparatus, described in the 1280s texts, has

1 Emmanuel Poulle, "Bernard de Verdun et le turquet", *Isis* 55 (1964), pp. 200–208: 204.

 | DOI:10.1163/9789004412842_006

a well-documented provenance (*Fig. 5.1*). It was made in Nuremberg and purchased there in September of 1444 by Nicholas of Cusa. For 38 Rhenish Gulden, the philosopher acquired not only a turketum but also an astrolabe, large wooden celestial globe, and a copy of Geber's *Flores ex almagesto*.[2] In addition to Nicholas's instrument, still preserved in the St. Nikolaus-Hospital in Bernkastel-Kues on the Mosel River, at least seven torqueti have survived from the fifteenth and sixteenth centuries.[3] The most famous early depiction appears in the background of Hans Holbein the Younger's 1533 painting, *The Ambassadors*, now in the National Gallery in London (*Fig. 5.2*).[4] The earliest printed image can be found in a Strasbourg 1512 edition of Gregor Reisch's *Margarita philosophica nova*.[5] The best known printed images appear

---

2 J. Hartmann, *Die astronomischen Instrumente des Kardinals Nikolaus Cusanus*, Weidmannsche Buchhandlung, Berlin, 1919, p. 8. Nicholas described this purchase in one of his manuscripts: Bernkastel-Kues, Bibliothek des St. Nikolaus-Hospitals/Cusanusstift, Cues 211, fol. 1r.

3 These include: 1. Hans Dorn, Vienna, constructed 1487 for Martin Bylica of Olkusz, now University of Krakow; Aleksander Birkenmajer, "Le torquetum de Marcin Bylica d'Olkusz à la lumière de la récente découverte du torquetum de Nicolas de Cusa (1922)", *Studia Copernicana* 4 (1972), pp. 537–540; Jay A. Levenson (ed.), *Circa 1492: Art in the Age of Exploration*, National Gallery of Art, Washington D.C., 1991, pp. 223–224, with colour plate; 2. signed LMN[orembergensis], ca. 1590, Bamberg Historisches Museum; *Kurt Pilz, 600 Jahre Astronomie in Nürnberg*, Verlag Hans Carl, Nuremberg, 1977, p. 259, Abb. 33; 3. Johann Praetorius, dated 1568, Nuremberg, Germanisches Nationalmuseum, Inv. No. WI 33; Ernst Zinner, *Deutsche und niederländische astronomische Instrumente des 11.-18. Jahrhunderts*, Beck, Munich, 1967², p. 472, pl. 11; 4. Jakob Cuno, sent 1561 to Duke Albrecht of Prussia, Kaliningrad, University Library; Zinner, op. cit., pp. 181 and 286; and three instruments attributed to Erasmus Habermel, 5. now at Hamburg, Museum für Kunst und Gewerbe, Inv. No. 1912.1435; Wolfgang Eckhardt, "Erasmus und Josua Habermel: Kunstgeschichtliche Anmerkungen zu den Werken der beiden Instrumentenmacher", *Jahrbuch der Hamburger Kunstsammlungen* 22 (1977), pp. 13–74: 32–33; 6. Munich, Deutsches Museum, Inv. No. 1687; Volker Bialas, *Gesetz und Harmonie: Sonderausstellung im Deutschen Museum aus Anlaß des 400. Geburtstages von Johannes Kepler*, Deutsches Museum, Munich, 1971, p. 11; and 7. ca. 1590, Kassel, Hessisches Landesmuseum, Inv. No. A 47; Karsten Gaulke, *Der Ptolemäus von Kassel*, Museumlandschaft Hessen Kassel, Kassel, 2007, pp. 240–241. An unsigned example in the Mathematisch-Physikalischer Salon, Dresden was lost in World War II but is reproduced in Ernst von Bassermann-Jordan, *Uhren: Ein Handbuch für Sammler und Liebhaber*, Richard Carl Schmidt & Co., Berlin, 1914, p. 47. For a recent study of the mathematics of the instrument, see Brian Albinson, Lennart Berggren, "The Heliodon and the Torquetum", *Compendium: A Publication of the North American Sundial Society* 26/3 (2019), pp. 27–35.

4 Elly Dekker, Kristen Lippincott, "The Scientific Instruments in Holbein's Ambassadors: A Re-Examination", *Journal of the Warburg and Courtauld Institutes* 62 (1999), pp. 93–125; J.D. North, *The Ambassadors' Secret: Holbein and the World of the Renaissance*, Hambledon and London, London, 2002, pp. 108–114.

5 Gregor Reisch, *Margarita philosophica nova*, Johann Grüninger, Strasbourg, 1512, sig. A2r.

FIGURE 5.1 Nicholas of Cusa's turketum, 1444. From J. Hartmann, *Die astronomischen Instrumente des Kardinals Nikolaus Cusanus*, Berlin, 1919, pl. 2

FIGURE 5.2 The turketum as depicted in Hans Holbein the Younger, *The Ambassadors*, 1533
LONDON, NATIONAL GALLERY, NG 1314

in Peter Apianus's 1533 book on cosmography[6] and in his magnificent 1540 *Astronomicum Caesareum* (*Fig. 5.3*).[7]

As can be seen from these images, the turketum was designed to sit on a table and to represent the essential circles of spherical astronomy with flat circular disks or semi-circles of thin brass or wood. The turketum replaced the complexities of the armillary sphere, with its several concentric rings mounted on various axes that (due to weight and friction) invariably flexed out of their specified planes, with much lighter and simpler plates.

An early text from the 1280s (see below) describes and names the parts of the instrument.[8] The base (missing in *Fig. 5.1*) represents the local horizon (*tabula horizontis*). Moving upward, the first circle rests in the plane of the equator (*tabula aequinoctialis*) and, as it rotates, represents the daily motion of the celestial sphere. The third circle is set to the plane of the ecliptic and gives celestial longitudes (*tabula orbis signorum*). The vertical circle (*christa*, comb of a chicken), attached to the pillar rising above the plane of the ecliptic, includes an alidade (*alidada circuli magni*) that marks latitudes with respect to the ecliptic plane (*regula latitudinum*). A pivoting vertical half circle (*semis*) with an attached bob (*perpendiculum*) marks altitudes with respect to the horizon. The device, in addition to demonstrating relations among the key circles of spherical astronomy, could be used to observe in altitude-azimuth, equatorial or ecliptic coordinates, or to convert among these coordinate systems.

As the circles are rotated around the equatorial axis in daily motion, however, the inclination of the vertical axis of the device against the horizon shifts by twice the obliquity. The several sixteenth-century exemplars I personally have manipulated, when set to central European latitudes around 50°, become top-heavy and easily tip onto their sides. Only if the base were firmly attached to a heavy table could users easily rotate the circles to represent positions of the celestial sphere at any date in the year or any time in the day. Making quantitative observations of objects in the night-time sky with the turketum, something I have not tried, would be exceedingly awkward. As instruments historian Anthony J. Turner has suggested, the turketum "may rather be considered an example of conspicuous intellectual consumption than a much-used instrument".[9]

---

6 P. Apianus, *Introductio geographica*, [Apianus], [Ingolstadt], 1533, sig. L4v.

7 P. Apianus, *Astronomicum Caesareum*, in aedibus nostris, Ingolstadt, 1540, sig. O4v.

8 Robert T. Gunther, *Early Science in Oxford*, Clarendon Press, Oxford, 1923–1945, 15 vols.: v. 2 (1923), pp. 370–375.

9 Anthony J. Turner, *Early Scientific Instruments: Europe, 1400–1800*, Sotheby's Publications, London, 1987, p. 18.

FIGURE 5.3 The turketum as depicted in P. Apianus, *Astronomicum Caesareum*, Ingolstadt, 1540, sig. O IIIIv

3 The Early Manuscripts and Users

Since the mid-nineteenth century, scholars have found five medieval Latin texts associated with our instrument. We begin by reviewing the names these texts assign to the device. Only one of these texts explicitly explains its choice for a name. And this is the only text (authored by Regiomontanus) that proposes an inventor for the device.

The most widely preserved of these texts (*incipit*: "De nominibus partium instrumenti quod turketus dicitur") was transcribed from a single manuscript (O1, see *Appendix*) and published in 1923 by Gunther.[10] Known in 35 manuscripts, it was occasionally copied with an *explicit* dated 1284.[11] Four late manuscripts assign an author to the text: "Magister Franchone de polonia parisiensis", an attribution first noticed in 1945 by Thorndike and since then generally accepted. No other texts have been firmly linked to this Franco and nothing further is known about him.[12] The nominative form of the name appears only

10 Gunther, op. cit. (n. 8), v. 2, pp. 370–375.

11 Hippolyte Aubert, "Notice sur les manuscrits Petau conservés à la Bibliothèque de Genève", *Bibliothèque de l'École des Chartes* 70 (1909), pp. 247–302: 291; Lynn Thorndike, "Franco de Polonia and the Turquet", *Isis* 36 (1945), p. 7. See *Appendix*. Explicits in E1, fol. 161v; V1, fol. 118v; and G, fol. 24r read: "Factum anno Domini 1284 secunda die Iulii in die dominica sole existente 17 Cancri, Luna 20 Aquarii, Saturno 20 Capricorni, Jove 21 Sagittarii, Marte 13 Virginis, Venere 16 Leonis, Mercurio 28 Gemini, Capito 12 Capricorni ante meridiem". These positions are correctly computed, to the nearest degree, from the Toledan Tables for 11 a.m. on 2 July 1284 for the meridian of that city. The Parisian Alfonsine Tables, assembled in the 1320s, give slightly different positions for Saturn and Jupiter. This use of the Toledan Tables supports the dating of this text as 1284; the text itself also refers to the Toledan Tables. See Gunther, op. cit. (n. 8), v. 2, p. 375.

12 *Ibid.*, p. 35. Gunther, who knew the description in the Halliwell manuscript that mentioned "Franco" (see n. 53 below), attributed the text to Franco of Lüttich (Liège), c. 1060, known to have composed treatises on music, squaring the circle, and theology. See Menso Folkerts, A.J.E.M. Smeur, "A Treatise on the Squaring of the Circle by Franco of Liège, of About 1050", *Archives internationales d'histoire des sciences* 26 (1976), pp. 59–105 and 225–253: 230–231. A manuscript in the Bodleian Library, Oxford, Can. misc. 556, contains a copy of the canon to the Toledan Tables with a gloss (fol. 16v) by a "Franco" that Poulle and Birkenmajer linked to Franco de Polonia. Fritz S. Pedersen, *The Toledan Tables: A Review of the Manuscripts and the Textual Versions with an Edition*, C.A. Reitzels Forlag, Copenhagen, 2002, pp. 142 and 336, rejects this link on chronological grounds. See Aleksander Birkenmajer, "Les astronomes et les astrologues silésiens au moyen-âge [1937]", *Studia Copernicana* 4 (1972), pp. 437–468: 444–448; *Die deutsche Literatur des Mittelalters, Verfasserlexikon*, de Guyter, Berlin, 1977–2008², s.v. "Franco de Polonia" by E. Poulle; Jarosław Włodarczyk, "Franko z Polski", in *Polski Wkład W Przyrodoznawstwo I Technikę. Słownik Polskich I Związanych Z Polską Odkrywców, Wynalazców Oraz Pionierów Nauk Matematyczno-Przyrodniczych I Technik*, ed. B. Orłowski, IHN PAN and IPN, Warsaw, 2015, p. 422. According to Pedersen, op. cit., p. 167, a "Franco" in 1285 glossed a Sorbonne manuscript copy of the Toledan

once in Franco's text. Fourteen manuscripts I have seen use the second declension, masculine form "turketus". Five manuscripts use the second declension, neuter form "turketum", the form I shall henceforth use since several important fourteenth-century authors (see below) referred to that form. In any case, it appears as if fourteenth- and fifteenth-century scribes were unsure about the grammatical form of the term, which presumably was completely new to them (see *Appendix* for the variant orthographies).

Franco's text provides no clues about the etymology of the name. In assigning only Latinate names to the various components of the instrument, Franco's text does not point toward Arabic origins of the turketum (unlike the astrolabe, e.g., with its Arabic-derived "almucantar" and "azimuth" lines).[13] Yet, as noted by Birkenmajer, the text offers no instructions about how to construct the instrument and we might guess that Franco did not consider himself as its inventor.[14] His text is divided into two parts. The first describes the various components of the turketum in very general terms. The second briefly explains seven problems that can be solved with the device, such as finding the ascendant or verifying stellar and planetary longitudes and latitudes after aligning the disks to the predicted lunar location. This latter exercise would enable users "to know by how much the Toledan Tables and tables for other cities [the Toulouse Tables?] err from the true places [...]. Elsewhere, I will explain" Franco concluded, "how to verify predictions of tables".[15] Apparently, Franco

Tables, which might support the attribution in P2 (see *Appendix*) to Francone parisiensis. Cf. Laure Miolo, "Le fonds scientifique d'un collège de théologie: le cas de la bibliothèque de Sorbonne, 1257–1500", PhD dissertation, Université Lumière Lyon II, 2017, pp. 395–396. For other references to a "Franco" at the end of the thirteenth century, see J. Handschin, "Aus der alten Musiktheorie, V: Zur Instrumentenkunde", *Acta musicologica* 16–17 (1944–1945), pp. 1–10: 4–5; Berend Wispelwey, *Biographical Index of the Middle Ages*, K.G. Saur Verlag, Munich, 2008, p. 389; C.P.E. Nothaft, "Criticism of Trepidation Models and Advocacy of Uniform Precession in Medieval Latin Astronomy", *Archive for History of Exact Sciences* 71 (2017), pp. 211–244: 219–220. The attribution of the turketum text to the Parisian music theorist Franco de Colonia, who died around 1240, must be rejected on chronological grounds. See Georges le Cerf, Edmond-René Labande, *Les traités d'Henri-Arnaut de Zwolle et de divers anonymes (Paris: Bibliothèque nationale, MS. Latin 7295)*, Bärenreiter, Kassel, 1972, p. viii; Hans Jürgen Rieckenberg, "Zur Biographie des Musiktheoretikers Franco von Köln", *Archiv für Kulturgeschichte* 42 (1960), pp. 280–293.

13 The term "alidade", of course, does derive from Arabic roots. However, this term had appeared since the eleventh century in Latin texts on the astrolabe and the quadrant and Franco might well have adopted the term from these earlier Latin texts. See Arno Borst, *Astrolab und Klosterreform an der Jahrtausendwende*, Carl Winter, Heidelberg, 1989, and the extensive literature cited therein.

14 Birkenmajer, op. cit. (n. 12), p. 447.

15 Gunther, op. cit. (n. 8), v. 2, p. 375. Another late thirteenth-century Paris astronomer, William Saint Cloud, exhibited a similar desire to compare predicted against observed

was more interested in testing received astronomical tables than in exploring the origins of the turketum or its name.

A second text on our instrument (*incipit*: "Sic facies instrumentum ad modum truqueti/turketi") was discovered in 1847 by the authors of the *Histoire littéraire de la France* and edited in 1961 from two manuscripts (P1 and P5) with a short commentary by Polykarp Hartmann.[16] Attributed to a little-known Franciscan friar, Bernard de Verdun, this text appears in a chapter on instruments embedded in a much longer *Tractatus super totam astrologiam* that treats many questions of natural philosophy, astrology, and a bit of mathematical astronomy. Bernard described how to make and use our instrument, introducing some, but not all, of Franco's names for the parts ("crista" does not appear but "tabula horizontis", "tabula aequinoctalis", "regula latitudinis" and "semis" do). Poulle guessed that, because of his less technical terminology, Bernard wrote as the inventor, seeing no need to explain what appeared self-evident. Poulle thus dates Bernard's text earlier than Franco's, arguing that Franco, upon reading Bernard, appreciated the sophistication of the new instrument and its need for a "more precise nomenclature" and thus added "crista" to Bernard's names. Such psychological speculation far exceeds my expertise and I offer no opinion as to the temporal order in which these two texts were composed. More important for my story is that Bernard also used the second declension "turketi" for the device (other scribes wrote "trunqueti", "truqueti", "torqueti" or "tarqueti"). Bernard used the term only in the genitive form. Like Franco, he offered no explanation for the origin of name.

A third anonymous text (*incipit*: "Si volueris componere turchetum accipe laminam") was transcribed from a single manuscript (Vi2) and published in 1964 by Poulle.[17] It provides more detailed instructions for constructing the device, dividing the circles, and setting the obliquity of the ecliptic to 23° 33′ 30″ (the value of the Toledan Tables).[18] Other manuscripts spell the term "turketum" or, in the sixteenth century, "torquetum". Again this text offers

positions. See Fritz S. Pedersen, "William of Saint-Cloud: Almanach Planetarum: An Edition of the Canons, a few Samples from the Tables and a Foray into the Numbers", *Cahiers de l'Institut du Moyen Âge grec et latin* 83 (2014), pp. 1–133; Richard L. Kremer, "On 'Disagreeing with the Common Tables': William of Saint Cloud's Almanac of 1292–1312", forthcoming.

16 Émile Littré, "Tractatus optimus super totam astrologiam", *Histoire littéraire de la France* 21 (1847), pp. 317–320; Polykarp Hartmann, *Tractatus super totam astrologiam/Bernardus de Virduno. Ediert und mit einem Kommentar versehen*, Dietrich-Coelde-Verlag, Werl, 1961, pp. 140–142; Poulle, op. cit (n. 1), p. 205.

17 Poulle, op. cit. (n. 1), pp. 207–208.

18 Pedersen, op. cit. (n. 12), pp. 961–966; José Chabás, B.R. Goldstein, *A Survey of European Astronomical Tables in the Middle Ages*, Brill, Leiden, 2012, p. 23.

no explanation for the name. An abbreviated fifteenth-century paraphrase of this text (*incipit*: "Si quis vult componere turchetum accipe laminam") is also found in copies using the terms "turketum" and "torquetum".

The fourth text (*incipit*: "Quia astronomicis student exercitiis"), the first for which we have a well-attributed author, is Regiomontanus's *De torqueto*, composed in 1469 and dedicated to the Archbishop of Esztergom, János Vitez, Regiomontanus's patron in those years. Published in 1544 with a canon describing how to solve twenty-one astronomical problems with the instrument, Regiomontanus's text was the first to historicize the instrument and to examine the etymology of its name.[19]

> [...] to this remarkable machine the name has been now for a long time "Torquetum" from turning, as I think, since the celestial form is partially represented by this likeness and is seen to be stretched all around into the form of a globe, as is the case in a ringed astrolabe but with the rings constrained and twisted [note: *contortus* comes from *contorqueo*] into some flat shape. For from the term torquibus [collars/rings] one can infer nothing precise of the ambiguity [*paranomasia*; a pun, different form, same sound] since one can examine adequately neither the appearance of the thing nor the derivation of the word. For perhaps it ought to be called "Torquatus", if one were to give place to the grammatical rule. But it does not matter to us whether or not "torquetum" was properly formed from "twisting", since for present purposes we do not take the mantle of the grammarian. We therefore will use the ancient designation as a kind of conventional name, lest we seem to have disdained the generative antiquity [ancient progenitors?] of the liberal arts.[20]

19 Johannes Schöner, *Scripta clarissimi mathematici M. Ioanis Regiomontani*, Johann Montanus & Ulrich Neuber, Nuremberg, 1544, sig. a4v–e1v; Ernst Zinner, *Leben und Wirken des Joh. Müller von Königsberg genannt Regiomontanus*, Zeller Verlag, Osnabrück, 1968², pp. 157–158 and 317. Regiomontanus's autograph draft is preserved in Vi3.

20 "Huic igitur spectabili machinae Torqueto nomen iam pridem fuit, a torquendo, ut arbitror, quoniam coelestis figura, quae hoc simulachro representatur non integra, atque in globi formam circumquaque distenta cernitur, veluti in Astrolabio annulari, verum coactis, ac in planam quandam speciem contortis circulis constat. Nam a torquibus nihil prorsus agnominationis mutuari potest, cum neque similitudinem rei, neque derivationem nominis satis tueatur. Torquatus enim vocari fortasse debuit, si formule grammaticae locus daretur. Caeterum torquetus a torquendo legitime ne fictus sit an non, parum curae nobis est, cum in praesentiarum grammatici munus non profiteamur. Prisco igitur cognomento perinde quasi fortuito nomine utemur, ne vetustatem bonarum artium parentem contempsisse videamur"; Schöner, op. cit. (n. 19), sig. b1v.

Regiomontanus's grammatical analysis is exceedingly interesting for our study. He assumes that the long-standing name of the device is not "turketum", as we find in most of the manuscripts of the Franco and Bernard texts, but rather "torquetum", from the Latin transitive verb *torquere*, to turn, twist, whirl, bend, even torture. He notes that the proper term, according to Latin grammar, should be *torquatus -a -um*, from the noun *torquis*. Regiomontanus says, however, that he will privilege history over grammar, i.e., he will retain the "conventional", if grammatically incorrect, name "torquetum". For Regiomontanus, "turketum" was thus a misspelling, the origin of which, he reluctantly concludes, he cannot explain.

I have found one early text that does use the spelling "torquetum" to describe our instrument, a usage that confirms Regiomontanus's claim that "torquetum" had long been used. The otherwise unknown John of Sicily composed a commentary on the canons to the Toledan Tables, dated by its modern editor to Paris, 1290s. At one point, John described measuring solar altitudes at the equinoxes "per astrolabium vel torquetum". Before Regiomontanus, however, this spelling apparently did not find usage beyond John's text.[21]

And Regiomontanus specified an inventor, "Geber the Spaniard" or Jābir ibn Aflaḥ, who had lived in Seville during the first half of the twelfth century. In the late 1450s, Regiomontanus himself had copied Geber's *Flores ex almagesto* (translated by Gerard of Cremona) into one of his notebooks.[22] This

---

21 See Fritz S. Pedersen, *Scriptum Johannes de Sicilia super canones Azarchilis de tabulis Toletanis*, Erik Paludan, Copenhagen, 1986, 2 vols.: v. 2, p. 257. Pedersen found this text in twelve manuscripts, ten of which include the passage referring to our instrument. Of the seven manuscripts I have inspected (Pedersen's P, M, V, N, H, F, J), dating from the early fourteenth through the fifteenth centuries, six use "torquetum". Only the late N uses "turquetum". Cf. *ibid.*, v. 1, pp. 20–37.

22 Regiomontanus's library, as inventoried in the sixteenth century, included ten of the thirteen Arabic astronomical/astrological authors whose works, in Latin translation, were printed before 1550. See Richard L. Kremer, "Text to Trophy: Shifting Representations of Regiomontanus's Library", in James Raven (ed.), *Lost Libraries: The Destruction of Great Book Collections since Antiquity*, Palgrave Macmillan, London, 2004, pp. 75–90: 79; Francis J. Carmody, *Arabic Astronomical and Astrological Sciences in Latin Translation: A Critical Bibliography*, University of California Press, Berkeley, 1956; Eberhard Knobloch, "Zur Rezeption der arabischen Astronomie im 15. und 16. Jahrhundert", in Joseph W. Dauben *et al.* (eds.), *History of Mathematics: States of the Art. Flores Quadrivii—Studies in Honor of Christoph J. Scriba*, Academic Press, San Diego, 1996, pp. 237–261; Charles Burnett, David Juste, "A New Catalogue of Medieval Translations into Latin of Texts on Astronomy and Astrology", in Faith Wallis, Robert Wisnovsky (eds.), *Medieval Textual Cultures*, de Gruyter, Berlin, 2016, pp. 63–76; Dag Nikolaus Hasse, *Success and Suppression: Arabic Sciences and Philosophy in the Renaissance*, Harvard University Press, Cambridge, 2016. In addition to Geber, Regiomontanus's Arabic-writing authors included Alcabitius, Alhazen,

text describes an awkward instrument (see below) comprised of a calibrated disk and two hinged rules that could be used to measure altitudes and, if the position of the Moon were known, to find the longitude and latitude of stars or planets. Its modern editor, Lorch, decided that Geber's instrument, in its Latin description, differs significantly from the turketum. But Regiomontanus obviously considered the two devices more closely related.[23] Regiomontanus thus pointed toward two approaches to discover the origin of our device: trace similarities in physical design of instruments, or trace the origin of words to explain their later meanings (this approach failed him).

A fifth text (*incipit*: "Vtilitates Turketi pro serenissimo et illustrissimo principe domino Matthia, dei gratia Rege Hungarie ...") that I recently found in a single manuscript copy (L2) was composed in the 1480s by a little-known Viennese astronomer, Johann de Epperies, and dedicated to King Matthias Corvinus of Hungary.[24] It tersely describes the components of the device and offers a canon of twelve problems that can be solved with the instrument. Although some of these overlap with those in Regiomontanus's text (e.g., finding right and oblique ascensions, the ascendant and the twelve astrological houses), others do not. This text seems to be independent of Regiomontanus's, especially since Johann used the term "turketum" and not Regiomontanus's "torquetum". Since King Matthias died in 1490, Johann undoubtedly wrote his text before that date and presumably did not know Regiomontanus's text.

Albategnius, Alpetragius, Alfraganus, Arzarchel, Abenezra, Haly filius Abenragel, Albumasar, Messahalah, Averroes, al-Khwārizimī, and pseudo-Thebit.

23 Ingeborg Neske, *Die Handschriften der Stadtbibliothek Nürnberg, Bd. V: Die lateinischen mittelalterlichen Handschriften, Varia, 13.-15. und 16.-18. Jahrhundert*, Harrossowitz Verlag, Wiesbaden, 1997, pp. 23–24; Richard Lorch, "The Astronomical Instruments of Jābir ibn Aflaḥ and the Torquetum", *Centaurus* 20 (1976), pp. 11–34: 23–30.

24 Although Johann de Epperies had matriculated in the Hungarian nation, i.e. as a "Hungarian", at the University of Vienna in 1492 and is included in Georg Tanstetteter's 1514 "viri mathematici" as an "outstanding man who, in this time, taught mathematics to the amazement of many", no archival records suggest that Johann ever taught at that university. He was associated, however, with the humanist circle around Konrad Celtis and in 1520 added marginal comments and a sketch of a drum clock to an astronomical manuscript now in Vienna. See Willy Szaivert, Franz Gall, *Die Matrikel der Universität Wien*, Böhlau, Vienna, 1954–1975, 5 vols.: v. 2, p. 227; Georg Tanstetter Collimitius (ed.), *Tabulae eclypsium Magistri Georgij Peurbachij, Tabula primi mobilis Joannis de Monte Regio*, Johann Winterburger, Vienna, 1514, sig. aa6r and aa8r–aa8v; Hans Rupprich (ed.), *Der Briefwechsel des Konrad Celtis*, Beck, Munich, 1934, pp. 93–95 and 387; Christoph Schöner, *Mathematik und Astronomie an der Universität Ingolstadt im 15. und 16. Jahrhundert*, Duncker & Humblot, Berlin, 1994, p. 111; Beatriz Porres de Mateo, "Les tables astronomiques de Jean de Gmunden: édition et étude comparative", PhD dissertation, Paris, École pratique des hautes études, 2003, pp. 65, 118–119 and 548; Jozsef Szinnyei, *Magyar irók: Élete és munkái*, Kiadja hornyanszky viktor Könyvkereskedése, Budapest, 1891–1914, s.v. "Eperjesi".

Johann's names for the parts also differ somewhat from Franco's (i.e., "tabula infima", "falanga altitudinis", "radix orbis signorum") and provides even more detail about dividing the circles than does "Si volueris componere". And Johann, unlike Regiomontanus, did not seek to explain the name.

The name turketum/turquetum apparently established itself rather quickly in Latin astronomy. In 1299 and 1301, Peter of Limoges, canon of Evreux with close ties to the Collège de Sorbonne, measured the longitude and latitude of several comets "per turquetum". Several years later, the astronomer John of Murs, Master of Arts and also closely related to the Sorbonne, praised the "turquetum" and astrolabe as especially reliable observational instruments yielding "truth". In 1326, Richard of Wallingford, astronomer and abbot of St. Albans, invented a simpler form of our device that he called the "rectangulus". Composed of rods rather than disks or spheres, it would enable users, he asserted, "to determine the paths and places of the planets and fixed stars [...] together with all the other problems which may be solved by the armillary, the astrolabe or the turketum" (scribes in two mid fourteenth-century manuscripts wrote "torketum").[25]

We may thus conclude that our instrument, by the mid fourteenth century, had an established name in Latin astronomy, the turketum. By the 1470s, thanks to Regiomontanus, its second name, the torquetum, came into broader use. The latter name, as Regiomontanus had explained, refers to the twisting or wobbling motion experienced by the vertical axis of the instrument as it rotates around the equatorial axis. This name derives from a visual account of the instrument in action. But the etymology of the earlier name remained a mystery.

4 Locating Origins of the "Turketum"

Since Regiomontanus, searches for an "origins story" usually have followed one of two paths. Look for precursor instruments that share all, or most, of the features described by Franco in 1284 and assume a kind of "natural evolution" from

25 Lynn Thorndike, "Peter of Limoges on the Comet of 1299", *Isis* 36 (1945), pp. 3–6; Jane L. Jervis, *Cometary Theory in Fifteenth-Century Europe*, Reidel, Dordrecht, 1985, pp. 29–31; Lynn Thorndike, *A History of Magic and Experimental Science*, Macmillan, New York, 1923–1958, 8 vols.: v. 3, pp. 295–296; Frank Hentschel, *Sinnlichkeit und Vernunft in der mittelalterlichen Musiktheorie: Strategien der Konsonanzwertung und der Gegenstand der Musica Sonora um 1300*, Steiner, Stuttgart, 2000, pp. 247–250; J.D. North (ed.), *Richard of Wallingford: An Edition of His Writings, with Introductions, English Translation and Commentary*, Clarendon Press, Oxford, 1976, 3 vols.: v. 1, pp. 406–432, v. 2, pp. 287–301.

the precursor device to the turketum. Or look for precursor words that sound like "turketum" and in their linguistic evolution might explain how that word came to be associated, semantically, with our instrument. I cannot, of course, systematically survey all these searches since 1469. Rather, I have selected a limited number of cases to illustrate the two approaches outlined above.

We begin with the search for precursor instruments. Tycho Brahe, in his 1598 book on astronomical instruments, mentioned the "torquetum", an instrument "which was in my opinion invented by the Arabians or the Chaldeans, and used by them". He elaborated neither the form of their device nor the sources that describe it, noting only that he had not constructed one "because, if it is to have the required size [for making observations], it will be oppressed by its own weight and be difficult to handle".[26] Did Tycho assume, therefore, that the "Arabians and Chaldeans" had less demanding standards than he did for their observational precision? Or that they would have used the turketum for pedagogical rather than observational purposes?

In the eighteenth and nineteenth centuries, the turketum's history became shorter. Leading historians of astronomy, such as Jean Bailly in France or Rudolf Wolf in Germany, were convinced that Regiomontanus himself had invented the device in the fifteenth century.[27] Franco's text was not yet known and their histories of astronomy were overwhelmingly European in focus.

By the twentieth century, European historians began to learn about earlier Arabic observational activities, especially Nasīr al-Dīn al-Ṭūsī's thirteenth-century observatory at Marāgha. In 1928, a German translation was published of a list of instruments at the early Marāgha observatory, including one with two graduated, orthogonal circles. Citing this list, George Sarton in 1931 noted that some (he specified no names) had named al-Ṭūsī as the inventor of the turketum (Gunther in 1923 had asserted this claim).[28] Sarton, however, preferred Regiomontanus's suggestion of Geber, but cited no sources to support his speculation. In 1944, the French engineer and historian of instruments Henri Michel rejected Regiomontanus's suggestion. Looking at the 1533 publication

26 Hans Raeder, Elis Strömgren, Bengt Strömgren (eds.), *Tycho Brahe's Description of his Instruments and Scientific Work as given in Astronomiae instauratae mechanica* (*Wandesburgi 1598*), Ejnar Munksgaard, Copenhagen, 1946, p. 53.

27 Jean Silvain Bailly, *Histoire de l'astronomie moderne depuis la fondation de l'école d'Alexandrie jusqu'à l'époque de M.D.CC.XXX*, Chez les frères de Bure, Paris, 1779–1782, v. 1, pp. 687–689; Rudolf Wolf, *Geschichte der Astronomie*, Oldenbourg, Munich, 1877, pp. 161–162.

28 Hugo J. Seemann, "Die Instrumente der Sternwarte zu Marâgha nach den Mitteilungen von Al 'Urḍî", *Sitzungsberichte der physikalisch-medizinischen Sozietät zu Erlangen* 60 (1928), pp. 15–126; George Sarton, *Introduction to the History of Science*, Williams & Wilkens, Baltimore, 1927–1948, 3 vols.: v. 2, p. 1005.

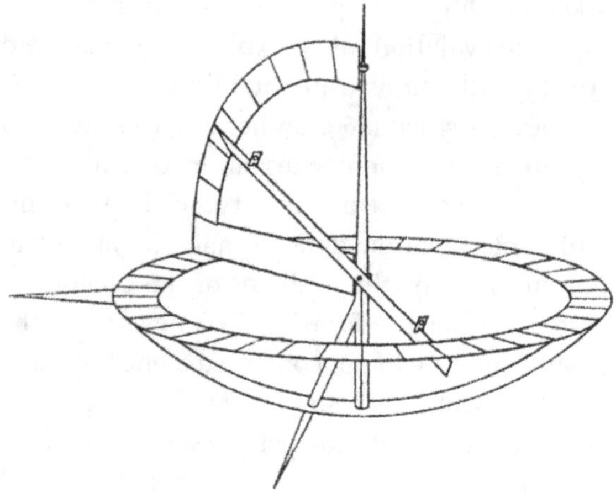

FIGURE 5.4
Lorch's reconstruction of the Arabic description of Geber's instrument. From Richard Lorch, "The Astronomical Instruments of Jābir ibn Aflaḥ and the Torquetum", *Centaurus* 20 (1976), p. 23

of a Latin version of Geber's text, Michel asserted, without providing any evidence, that Geber's device "differed sensibly" from the turketum described by Franco.[29]

Richard Lorch in a 1976 article fully investigated Geber's instrument and its relation to Franco's turketum.[30] Lorch collated five manuscripts (the earliest is dated 1229) of the Arabic *Iṣlāh al-Majisṭī*, translated the text, and tried to reconstruct a diagram of the instrument (the manuscripts contain no sketches). Geber did not name the device that, he wrote, consists of one "armilla" (metal hoop) in the ecliptic plane, one quadrant perpendicular to that and a pivoting bar with two sights (*Fig. 5.4*). This construction pivots on an equatorial axis (but lacks a circle in the equatorial plane). Like Ptolemy's armillary sphere, Geber's device can measure longitude and latitude if the user can align it to the ecliptic by sighting a celestial body, like the Moon, of known position.

Lorch also translated the Latin version of Geber's text, which likewise offers no name for the device, and reconstructed a sketch of the device it presents (*Fig. 5.5*). Consisting of an armilla for the ecliptic plane, this version uses two hinged rods (like the Ptolemaic ruler) to measure latitudes rather than a quadrant. As such, it is even more distant from Franco's turketum, structurally, than the device presented in the Arabic version. After these reconstructions, Lorch asserted that there is "no hint" of Geber's instruments in the list of Marāgha

29 Henri Michel, "Le Rectangulus de Wallingford précédé d'une note sur le Torquetum", *Ciel et terre* 60 (1944), pp. 196–204: 196; Jābir ibn Aflaḥ, "Liber Geber super almagesti", in P. Apianus (ed.), *Instrumentum primi mobilis*, Johann Petreius, Nuremberg, 1534.

30 Lorch, op. cit. (n. 23).

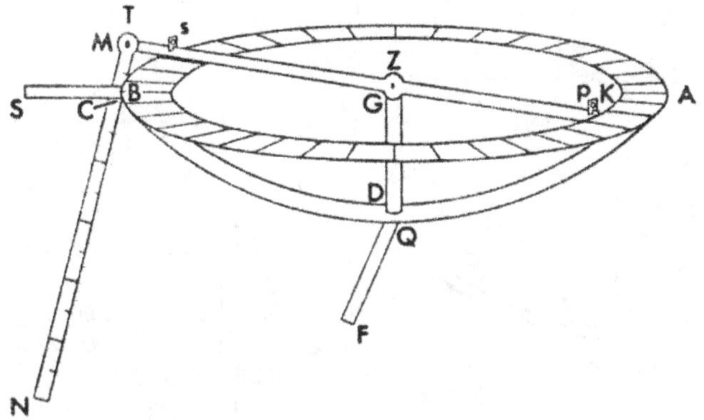

FIGURE 5.5 Lorch's reconstruction of the Latin description of Geber's instrument. From Richard Lorch, "The Astronomical Instruments of Jābir ibn Aflaḥ and the Torquetum", *Centaurus* 20 (1976), p. 27

instruments, that "in the east" the turketum was "apparently not known", that the "torquetum" was first mentioned by a late thirteenth-century Latin writer, and that in the Latin West Regiomontanus was the first to explicitly mention Geber's instrument. He concluded cautiously:

> If, therefore, Jābir did inspire the invention of the torquetum—an uncertain premise—it is likely that the inventor lived in the West in the thirteenth (or late twelfth) century, and that he read Arabic. Until new evidence appears, it would perhaps be unwise to speculate further.[31]

A final Arabic instrument that has been compared to the turketum is the *da'ire-yi mu'addel* or equatorial circle (literally, the day length circle). This device consists of a horizon plate to which is attached a hinged equatorial semi-circle that can be adjusted to given geographical latitudes by means of a notched arch. An alidade pivots around the axis of this semi-circle, on which is fixed orthogonally a semi-circular arc with a thin slit along its arc. A magnetic compass needle (magnetic declination of 7° is marked) is placed at the centre of the horizon plate, allowing the device to be oriented to the user's meridian.

31 *Ibid.*, p. 33. For a recent survey that finds no equatorially mounted instruments in Marāgha, see Mohammed Mozzafari, Georg Zotti, "The Observational Instruments at Marāgha Observatory after 1300", *Suhayl* 12 (2013), pp. 45–179.

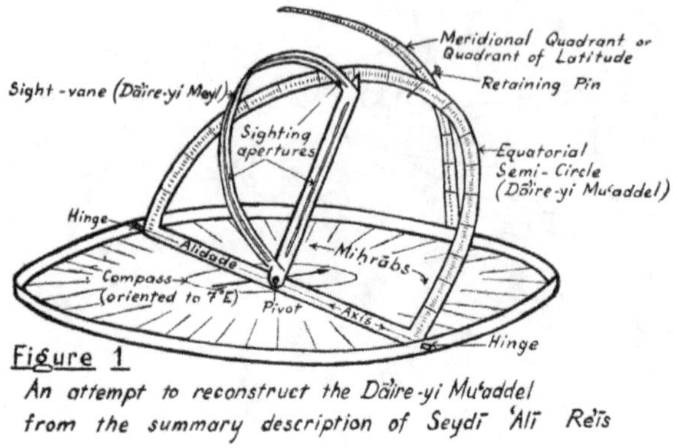

FIGURE 5.6
Reconstruction of the Arabic equatorial circle. From William Brice, Colin Imber, Richard Lorch, *The Dāʾire-yī Muʿaddel of Seydī ʿAli Reʾīs*, Manchester, 1976, p. 5

The slotted alidade enables the device to function as a sundial and to mark hour angles. The ecliptic is not represented on this instrument (*Fig. 5.6*).[32]

Apparently, this equatorial circle was widely used in eastern Islam. In 1961, Sevim Tekeli published Turkish and English translations of a late fifteenth-century Arabic text by the Egyptian author Muhammad al-Wafai that briefly describes the instrument and its use. In 1976, Brice, Imber and Lorch edited and translated a later and shorter Arabic text by the well-known sixteenth-century Ottoman admiral and geographer, Seydi ʿAli Reis. Tekeli suggested that the equatorial circle "has the characteristic[s] of the torquetum", asserting (without evidence) that the ecliptic circle "was added later" to Franco's device.[33] Brice, Imber and Lorch do not mention the turketum. In any case, the Arabic texts were written much later than Franco's and they do not help us answer our question about how the turketum acquired its name.

Another precursor, that Needham called an "equatorial torquetum", appears in a list of seventeen instruments documented at the imperial Chinese observatory from 1276–1279 (Yuan Dynasty). Apparently devised by Kuo Shou-Ching around 1270, this large instrument had an equatorial mount on which circles

32 For an analysis of the geometry of this dial, see William Brice, Colin Imber, Richard Lorch, *The Dāʾire-yī Muʿaddel of Seydī ʿAli Reʾīs*, « Seminar on Early Islamic Science 1 », University of Manchester, Manchester, 1976, pp. 6–14, and the critical review by David A. King, "An Islamic Astronomical Instrument", *Journal for the History of Astronomy* 10 (1979), pp. 51–53.

33 Sevim Tekeli, "İzzüddin b. Muhammed Al-Vefaî'nin 'Ekvatör Halkası' Adlı Makalesi ve Torquetum ('Equatorial Armilla' of Iz Al-Din b. Muhammad Al-Wafai and Torquetum)", *Ankara Üniversitesi, Dil ve Tarih Coğrafya Fakültesi Dergisi* 18, 3–4 (1961), pp. 227–259: 241. For a photograph of an equatorial circle now at the Kandilli Observatory in Istanbul, see King, op. cit. (n. 32), p. 52; for a seventeenth-century equatorial circle once in the National Museum in Damascus, see Seyyed Hossein Nasr, *Islamic Science: An Illustrated Study*, World of Islam Festival Publishing Company Ltd., London, 1976, p. 44.

for the equator and local horizon turn, allowing direct conversion between equatorial and alt-azimuth coordinates. Needham speculated that the equatorial design was "invented" by Geber; he did not suggest, however, that information about Kuo's instrument had made its way to Paris by the 1280s.[34]

We turn, finally, to attempts to understand the linguistic origins of 'turketum', the mysterious term employed in the earliest Latin texts to describe our instrument. The most widely circulated etymological 'origins story' for this term makes the linguistic assumption that words sounding similar (phonetically similar) must have shared semantic roots (rather than emerging independently). Hence, since the medieval Latin words for 'Turkey' and 'Turks' ("turchia", "turquia", "turcus") sound similar to 'turketum', our device must have Turkish roots.

The earliest example I have found of a scholar positing a 'Turkish' origins story appears in the description of the oldest extant brass example of our device, purchased in 1444 in Nuremberg by Nicholas of Cusa (see above). In 1919, the German historian of instruments Hartmann reported finding in Nicholas's fifteenth-century manuscripts the term "turketum". Discounting earlier manuscripts bearing the term as not referring to this instrument,[35] Hartmann wondered who, in early fifteenth-century Nuremberg, could have constructed Nicholas's instrument. Finding in the same manuscript (Cues 211)[36] an astronomical table for computing times of true syzygy (astronomical conjunctions or oppositions) "per Nycholaum dictum de Heybech", Hartmann guessed that this Nicholas had authored other tables in Cues 211 that are computed for the meridian of Nuremberg. A Nicholas also appears several times in protocols of the Nuremberg City Council; in 1431, they had paid "pro m[agistr]o Nicolau" 50 florens for a clock and "pro Turcis, hospiti ipsorum" 40 florens. Hartmann eagerly speculated:

> If this Nicholas were identical with Nicholas de Heybech, we could see in him the inventor of the turketum. We might assume that "Turcis" refers

34 Joseph Needham, *Science and Civilization in China*, Cambridge University Press, Cambridge, 1954–2016, 7 vols.: v. 3, pp. 370–378.

35 Siegmund Günther had mentioned, as an early reference to the "torquetum", the canon to John Lignères' *Tabulas primi mobili* (dated 1322). However, J. Hartmann, op. cit. (n. 2), p. 9, correctly realized that this passage referred to the Ptolemaic ruler, constructed from three rigid bars "vel ferro vel cupro fortes et rigidas, ut de livi torqueri non possint, quadrilatere superficiei". See Maximilian Curtze, "Urkunden zur Geschichte der Trigonometrie im christlichen Mittelalter, No. 7, Die 'Canones tabularum primi mobilis' des Ioannes de Lineriis", *Biblioteca mathematica* 3, Ser. 1 (1900), pp. 390–413: 397. Hartmann knew neither Franco's nor Bernard's text.

36 See n. 2.

> to his assistant or at least accomplice, which would give us an explanation of the hitherto completely unknown designation of the turketum as an 'instrument of the Turks' [*Türkeninstrument*]. We could guess that Heybech, during his long stay in 1431, worked for a Nuremberg Bürger; supported by that town's well-known skilled metal workers, he built a turketum that was named after Heybech's [Turkish] assistant. Later it would assume the name torquetum.[37]

Hartmann's story is very specific; in 1431 a "Turk" in Nuremberg had constructed a device that henceforth would be called the "instrument of the Turks".

The leading instruments historian of his generation, Ernst Zinner, also accepted this word association, coining the term "instrument of the Turks" or *Türkengerät* in his 1922 review of Hartmann's book. Already in this piece Zinner brought Franco's 1284 text into connection with Nicholas of Cusa's instrument; in his 1925 inventory of astronomical manuscripts, Zinner listed fourteen manuscripts of Franco's text. By 1931, Zinner read "Turkish" to mean "Arabic" and, silently following Regiomontanus, named Geber's as the model instrument that had prompted Franco's "invention".[38] In his 1967 book on astronomical instruments, Zinner revised this claim, offering a curious argument about the technical terms for the instrument's components:

> The *Türkengerät* belongs to those medieval instruments that were not borrowed from the Arabs or antiquity, as follows from the lack of Arabic or Greek technical terms in its description. Already in the earliest text, by Franco, it is called a '*turketum*', which means '*Türkengerät*'.[39]

---

37 Hartmann, op. cit. (n. 2), p. 14. For a more recent analysis of Nicholas's astronomical manuscripts that judges Hartmann's conclusions as "not acceptable", see Alois Krchnák, "Die Herkunft der astronomischen Handschriften und Instrumente des Nikolaus von Kues", *Mitteilungen und Forschungsbeiträge der Cusanus-Gesellschaft* 3 (1963), pp. 109–180: 167–168.

38 Ernst Zinner, [Review of Hartmann, op. cit. (n. 2)], *Vierteljahresschrift der Astronomischen Gesellschaft* 57 (1922), pp. 9–19: 14; Zinner, *Verzeichnis der astronomischen Handschriften des deutschen Kulturgebietes*, Beck, Munich, 1925, p. 99; Zinner, *Geschichte der Sternkunde von den ersten Anfängen bis zur Gegenwart*, Springer, Berlin, 1931, p. 394.

39 Zinner, *Deutsche und niederländische astronomische Instrumente*, cit. (n. 3), p. 177: "Das Türkengerät gehört zu den mittelalterlichen Geräten, die nicht von den Arabern oder vom Altertum übernommen wurden, wie aus dem Fehlen besonderer arabischer und griechischer Fachwörter in ihren Beschreibungen hervorgeht. Bereits in der frühesten Arbeit, nämlich des Franco, wird es Turketum genannt, was Türkengerät bedeutet." Zinner did not comment on the term "alidade" used by Franco. See n. 13.

If no Arabic terms means the instrument was not borrowed from the Arabs, why assume that an instrument given Latin terms was borrowed from the Turks? The logic seems flawed; nonetheless, Zinner gave preference to the name "turketum" in his origin story.

Since Zinner, a non-European heritage of the name, if not the instrument, has remained firmly fixed in the historiography. Hence, we read in the 1998 *Instruments of Science: An Historical Encyclopedia* that the early manuscripts describing our instrument use the term "turquetus" or "turketus" which "must have meant the 'Muslim instrument', *turkus* being a medieval Latin word covering a more general area than Turkey or Turkmenistan alone. It is this original meaning that is reflected in the English 'turquet' and German *Türkengerät*".[40]

At least one other word association has been proposed to explain our name. In his 1964 article on the instrument, Poulle noticed that one of the Bernard manuscripts (P1) employs the spelling "truquetus" (this term appears only once in all the manuscript copies of the early texts I have examined). This word reminded Poulle of the Old French "troche", a word he found defined in Godefroy's *Dictionnaire de l'ancienne langue française* as "assemblage, faisceau, masse, troupe, quantité". This is "without doubt", concluded Poulle, "the same idea Bernard of Verdun had when, wishing to evoke with a word the instrument he was about to describe, he was struck by its strange silhouette".[41]

Interestingly, Poulle did not ask about the origins of "troche", surely related to the Latin "trochus" (metal hoop) and the Greek "τροχός" (wheel). In ancient Greek, the term usually appears in technological contexts, referring to a cartwheel, block and tackle, parts of a ship or round stones shaped by water. Early Latin writers, however, used the term in poetic texts, referring to hoops rolled by playing children or exercising youth.[42] In several of these poems, the verb "ludere" (to play/tease/trick) is linked to the "trochus". Such a word association would make the turketum a tool of delight, entertainment and imagination, an image that I have found associated with other early modern astronomical instruments.[43] But based on an ambiguous reading of a single manuscript, such an association might be only my fantasy.

---

40 Giles M. Hudson, "Torquetum", in Robert Bud, Deborah Jean Warner (eds.), *Instruments of Science: An Historical Encyclopedia*, Science Museum, London, 1998, pp. 623–626: 624.

41 Frédéric Godefroy, *Dictionnaire de l'ancienne langue française*, Ministère de l'instruction publique, Paris, 1881–1902, 10 vols.: v. 8, p. 80; Poulle, op. cit. (n. 1), pp. 204–205.

42 R.S.P. Beekes, *Etymological Dictionary of Greek*, Brill, Leiden, 2010, pp. 1506–1507; P.G.W. Glare (ed.), *Oxford Latin Dictionary*, Oxford University Press, Oxford, 2012^2, p. 2182. Cf. Horace, *Carmina* 3.24; Ovid, *Ars amatoria* 3.380; Ovid, *Tristia* 3.12.19–22; Martial, *Epigrams* 14.168:2; or Sextus Propertius, *Elegies* 13.14:6.

43 Richard L. Kremer, "Playing with Geometrical Tools: Johannes Stabius's *Astrolabium Imperatorium* (1515) and Its Successors", *Centaurus* 58 (2016), pp. 104–134.

5 From 'Origins Stories' to History

I have not been able to identify any single text that answers our question about how the turketum (or torquetum) got its name. If, however, we turn from the etymological (word associations) and the morphological (structural or design associations) to the historical, perhaps we could reframe our question, at least about the "Turkish" association. In the 1280s, what could Franco or Bernard, two authors working in or near Paris, have known about "the Turk"?

It is widely known that, after Mehmet II's conquest of Constantinople in 1453, all things Turkish quickly flooded Western Europe. The new printing press, already in 1454, issued a so-called "Turkish Calendar", an astronomical rhyme calling for a crusade "widd die durken" (against the Turks).[44] Popes sold indulgences to raise funds for this crusade. The 1454 German Reichstag debated this topic. The greatest European orator of the time, the Sienese Enea Silvio Piccolomini (later Pope Pius II) preached crusade. The "Turk" was the enemy. Yet with fear came curiosity and orientalizing, sparked by a burst of travel and trade between Venice and the Ottoman East. "Turkish" carpets and tapestries, etched "turkish" armour, "turkish" spices, paintings of "Turks", even "turkish women's headdresses" (*Fig. 5.7*) suddenly were in great demand across Europe.[45]

However, 1453 was more than one hundred fifty years after Franco and Bernard deployed the term "turketum". Might Latin words for "Turks" have been circulating in late thirteenth-century Europe? Fully investigating this topic, of course, would far exceed the bounds of this chapter. We can find, however, some scattered hints. In *Histories*, IV, 22, Herodotus had mentioned an ancient nation, living on the north-east trade route beyond the Thyssagetae (north of the Caspian Sea), that he called the "Iyrcae". In Pomponius Mela's *Geography*, I, 116, these people became the "Turcae"; in Pliny's *Natural History*, VI, 7, the

---

44 Issued by the "Printer of the 36-line Bible" (probably Gutenberg), this edition is the oldest European book printed with moveable type that is preserved as a complete copy. See Eckehard Simon, *The* Türkenkalender *(1454) Attributed to Gutenberg and the Strasbourg Lunation Tracts*, Medieval Academy of America, Cambridge, 1988.

45 For recent examples from this massive literature, see James Harper (ed.), *The Turk and Islam in the Western Eye, 1450–1750*, Ashgate, Farnham, 2011; Andrei Pippidi, *Visions of the Ottoman World in Renaissance Europe*, Hurst, London, 2012; Karoline Dominika Döring, *Türkenkrieg und Medienwandel im 15. Jahrhundert, mit einem Katalog der europäischen Türkendrucke bis 1500*, Matthiesen, Hsum, 2013; Charlotte Colding Smith, *Images of Islam, 1453–1600: Turks in Germany and Central Europe*, Pickering & Chatto, London, 2014; Sabine Haag, Guido Messling (eds.), *Occident and Crescent Moon: The Ottoman Orient in Renaissance Art*, Kunsthistorisches Museum, Vienna, 2015.

FIGURE 5.7 "Balzo alla turca" (Turkish-style headdress). Both its name and turban form point to European fascination with "Turkey". Parmigianino, *Portrait of a Young Woman*, c. 1530
VIENNA, KUNSTHISTORISCHES MUSEUM, INV. GG 327

"Tyrcae". Such geographical texts would have supplied European readers with a name for a people located somewhere north of Anatolia.[46]

By the eleventh century, as the Seljuk Turks began to expand their influence (taking control of Baghdad in 1055, of Damascus and Aleppo in 1076 and 1086, respectively), Byzantines devoted increasing attention to these

46 Antonín Salač, "Über den Namen der Türken im Griechischen und Lateinischen", *Listy filologické/Folia philologica* 80, 2 (1957), pp. 50–55: 54.

invaders and began communicating images of "Turks" to European ambassadors, mercenaries, merchants and pilgrims. As Morton has argued, "the Greeks appear to have exerted a profound influence over early Western European ideas concerning the Turks".[47] But it was the historical chronicles written after the First Crusade that spread representations of "Turcus/Turci" widely throughout Europe. Baldric of Bourgueil's *Historia Hierosolimytana* (c. 1105), Guibert of Nogent's *Dei gesta per Francos* (ca. 1108), or Fulcher of Chartres's *Historia Hierosolymitana* (1099–1127) list many different "pagans" encountered by the Christian crusaders but the term "Turks" invariably dominates these narratives. By early in the thirteenth century, Latin readers would have known the "Turks" as infidels, clever military tacticians, brutal warriors, and a people who thought they "shared an ancestry with the Franks, and that the highest military prowess belonged particularly to the Turks and Franks, above all other people".[48]

As far as I know, however, none of these crusade histories mention scholarly or astronomical activities among the Seljuk Turks. A recent study of scholars in the early Ottoman Empire finds several *madrasas* established in the 1330s in Bursa and İznik; but these educational centres would begin to expand only in the early fifteenth century.[49] Likewise, immigrant scholars from Egypt, Syria, Iraq and Yemen began to arrive in Anatolia by 1400. Given this context, it seems quite unlikely that an astronomical text of "Turkish" origins could have reached Paris by the 1280s.

Franco and Bernard surely would have known about "Turks". Perhaps they would have localized these people to Anatolia. Perhaps they understood "Turk" as a more general term for Arabs or Muslims. Perhaps they had encountered an Arabic text about an instrument similar to ours. But would Latin readers have assigned a "Turkish" provenance to an Arabic text describing an astronomical instrument that simplifies the armillary sphere? Until we find clues about such

47 Nicholas Morton, "Encountering the Turks: The First Crusaders' Foreknowledge of Their Enemy: Some Preliminary Findings", in Simon John, Nicholas Morton (eds.), *Crusading and Warfare in the Middle Ages*, Ashgate, Farnham, 2014, pp. 47–68: 67.

48 Guibert of Nogent, *The Deeds of God through the Franks, a Translation of Guibert De Nogent's 'Gesta dei per francos'* (trans. by Robert Levine), Boydell Press, Woodbridge, 1997, p. 61.

49 Abdurrahman Atçil, *Scholars and Sultans in the Early Modern Ottoman Empire*, Cambridge University Press, Cambridge, 2017, p. 31. Cf. Ahmet Yaşar Ocak, "Social, Cultural and Intellectual Life, 1071–1453", in Kate Fleet (ed.), *The Cambridge History of Turkey, Volume 1: Byzantium to Turkey, 1071–1453*, Cambridge University Press, Cambridge, 2009, pp. 353–422.

an imagined Arabic text, such historical speculation cannot provide a satisfactory answer to the question in our title.[50]

6 Conclusion

In 1997, in his final publication on our instrument, Lorch concluded: "The origin of the instrument and also its name is unknown. We do not know whether it was invented by the Arabs (the universal instrument of Geber is different)".[51] Alas, as indicated by my review of the scholarship over the past twenty years, Lorch's conclusion remains intact. The sources hitherto brought into play do not provide an unambiguous 'origins story' for the turketum or its name. I have located one previously unknown source, a late fifteenth-century text by Johann de Epperies, but it offers nothing relevant for the question of provenance.

Perhaps, therefore, we should narrow our question. Rather than seeking to trace the origin of the instrument, i.e., looking for texts firmly dated before 1284 that describe an instrument that we judge to be similar in design and function to the turketum, we should focus on the Latin name "turketum" that we find in nearly all manuscript copies of the earliest known texts. If Franco or Bernard did not invent this word "out of thin air", how did they come to employ it as a name for their instrument? If this question allows an answer, surely it will come from a study of the circulation and transformation of people, codices and words in late thirteenth-century Paris, the place where the earliest Latin texts emerged. We have well documented accounts of how early astrolabes traveled from Arabic Spain to Latin Europe.[52] Since, however, we lack both textual sources and brass artefacts for the early history of our instrument, we might focus instead on words deployed in other contexts that late thirteenth-century Parisian readers might have encountered.

50 A rich scholarship exploring issues of transmission, translation, and transformation in medieval interactions between Islamic and Latin cultures has emerged in the past several decades. See F.J. Ragep, Sally P. Ragep (eds.), *Tradition, Transmission, Transformation: Proceedings of Two Conferences on Pre-Modern Science Held at the University of Oklahoma*, Brill, Leiden, 1996; Mohammed Abattouy, Jürgen Renn, Paul Weinig (eds.), "Experience and Knowledge Structures in Arabic and Latin Sciences", *Science in Context* 14, 1–2 (2001), pp. 1–12; Robert Wisnovsky *et al.* (eds.), *Vehicles of Transmission, Translation, and Transformation in Medieval Textual Culture*, Brepols, Turnhout, 2011; Hasse, op. cit. (n. 22).

51 Richard Lorch, "Turquetum", in *Lexikon des Mittelalters*, Artemis-Verlag, Munich, 1997, 9 vols.: v. 8, p. 1120.

52 Cf. Borst, op. cit. (n. 13); Petra Schmidl, "Knowledge in Motion: An Early European Astrolabe and its Possible Medieval Itinerary", *Medieval Encounters* 23 (2017), pp. 149–197.

Acknowledgments

For their suggestions and advice, I thank David Abosso, Silke Ackermann, Gaye Danışan, Oliver Duntze, Falk Eisermann, Péter Ekler, Cecilia Gaposchkin, Alena Hadravová, David Juste, Christoph Mackert, Laure Miolo, C.E.P. Nothaft, Michael Shank, Robert Stewart, Edith Sylla, Jarosław Włodarczyk, and especially the late Inge Elmqvist Söderlund who organized the session where this chapter was originally presented.

Appendix: Manuscript Copies of the Early Turketum Texts

This preliminary and incomplete list includes only manuscripts that I have inspected or that have been described by printed catalogues. Dates are taken from the catalogues, incipits from Lynn Thorndike and Pearl Kibre, *Catalogue of Incipits of Mediaeval Scientific Writings* (rev. and augmented ed.), Mediaeval Academy of America, Cambridge MA, 1963 (abbreviated to Th/K, below), and the terms listed are found in the manuscripts.

	Library Siglum
B	Berlin, Staatsbibliothek
E	Erfurt, Bibliotheca Amploniana, now incorporated in the Universitätsbibliothek Erfurt
F	Florence, Biblioteca Nazionale Centrale
G	Geneva, Bibliothèque de Genève
I	Innsbruck, Bibliothek des Tiroler Landesmuseums Ferdinandeum
J	Jena,Thüringer Universitäts und Landesbibliothek
L	Leipzig, Universitätsbibliothek
Lo	London, British Library
Mi	Milan, Biblioteca Ambrosiana
M	Munich, Bayerische Staatsbibliothek
MU	Munich, Universitätsbibliothek der Ludwig Maximilians Universität
O	Oxford, Bodleian Library
P	Paris, Bibliothèque nationale de France
Po	Pommersfeld, Schlossbibliothek, Schloss Weißenstein
Pr	Prague, National Library of the Czech Republic
S	Seville, Biblioteca Capitular y Colombina
U	Uppsala, Universitetsbibliotek
V	Vatican City, Biblioteca Apostolica Vaticana
VeM	Venice, Biblioteca Nazionale Marciana
VeC	Venice, Museo Civico Correr

Vi	Vienna, Österreichische Nationalbibliothek
W	Würzburg, Universitätsbibliothek
Y	New Haven, Yale University Medical Library

Manuscript Lists

Franco de Polonia, De nominibus partium instrumenti quod turketus dicitur, Th/K 383

E1	1300	CA 4°349, fols. 160r–161v	turketus, explicit dated 1284
P1	1300	Lat. 7333, fols. 49r–52v	turketus, explicit dated 1284
E2	1325	CA 4°369, fols. 174r–v	turketus
E3	1350	CA 4°386, fols. 105v–107v	turketus
Lo	1350	Royal 12.C.XVIII, fols. 88–91	turketum
VeM	1350	Lat. XI.165(4254), fols. 57r–62r	turketum
Po	1350	Ms 129, fols. 8–9	torketum (Zinner 1925)
Vi1	1360	Cod. 5311, fols. 64v–66v	targeti, torqueti, turgeti, explicit dated 1284
O2	1400	Digby 167, fols. 62–63	turketus, explicit dated 1284
L1	1400	Ms 1469, fols. 211r–214r	turquetus
U	1400	C 653, fols. 59v–61r	turketum
V1	1400	Pal. lat. 1122, fols. 117r–118v	turketus, explicit dated 1284
Mi2	1400?	H 75 sup., fols. 65r–66v	turketus, magister Franchone de polonia
Mi3	1400?	H 75 sup., fols. 70r–73r	turketum, explicit dated 1284
O1	1425	Ashmole 1522, fols. 176r–178v	turketus (ed Gunther 1923)
Pr	1425	III.C.2, fols. 9r–11r	turgeti
Y1	1433	Ms 25, fols. 57r–62v	torquetum (additional text plus "De torqueti correctione")
Mi1	1450	C 299, fols. 183–184	tordetus/torquatus
VeC	1450	Cic. 3747(2712), fols. 167r–168v	turcetum
B	1450	Lat. F.246, fols. 81v–82v	torquetus (rubric), turketus (incipit dated 1284), invento in regia urbe Pragensi [instead of Parisiensis?]
P2	1450	Lat. 7197, fols. 15r–16v	torketus, explicit dated 1284, magister francone de polonia parisiensis
P3	1450	Lat. 7295, fols. 14–18	turketus
P4	1450	Lat. 14068, fols. 2r–5r	turketus
V2	1450	Pal. lat. 1340, fols. 28v–30v	turketus
Vi2	1450	Cod. 5258, fols. 18r–21v	turketus

M1	1480	Clm 25004, fols. 140–142	turceti
Y2	1480	Ms 24, pp. 421–426	turchetus
M2	1486	Clm 125, fol. 286	turketum (Zinner 1925)
L2	1490	Ms 1469a, ff, 2r–3r (incomplete)	torquetus
H	1494	Halliwell Ms 9[53]	torquatem, Magister Franco
G	1500	Lat. 80, fols. 22r–24r	turketus, explicit dated 1284, magister franchone de polonia
Vi3	1520	Cod. 5228, fols. 30v–33v	turketus (same additional text as Y1, Johannes Schöner autograph)
W	1525	I.t.q.236, fols. 75r–80r	torquetum (same additional text as Y1)
S	1530	Ms 7-7-12, fols. 92r–94v	turchetus
J1	1536	El. f. 73, 196r–197r	? (fragment, lacks first sentence)
M3	1550	Clm 92, fols. 132–142	turthetum

Bernard de Verdun, Si facies instrumentum ad modum turketi, Th/K 1479

P1	1300	Lat. 7333, fols. 46v–48r	truqueti[54]
P5	1350	Lat. 7334, fols. 24r–26r	turketi
E3	1350	CA 4°386, fols. 23v–24r	turketi
M4	1436	Clm 10662, fols. 224r–v	torqueti
Vi2	1450	Cod. 5258, fols. 134r–v	tarqueti
V3	1450	Lat. 3097, fol. 72v	trunqueti

Anonymous, Si volueris componere turchetum accipe laminam, Th/K 1476

MU	1450	4°738, fols. 84r–87v	turchetum
Vi2	1450	Cod. 5258, fols. 16r–18r	turchetum (ed. Poulle 1964)
Y2	1480	Ms 25, pp. 417–426	turchetum (same additional text as Y1)
L2	1485	Ms 1469a, fols. 1r–2r	turketum

53 Samuel Leigh Sotheby, *A Catalogue of a Collection of Scientific and Historical Manuscripts [...] Forming a Portion of the Very Valuable Collection of Manuscripts of James Orchard Halliwell*, Compten and Ritchie, London, 1840, p. 9. The current location of this codex is unknown.

54 Polykarp Hartmann, op. cit. (n. 16), p. 140, expanded the abbreviated term "truqti" as "trunqueti", Poulle, op. cit. (n. 1), p. 204 as "truqueti". Since the scribe wrote only one tilde above the "q", Poulle's reading is preferred. Note, however, that in the later V3 the scribe added a tilde above both the "u" and "q", which I expand as "trunqueti."

Vi3	1520	Cod. 5228, fols. 155r–156v	torquetum (Johannes Schöner autograph)
W	1525	I.t.q.235, fols. 74r–75v	torquetum (same additional text as Y1)

Anonymous, Si quis vult componere turchetum accipe laminam, Th/K 1463 (abbreviated paraphrase of Th/K 1476)

Y1	1433	Ms 25, fols. 56r–57r	torquetum
Vi4	1450	Cod. 5296, fols. 168r–173v	torquetum
I	1475	W.3277, fols. 24r–26v	torquetum
M6	1475	Clm 27, fols. 132–136	turketum
J1	1536	El. 2. 73, fols. 195r–196r	torquetum
Vi5	1550	Cod. 5277, fols. 118r–123v	torquetum
M7	1550	Clm 19689, fols. 1–7	?

Regiomontanus, Quia astronomicis student exercitiis, Th/K 1204

Vi3	1469	Cod. 5496, fols. 14r–30v	torquetus (Regiomontanus autograph)
F	1470	Magliabech. XI.144, fols. 109r–121v	torqueti (rubric)
J2	1501	El. f. 72, fols. 86r–86v	torquetus
M4	1527	Clm 28144, fols. 11r–12v	torquetus (Jakob Ziegler paraphrase)
J1	1536	El. f. 73, fols. 191r–194v	torquetus
M5	1550	Clm 19689, fols. 13–27	torquetus
Vi4	1500	Cod. 5292, fols. 95r–103v	torquetus

Johann de Epperies, Vtilitates Turketi pro serenissimo et illustrissimo principe domino Matthia, dei gratia Rege Hungarie

L2	1490	Ms 1469a, fols. 3v–8r	turketum

CHAPTER 6

A Mingling of Traditions

Aspects of Dialling in Islam

Anthony Turner

1 Islamic Dials in the Fourteenth and Fifteenth Centuries

> Among instruments he cut the plates of sundials in such a way that one would stay standing to admire them, because of their elegance, because of the beauty of the script, because of the way in which the work was balanced and correctly laid out. In this subject the maker reached a very high rank and attained a level well above that of many scholars of previous ages. He left far behind him the *khamā'irrīyat*[1] and the *saffārīyat*,[2] just as he did all the other instruments made by serious makers. People competed to show their zeal to buy them at the highest price.[3]

Thus, in his *Inhāta,* the fourteenth-century biographer, Ibn al-Khatīb, described Ibn Baso (d. 709 H/1309–1310 CE). Ibn Baso was responsible for the times of the prayers in the great mosque of Granada, acting in effect as a *muwaqqit* although the term itself is not used, and was famous for his construction of astronomical instruments. *Muwaqqits* are a class of astronomers who first appear as a distinct, so named, group in thirteenth-century Egypt. Specialists in

1 Instruments made by Muhammad b. Fattūh al-Khamā'irī, a noted maker in Seville from whom some fourteen instruments have survived. See under this name in Alain Brieux and Francis Maddison, *Répertoire des facteurs d'astrolabes et de leurs œuvres,* in preparation. Pending the appearance of this, see the summary listing given by David A. King, "A Checklist of Islamic astronomical instruments to *ca.* 1500, ordered chronologically by region", in D.A. King, *In Synchrony with the Heavens. Studies in Astronomical Timekeeping and Instrumentation in Medieval Islamic Civilization*, Brill, Leiden and Boston, 2004–2005, 2 vols.: v. 2, "*Instruments of Mass Calculation*" (2005), pp. 993–1020: 1010, entry 6.2.

2 Instruments made by Ibn al-Sāffar, an equally noted maker. See under this name in Brieux and Maddison, op. cit. (n. 1), and King, op. cit. (n. 1), p. 1006, entry 3.3.

3 Translated from H.J.P. Renaud, "Notes critiques d'histoire des sciences chez les Musulmans I: les Ibn Bāso", *Hesperia* 24 (1937), pp. 1–12: 7. On Ibn Baso see now also Emilia Calvo Labarta, *Abū 'Alī al-Husayn ibn Baso (m. 716/1316) Risalāt al-Safīha al-Yāmi'a li-Yamī' al-'urūd (Tratado sobre la lamna general para todas las latitudes. Edicion critica, traccion y estudio,* « Fuentes Arabico-Hispanas 13 », Consejo superior de investigaciones cientificas, Instituto de Cooperación con el Mundo Arabe, Madrid, 1993.

 | DOI:10.1163/9789004412842_007

ʿilm al-mīqāt (astronomical time reckoning for religious purposes), they had a formal attachment to a mosque, as other specialists of this type could have, although a parallel group, the *mīqāti*, specialists in spherical astronomy and astronomical timekeeping, were not generally specifically associated with a religious institution. Both groups occupied themselves with instruments, in particular astrolabes and sundials.[4]

By the fourteenth and fifteenth centuries, sundials had an accepted place in Arab-Islamic society. Not only did they have an intrinsic mathematical interest, but they were useful to *muwaqqits* and *mīqāti*. Their integrated place in society and culture emerges clearly from a mid-fifteenth-century work by al-Qalqashandi describing the structure and personnel of the state chancery. In it he gives a classified bibliography of the most useful books in the fifty-five sciences with which he considered a competent, professional chancery scribe should be acquainted. Timekeeping and sundials find their natural place in this scheme being, respectively, the second and fifth of the five sections of the fifth chapter which deals with astronomy. Of the two works recommended for the determination of hours, one was the treatise of al-Marrākushī,[5] the other an anonymous hand-book for *muwaqqits*. For sundials al-Qalqashandi says that a certain number of convincing books are known and he gives an interesting illustration of their longevity by mentioning only the tenth-century work on the subject by Ibrahim b. Sinan al-Harranī.[6]

Erected in mosques and *madrasas*, sundials, apart from their immediate utility in prayer-time determination,[7] were also public indicators of the skill in mathematics of the person who designed them. A complex dial was therefore an appropriate object for a ruler to commission from a favoured mathematical

4 For details see D.A. King, "On the role of the *muezzin* and *muwaqqit* in medieval Islamic societies", in King, *In Synchrony with the Heavens*, cit. (n. 1), v. 1, pp. 623–677.

5 Abū ʿAlī al-Hasan ibn ʿAlī ibn ʿUmar al-Marrākushī, *Jāmiʿ al-mabādī wa-l-jāyāt* (All the Beginnings [Principles] and Ends [Results]), a treatise on the astronomical instruments used in Islam with an extensive treatment of sundials.

6 Gaston Wiet, "Les classiques du scribe Egyptien au XVᵉ siècle", *Studia Islamica* 18 (1962), pp. 41–80: 55. Ibrahim b. Sinan is Abū Ïshāq Ibrāhīm ibn Sinān ibn Thābit (*c.* 908–*c.* 946), grandson of the famous Thābīt ibn Qurrah. In the *Fihrist*, al-Nadīm says of him that "although he died at a youthful age, he excelled and was pre-eminent in the science of geometry. During his time no one appeared who was more brilliant than he was". Bayard Dodge (ed. and tr.), '*The Fihrist of al-Nadīm*': *a Tenth Century Survey of Muslim Culture*, Columbia University Press, New York, 1970, 2 vols.: v. 2, pp. 648–649.

7 "For the daytime prayers, a sundial marked with curves for the prayer-times obviated the use of any tables; indeed a single sundial in a mosque could actually replace the services of a *muwaqqit* during the daytime provided the sun was shining." King, "On the role of the *muezzin* ...", cit. (n. 4), p. 654.

servant. Conversely it could also be an appropriate object for a scholar summoned to a new, and lucrative, appointment to erect. Such was perhaps the case with a dial by 'Alā ad-Din 'Alī bin Muhammad al-Qawshajī or Qūshjī (d. 879 H/1474 CE), a noted astronomer, mathematician and grammarian. The son of a falconer in the service of Ulugh Bey, 'Alī studied at Samarkand and at Kirman. "At a tender age and at the prime of his youth [he] carried away the reed of pre-eminence in the vast realm of the sciences".[8] Returning to Samarkand, 'Alī worked in Ulugh Bey's observatory where he completed the *zīj al-jadīd*. After the death of Ulugh Bey, 'Alī took service with the Sultan Uzun Hasan at Tabriz, by whom he was sent to Istanbul as ambassador to the Ottoman Sultan Mehmet II. Mehmed conceived a great affection for the learned ambassador and on the completion of his mission attached him to his service to teach mathematics and astronomy first at the *madrasa* attached to the Hagia Sophia mosque, later at the Fatih mosque and *kulliye* (place of higher learning) founded about 1463. In 878 H (1473–1474 CE), shortly after its completion, 'Alī adorned it with a vertical dial, but no other work of this kind by him is known although he wrote several treatises on mathematical and astronomical subjects.[9]

If 'Alī Qūshjī's dial supplied a model or a stimulus for the erection of dials in Istanbul in later centuries, then they may have been the one-off productions of teachers or mathematical practitioners. Surviving examples are often not signed, and when they are the makers are seldom known for any other instruments.[10] One dial, which predates that of 'Alī Qūshjī, was erected 812 H (1409–1410 CE) on the mosque of Hadji Hasan, Konya, by a goldsmith, Hasan al-Sā'igh.[11] Most of the dials known are found in mosques or *madrasas*, only

---

8 Introduction to Ulugh Bey, *al-Zīj al-jadīd*, quoted from Aydin Sayılı, *The Observatory in Islam and its Place in the General History of the Observatory*, « Publications of the Turkish Historical Society, ser. 7, no. 38 », Türk Tarih Kumuru Basimevi, Ankara, 1960; anastatic reprint: Arno Press, New York, 1981, p. 265.

9 For 'Alī see L.A. Mayer, *Islamic Astrolabists and Their Works*, Albert Kundig, Geneva, 1956, pp. 42–43; Brieux and Maddison, op. cit. (n. 1), who explain the transliteration of the name that has been followed here; C.A. Storey, *Persian Literature: A Bio-Bibliographical Survey* (reprinted edition), Luzac & Co., London, 1972, 2 vols.: v. 2, pp. 9–10 and 75–77, and references there given. For his work at Samarkand see Sayılı, op. cit. (n. 8), pp. 27ff.

10 For surviving dials in Istanbul and elsewhere in Turkey see A. Süheyl Ünver, "Sur les cadrans solaires horizontaux et verticaux en Turquie", *Archives Internationales d'Histoire des Sciences* 7 (1954), pp. 254–266. He lists 43 dials of which 11 are signed. Of 22 dials in Istanbul, 7 are signed.

11 Ünver, op. cit. (n. 10), p. 258; Mayer, op. cit. (n. 9), p. 47.

rarely in any other public space.[12] They spread chronologically from the fifteenth to the nineteenth centuries, but so few have survived—or at least been recorded—that no clear tradition or continuity is discernible in their making, and only occasionally can a local centre of manufacture be identified. Perhaps it was because fixed dials were thus hidden from infidel eyes in sacred buildings that Félix de Beaujour stated categorically that Turkey was "a country where the use of sundials and public clocks is unknown".[13] If dialling thus had a low profile this was in part because it was the private domain of the *mīqāti*. Time determination and distribution belonged to the *muwaqqit* and the *muezzin*. Ghislain de Busbecq noted in 1560 that it was thought that by placing public timepieces "the authority of their muezzins and their ancient rites would be thereby impaired", just as printing lessened the sacred nature of the scriptures.[14] Robert de Dreux in 1665, astounded to find a public clock in Serres (Macedonia), noted that in all his travels during four years in Ottoman domains he had never seen another, these being forbidden as "unnecessary".[15] As a subject auxiliary to mathematics and astronomy, with certain usefulness

12 One exception to this is the declining dial erected on a wall of the palace of Ulugh Bey at Samarkand where an aperture meridian was also perhaps constructed. In general however, localisation of sundials to mosques seems to have been general throughout Islam. Certainly, examination of photographic surveys of Islamic cities such as those by Wim Swaam and his co-authors (*Isfahan, Pearl of Persia*, 1966; *Morocco*, 1967; *Moorish Spain*, 1963, all Elek Books, London) shows a striking absence of dials.

13 "[…] un pays où l'on ne connaît pas l'usage de cadrans solaires et horloges publiques"; Félix de Beaujour, *Tableau du commerce de la Grèce formé d'après une année moyenne, depuis 1787 jusqu'en 1792*, Paris, 1800, 2 vols.: v. 2, p. 19. It was however exactly at the time de Beaujour was travelling in Turkey that the first tower-clocks were being erected. See Klaus Kreiser, "Les tours d'horloges ottomanes: inventaire préliminaire et remarques générales", in François Georgeon, Frédéric Hitzel (eds.), *Les ottomans et le temps*, Brill, Leiden and Boston, 2012, pp. 61–76.

14 Ogier Ghislain de Busbecq, *The Life and Letters of Ogier Ghislain de Busbecq* (trans. by C.T. Forster and F.H.B. Daniell), C. Kegan Paul & Co., London, 1881, 2 vols.: v. 1, p. 255.

15 Hubert Pernot (ed.), *Voyage en Turquie et en Grèce du R. P. Robert de Dreux, aumonier de l'Ambassadeur de France (1665–1669)*, Les Belles Lettres, Paris, 1925, pp. 93–94. In fact other clock towers did exist. Some eight or nine, either surviving from pre-Ottoman times, or constructed during the sixteenth and seventeenth centuries, are listed by Kreiser, op. cit. (n. 13), pp. 72–74, while Feza Günergun has pointed out to me that in the 1670s Evliya Çelebi counted thirty clock towers with bells while travelling through the commercially most important Balkan cities of this era. He also noted that *muezzins* would adjust the prayer times according to mechanical clocks. See Heath W. Lowry, *Evliya Çelebi'nin Ayak İzlerinde Bir Rehber Olarak Seyahatname* (*In the Footsteps of the Evliya Çelebi—The Seyahatname as Guidebook*), Bahçeşehir Üniversitesi, Istanbul, 2012, pp. 49–57. The concentration of these towers and usages in the Ottoman West, the Balkans neighbouring onto Europe, is probably significant, for the distribution of clocks, towers and dials no doubt varied in both time and place.

in religious life, the place of dialling in the Islamic world was thus recognised but unobtrusive, it was an arcane subject that added to the aura of learning of the *muwaqqit-muezzin* duo, and so reinforced their authority. It was not emphasised in teaching or daily life and only occasionally were there upsurges of interest leading to renewed activity and even innovation.

2 The Compass and a New Form of Sundial, the *Sandūq*

One such moment of innovation followed the introduction of the dry-mounted compass needle. The compass, at least as a floated needle on water, was known in Europe by the late twelfth century (it is noted by Alexander Nequam *c.* 1187).[16] In the course of the thirteenth century, the pivoted 'dry' needle was developed and both, by the end of the same century, were also known in Islamic lands. The 'wet' compass was described by the Yemenite Sultan al-Ashraf (*c.* 1290), and the 'dry' compass by the Cairene *muezzin* and astronomer Ibn Simʿun (*c.* 1300).[17] In both central Islam and Europe this stimulated the development in the fourteenth and fifteenth centuries of new forms of portable dials. Notable among these was the *sandūq al-yawāqīt* (casket of precious stones).[18] This is a small (120 mm × 120 mm) portable instrument which, according to one of the two inscriptions it carries, amongst other functions, offers the means "for the determination of the hours of prayer made and created for the first time by ʿAlī ibn Ibrahīm ibn al-Shātir".[19] Although the instrument is incomplete, it is

16 *De naturis rerum* and *De utensilibus*, both probably written *c.* 1187. See the mentions and discussion in W.E. May, *A History of Marine Navigation*, G.T. Foulis & Co., Henley-on-Thames, 1973, pp. 43 and 104–105. For Nequam (or Neckam) the fullest account is R.W. Hunt, *The Schools and the Cloister: The Life and Writings of Alexander Nequam (1157–1217)* (ed. and rev. by Margaret Gibson), Clarendon Press, Oxford and New York, 1984.

17 For a survey of early mentions of the compass and its probable origin in China see Joseph Needham, Wang Ling, Kenneth Girdwood Robinson, *Science and Civilization in China*, Cambridge University Press, Cambridge, 1954–2016, 7 vols.: v. 4, pt. 1 "Physics and physical technology: Physics" (1962), pp. 245ff. For al-Ashraf and Ibn Simʿun see Petra G. Schmidl, "Two Early Arabic Sources on the Magnetic Compass", *Journal of Arabic and Islamic Studies* 1 (1997–1998), pp. 81–132.

18 Louis Janin, David A. King, "Ibn al-Shātir's Sanduq al-Yawāqīt: An Astronomical 'Compendium'", *Journal for the History of Arabic Science* 1 (1977), pp. 187–256; reprinted in D.A. King, *Islamic Astronomical Instruments*, Variorum reprints, London, 1987, ch. XII, same p. nos. On p. 216 other notes on the compass can be found.

19 Francis Maddison, Anthony Turner, *Science and Technology in Islam* (Catalogue of an international loan exhibition held in the Science Museum, London, 6 April–31 August 1976, as part of the World of Islam Festival, London, 1976), unpublished typescript, [p. 160].

clear that originally the box contained a magnetic compass needle to orient the instrument when it was used to find equal hours or a *qibla* direction.

The *sandūq*, dated H 767 (1365–1366 CE) is the second oldest known Arab-Islamic portable dial, and perhaps the oldest dated direction dial known from any part of the world. It had some, but rather little, posterity. In H 847 (1443–1444 CE), ʿAbd al-ʿAziz ibn Muhammad al-Wafāʾī al-Miqātī (d. H 876, 1471–1472 CE) had a version of the instrument made, as an inscription on a surviving fragment reveals: "The casket of precious stones uniting the operations for the determination of prayer times, creation of Shaykh the imam, the wise, the very learned ʿAlāʾ al-Dīn b. al-Shātir. Developed by the master Muhammad al-Jawharī. Property of the servant avid for God the highest, ʿAbd al-ʿAziz ibn Muhammad al-Wafāʾī al-Miqātī. And this work was made in the year 847 of the Hijra".[20] A vertical dial diagram on one face of the surviving plate is drawn for latitude 30° and so locates the item in Cairo; it was also in Egypt that late in the same century Shams al-Dīn Muhammad ibn Abi al-Fath al-Sūfī wrote a short account of the instrument.[21] More important however is that following his work on the *sandūq*, al-Wafāʾī invented two further forms of equinoctial dials, the *muqawwar* and the *dāʾirat al-muʿaddil*. No example of the first is yet known, but the *dāʾirat* clearly had some popularity. Surviving examples date from the seventeenth to the nineteenth centuries and several descriptions of it were written.[22] Even so, the magnetic compass, although it continued to be deployed in the Arab-Islamic world, did not lead to an efflorescence of new types and styles of portable direction dials as it did in Renaissance Europe, although some elaborate instruments combining compass and gnomon in a *qibla* indicator are known. An example is that by Bayram b. Ilyās (990 H/1582–1583 CE) but, made of painted ivory and gilt-brass, this is a luxury object.[23] Whatever the vagaries that affect the survival of objects, such a relative scarcity

20 Brieux and Maddison, op. cit. (n. 1) under Muhammad al-Jawharī. The fragment, from the museum of the Kandilli Observatory, is described by Muammer Dizer, "Early Scientific Instruments Exhibited by the Museum of the History of Science, Kandilli", unpublished typescript, n.d., and listed in King, "A Checklist of Islamic Astronomical Instruments …", cit. (n. 1), p. 1019, entry 5.2, no. 8006.

21 Published and translated by David A. King in Janin and King, op. cit. (n. 18), pp. 196–198 and 248–250.

22 For which see Francis Maddison, Emilie Savage-Smith, *Science, Tools and Magic*, « *The Nasser D. Khalili Collection of Islamic Art* 12 », The Nour Foundation in association with Azimuth editions and Oxford University Press, London and Oxford, 1997, 2 vols.: v. 1 "Body and Spirit, Mapping the Universe", pp. 277–279, where two eighteenth-century examples are described and illustrated.

23 British Museum, inv. no. 1921,0625.1.

of portable direction dials during this period must mean that rather few of them were made.

Lack of an object in society implies a lack of need for it, which was the case for portable dials in Islam during the early Ottoman period. For the primary purpose of dials to indicate the times of prayer, fixed dials in mosques, astrolabes and quadrants already existed in some numbers. The portable direction dial, even adapted to this task, did not offer any great advantages. In a relatively static society with only a small bourgeois middle-class, the kind of consumer demand for novelty that underlay the proliferation of different kinds of portable dials in Europe during the sixteenth and seventeenth centuries—string-gnomon, diptych, compass, equinoctial dials and compendia of different forms and materials—was less powerful. Moreover, insofar as there was demand for portable timepieces, it was oriented towards European novelties rather than native products, in particular it was oriented towards self-sounding time-machines—that is towards 'clock-watches', watches incorporating a mechanism that automatically sounded the hour.

3 European Sundials and Clocks at the Ottoman Court

It was devices such as these that dominated the Ottoman court and market. In 1477, three years after ʿAlī Qūshjī had erected his sundial in the Fatih mosque for Mehmet II, that ruler arranged for a clock-maker to be sent from Venice to work in Istanbul.[24] Thereafter demand for striking clocks and watches at the Sultan's court was insatiable. Portable sundials, although they could be decorative, could not compete with the tinkling enchantment of chiming watches which might even be enhanced with music and calendar indications, nor with the whizzing fascination of their moving parts. Nonetheless, European commercial enterprise did make some efforts to develop a dialling market. In 1552,[25] Georg Hartmann in Nuremberg engraved and printed designs for a diptych dial with four hour-scales—a vertical equal hours scale (Nuremberg or Babylonian hours), common equal hours counted from midnight to midnight, an horizontal dial for equal hours at latitude 45°, equal 'Italian' hours counted from sunset—, and with his signature, like all the hour numerals, in Arabic.[26] Although it is not known what success attended this initiative it

24 Otto Kurz, *European Clocks and Watches in the Near East*, The Warburg Institute & Brill, London and Leiden, 1975, p. 21.

25 Following *ibid.*, p. 52, rather than Ernst Zinner, *Deutsch und niederländische astronomische Instrumente des 11.-18. Jahrhunderts*, C.H. Beck'sche Verlagsbuchhandlung, Munich, 1956, p. 363.

26 The instrument is discussed in some detail by Kurz, op. cit. (n. 24), pp. 51–53.

was not ill-founded for there were pockets of interest in astronomy and time-measurement in the Arab-Islamic, Persian, and Ottoman worlds. In general, this interest was allied to modernist attitudes that sought to introduce the products of European technology, which seemed culturally to be value-free, into Islamic realms.

Advanced European technology first arrived in the form of diplomatic gifts and tributes. The important place held by clocks and watches in imperial payments to the Ottoman state in the second half of the sixteenth century is well-known,[27] but they were also deployed by other European states. In 1514 the Serenissima Signoria of Venice intended to send a clock for the Mamluk Sultan Qansuh al-Ghurī; clocks were presented to the Ottoman Sultan from the French crown in 1543–1544 and 1547 and from the French ambassador in Tripoli in 1551.[28] Somewhat later (1583), William Harborne, English ambassador to the Sultan, would offer him a sumptuous silver automaton clock while Thomas Dallam's organ clock of 1598–1599 presented to Sultan Mehmet III is particularly well known.[29] Elsewhere instruments of learning were also seen to have a role to play. In 1600, Thomas Bernhere, an English factor at Marrakesh, wrote to his brother-in-law, the London mathematical practitioner Edward Wright (1561–1615),[30] that

> This King Muley Hamet [Mukai Ahmad al-Mansūr (1549–1603), first Saadian Sultan of Morocco] is much delighted in the studie of Astronomy and Astrologie, and valueth Instruments serving for the course of the Sunne and the Moone, that are of rare device, exceedingly. Wherefore your Spheare,[31] your Watch, your Mundane Diall,[32] and your sextans,

27 See the survey by Gottfried Mraz, "The Role of Clocks in the Imperial Honoraria for the Turks", in Klaus Maurice, Otto Mayr (eds.), *The Clockwork Universe, German Clocks and Automata 1550–1650*, Smithsonian Institution & Neale Watson Academic Publications, Washington D.C. and New York, 1980, pp. 37–48.

28 All these examples from Kurz, *op. cit.* (n. 24), pp. 24–27, where several others are given.

29 For the 1583 clock and that by Dallam see Ian White, *English Clocks for the Eastern Markets. English Clockmakers Trading in China and the Ottoman Empire 1580–1815*, The Antiquarian Horological Society, Ticehurst, 2012, pp. 47–52.

30 *Oxford Dictionary of National Biography*, Oxford University Press, Oxford, 2004, s.v. "Edward Wright" by A.J. Apt.

31 About 1600 Wright wrote *The Description and Use of the Sphaere*, which would be published by John Tap in 1613 with a second edition in 1627. He also designed, but probably later, a 'Celestial Automaton' of which a manuscript description can be found in British Library, Sloane MS 4405.

32 Wright's *The Art of Dialling; shewing how to make any kind of diall vpon a plaine superfivcie howsoever placed*, appeared in 1614 and had a second issue the same year under the variant title of *A Short Treatise of Dialling* ...; see the online *English Short Title Catalogue (1473–1800)*, 26023; http://estc.ucr.edu/ (accessed 18 February 2019).

> your new Magneticall Instrument for Declination,[33] or any Astrolabe that hath something extraordinarie in it will be accepted; and you might sell the same at good prices.[34]

In the context of a fascinating new technology a 'Mundane Diall' could pass as a novelty, that which stemmed from the indigenous tradition would command no attention, nor would that tradition develop except by the mingling into it of new European ideas.

Some of the ways in which this mingling took place are already indicated in Bernhere's letter. The bearer of it (Master Pate), he tells Wright, and the captain of the ship in which he is travelling, will bring the Moroccan ambassador to Queen Elizabeth I, "Abdala Wahed Anoone' [Abdul Wahid ibn Masʿud (1558–?)] who has some insight in such matters [astronomy]", to see Wright. He should show Abdul Wahid:

> all varietie of Instruments that you have either in your owne hands, or have sold and lent to others; that hee may choose some for the Kings use and his owne. You may shew them also the Draughts and Lineaments of whatsoever you have in Paper, all of which I know will make them admire and be desirous to have some that they can understand how to use. You may cause to be framed some Instruments in Brasse or Silver, leaving the spaces for the Arabique words and figures, and the words in Latine, or Spanish, which is farre better: there will be found here that can grave the same in Arabique upon the Instruments having some direction from you about the matter. Or Abdala Waked being a perfect pen-man, can set the Arabique Leters, figures and words down very faire; and so any of your Gravers can worke the same in Metall, having his Writing before them.[35]

Instruments produced in both ways, engraved in Europe in Arabic before being shipped, or shipped unfinished to be completed locally, have survived from later centuries, as have examples of purely European instruments adapted for

---

33 Wright had worked on magnetism and declination in conjunction with William Gilbert. According to the near contemporary writer Mark Ridley, Wright was the anonymous author of Chapter 12 of Book IV of Gilbert's *De Magnete* (1600) which indeed describes two declination instruments.

34 Samuel Purchas, *Hakluytus Posthumus or Purchas his Pilgrimes, containing a History of the World in Sea Voyages and Lande Travells by Englishmen and Others*, James MacLehose & Sons, Glasgow, 1905–1906, 20 vols.: v. 6, pp. 57–59, quoted from Kurz, op. cit. (n. 24), p. 66.

35 *Ibid.*

local use by having Arabic translations added to them.[36] No instruments of any kind, however, have survived in anything like the same numbers as have clocks and watches. Even as European novelties, portable sundials were of relatively minor importance.

4 Sundials at the Safavid Court in Persia, the *Qiblanuma*

But this said, there is of course an exception. At the end of his letter to Wright, Bernhere remarked that

> Your Magneticall Instrument of Declination would be commodious for a yeerely Voyage, which some make for the King over a Sandy Sea (wherein they must use a Needle and Compasse) to Gago.[37] If you question about the matter, and show them some instrument serving for this purpose, it will give great content.[38]

Orientation, either for the purposes of navigation or for praying, was a matter of high importance to travellers in Islam. It was in this context therefore that the dry-mounted magnetic compass made a stronger impact.

It did so particularly in Persia, at the Safavid court and capital at Isfahan (1598–1736). Imbued with an unusual curiosity, Shah 'Abbas I (the Great, 985–1038 H/1577–1629 CE), was also relatively tolerant of Christianity and anxious to encourage trade and exchange with Europe. Foreign goods, like foreign travellers were therefore not uncommon in his new capital where astronomy, astrology and time-measurement were actively pursued, the Shah governing his every action by the advice of his astrologers. Automaton clocks adorned the city and in the list of gifts that would please the Shah communicated to representatives of the English East India Company about to send their first legation to Persia were several technical items including "horizontal dials which

36 See below nn. 49 and 50.

37 Gao, the capital of the Songhai Empire in what is now eastern Mali. With Timbuktu it was conquered by Ahmad al-Mansur in 1591.

38 Purchas, op. cit. (n. 34), pp. 57–59, quoted from Kurz, op. cit. (n. 24), p. 67. The need for compasses is revealed in a letter from Laurence Madoc in August 1594 to the London merchant Anthony Dassel (printed in *The Classical Journal* 25–26, 1822, pp. 92–94), concerning the riches and the conquest of Timbuktu and Gago where he notes that of the 1,700 men who left on the expedition, a third perished while crossing the sandy sea for want of water.

may answer to the latitude of Ispahan".[39] What forms of dials may have been sent to Isfahan (if any) there to influence the local tradition in response to this is not known, but it is clear that during the seventeenth century, compasses specifically designed to aid in determining the direction of Mecca, instruments known as *qiblanuma*, came into use. Although no surviving examples of such instruments can be dated to this century with any security, the magnetic needle they employed appears as an image in the new 'Indian style' poetry, innovative poetry current from the late sixteenth century onwards, in which novel metaphors and images "many of them reflecting the meeting of Indo-Muslim civilization with new foreign influences" were employed.[40] So in a *qasīda* from his *Divān*, Abu Taleb Kalim of Kashan (d. 1651) wrote that:

> Only the *qiblanuma* can look toward the face of the beloved,
> And thus guide people who do not see Divine Truth.[41]

Sometimes the poets speak of the '*qibla*-showing bird'. Mirza 'Azima Iksir has "flies like a *qiblanuma* in his own house", and in his *Dīwan-i-farsī,* Mir David (d. 1785) claims to be the guide of the people who, like a *qibla*-showing bird, are pent in a cage.[42]

The poets spoke truth. Some surviving *qiblanumas*, such as that by Abd al-A'imma mentioned below, still retain their original compass-needles and these have the form of a bird with outspread wings. So too does a compass inset to the throne of an astrolabe by Muhammad Muqīm al-Yazdī, decorated by Muhammad Mahdī al-Yazdī of 1055 H (1645–1646 CE) and another in an astrolabe by Muhammad Mahdī al-Yazdī of 1070 H (1659–1660 CE).[43] For the form of the early dials in which such needles were incorporated no information is available except for two exceptional Iranian world maps at the centre of which the distance and direction of Mecca can be read. On both of these, an inscription beside the compass box tells us that "In this stone is significance on land and on sea: it moves towards the south pole like a blue-feathered

39 *Calendar of State Papers Colonial, East Indies, China and Japan*, v. 3, 1617–1621, p. 159, entry 339.

40 Annemarie Schimmel, *A Two-coloured Brocade: The Imagery of Persian Poetry*, University of North Carolina Press, Chapel Hill, 1992, pp. 2–3.

41 *Ibid.*, p. 90.

42 *Ibid.*, p. 90, n. 14, and p. 130.

43 Both are described by François Charrette in Koenraad van Cleempoel, *Astrolabes at Greenwich. A Catalogue of the Astrolabes in the National Maritime Museum, Greenwich*, Oxford University Press and the National Maritime Museum, Oxford, 2005, pp. 236–257.

bird [*murgh*]".[44] That several different types of *qibla* indicators employing magnetic compasses are known from later periods however suggests that there was also some variety earlier. A fine instrument by Abd al-A'imma offers no features that could not have been developed within the indigenous tradition of mathematical instrument-making.[45] Another instrument however is likely to have resulted from a fusion of a European-style instrument with the particular needs of Islamic life.

It consists of a cylindrical box, opening into two almost equal parts which are hinged together. The box contains a compass with a pivoted needle (missing) over which a cutaway plate, engraved with a semi-circular scale of degrees and carrying a folding gnomon for the dial, is placed. Tables giving the angular distance of Mecca from different towns are engraved on the inner and outer surfaces of the lid and the outside of the base. The instrument is signed "ʿAbd al-[illegible]" but if the remaining part of the inscription is to be supplied by 'A'imma', which would fit the space and for which the engraving of the instrument is of sufficiently high quality, then it must date from the first quarter of the eighteenth century, the period when Abd al-A'imma is known to have been working from dates on other pieces.[46] In this case the instrument becomes the earliest *qiblanuma* of this type so far known.

The cylindrical, box-type, *qiblanuma* is analogous to the compass-dial among European instruments. Both are contained in cylindrical boxes usually of less than 10 cm diameter. Both contain the compass in a box with a cutaway plate, carrying the hour scale and gnomon of the dial, set above it. Both may carry indications in the lid although the European instrument tends to have a plain external surface. In Europe compass-dials as an independent type go back to the sixteenth century and their origins may be traced back even earlier. That the Persian *qiblanuma*, which seems to be unknown before the end of the seventeenth century was developed from this simple form of European pocket dial carried to the Safavid realms, and perhaps elsewhere, by European visitors in the course of that century seems highly likely.

---

44 Translation quoted from D.A. King, *World Maps for Finding the Direction and Distance to Mecca: Innovation and Tradition in Islamic Science*, Islamic Heritage Foundation & Brill, London, Boston and Cologne, 1999, p. 205.

45 A second instrument with exactly the same indications but slightly differently arranged is preserved in the History of Science Museum, University of Oxford, inv. no. 48472. The engraving style of the two instruments is sufficiently close for the Oxford instrument to be ascribed to the workshop of Abd al-A'imma with some confidence. For a discussion of it see King, *World Maps ...*, cit. (n. 44), pp. 118–120.

46 For surviving works see Brieux and Maddison, op. cit. (n. 1). The instrument discussed here is illustrated in Harriet Wynter, Anthony Turner, *Scientific Instruments*, Studio Vista, London, 1975, p. 118.

Once introduced the shallow, cylinder *qiblanuma* became popular. A simplification was for the gnomon to act only as a meridian, thus negating the need to engrave a full hour scale, but still allowing the orientation shown by the compass to be checked when noon was determined. From Persia such instruments spread to other parts of the Islamic world and they could still be found in use in early nineteenth-century Egypt by the noted Arabist Edward William Lane (1801–1876) during his sojourns in Egypt in the 1820s and 1830s.

> The astrolabe and the quadrant are almost the only astronomical instruments used in Egypt. Telescopes are rarely seen here, and the magnetic needle is seldom employed, except to discover the direction of Mekeh; for which purpose convenient little compasses (called "kibleeyehs") shewing the direction of the Kibleh at various large towns in different countries, are constructed, mostly at Dimyat: many of these have a dial, which shows the time at noon, and also that of the ʿasr [the evening prayer] at different places and seasons.[47]

Qibla indicators existed in many forms. Another, although much rarer, form of the instrument also incorporates a sundial. This is the cartographic dial mentioned above of which two examples are known.[48] One of them, now incomplete, evidently derives from early eighteenth-century Isfahan, although the geographical coordinates underlying the recto-azimuthal map derive from the fifteenth-century *zij* of Ulugh Bey. While the solution found for indicating the azimuth of the *qibla* and the distance between two localities on the map derive directly from the traditions of Islamic cartography and mathematical astronomy, the dial with which the instrument is fitted supplies another example of the incorporation of European dialling ideas into the Islamic tradition. The dial on the map, signed by Muhammad Husayn, is an inclining dial, a type that first appears in European dialling towards the end of the seventeenth century. How knowledge of it arrived in Persia and whether first from French or English sources or elsewhere is unknown, although that it occurred in the context here sketched out of lively commercial and technical exchange between Isfahan and Europe is highly probable. Such dials moreover were not confined to use with cartographic *qiblas*. Examples of English inclining dials engraved

47 Edward William Lane, *Account of the Manners and Customs of the Modern Egyptians*, Charles Knight & Co., London, 1836, 2 vols., and several other editions, cited here from the edition included in « Everyman's Library 315 », J.M. Dent, London, 1908, re-print 1936, p. 223.

48 In private collections.

with both European and Arabic inscriptions are known,[49] and some clearly followed Bernhere's earlier prescription and were shipped not engraved, to be completed locally.[50] Such, however, was not the case for an Ottoman dial, written in Arabic script, based on a European model. This is a silver universal equinoctial ring dial signed "'ikhtira-i bārūn 1146" (Invention of, or 'Made by', Bārūn 1733–1734).[51] As yet little enough is known about it, but, like the foregoing examples, it shows that some European portable dials mingled, during the sixteenth to eighteenth centuries, with Arab-Islamic dialling there to modify it, and themselves be adapted.

---

49 For example, one by Culpeper London (perhaps *c.* 1730, it could be by either the father or the son of this name), present location unknown, see E. Weil, *Catalogue 29. Astronomy, Navigation, Horology, Mathematics, Globes & Instruments*, E. Weil, London, n.d., no. 129; reprinted in *Catalogue of Books, Manuscripts, Photographs and Scientific Instruments Fully Described and Offered for Sale by Ernst Weil 1943–1965: Original thirty-three catalogues …*, Jeff Weber, Glendale (CA), n.d., 2 vols.: v. 2, no. 129. One by John Gilbert, now in the Royal Museums of Greenwich, inv. no. D368/37-11c, is described in Hester Higton, *Sundials at Greenwich. A Catalogue of the Sundials, Nocturnals and Horary Quadrants in the National Maritime Museum, Greenwich*, Oxford University Press and the National Maritime Museum, Oxford, 2002, pp. 123–124.

50 For example, a dial in the History of Science Museum, University of Oxford, inv. no. 33518 (Billmeir Collection 57-84/210). See [Francis Maddison], *A Supplement to a Catalogue of Scientific Instruments in the Collection of J.A. Billmeir, Esq., C.B.E.: Exhibited by the Museum of the History of Science, Oxford*, Partridge & Sons, London, 1957, p. 57.

51 Sold at auction in Rodez, France, by Me Hervé Legroux, 1st December 2016, lot 429. I thank Dr. Taha Arslan for his reading of the signature on this so far unique piece. The recent publication by Feza Günergun, "La traduction de l'*Abrégé de la sphère* de Jacques Robbe, géographe du Roi de France par Petros Baronian, drogman à Istanbul: *Cem-nümâ fi fenn el-coğrafya*", *La Révolution française* 12 (2017), on line version http://lrf.revues.org/1841 (accessed 18 February 2019), which appeared after this chapter was written, reveals that Barūn is to be identified with the Armenian dragoman Petros Baronian, known both for the translation there discussed and for a form of *qibla* indicator. The signature formula on surviving examples of this shows that the translation 'Invention of' is to be preferred for the inscription on the ring dial.

CHAPTER 7

Kāshānī's Equatorium

Employing Different Plates for Determining Planetary Longitudes

Hamid Bohloul

Ghīyāth al-Dīn Jamshīd ibn Mas'ūd ibn Maḥmūd ibn Muḥammad Kāshānī (or al-Kāshī), the fifteenth-century Iranian astronomer and mathematician, is known nowadays for his precise calculation of π and Sin 1°, and for his contribution to Ulugh Beg's observatory at Samarkand. In 818 H (1416 CE), most probably in his home town of Kāshān, he invented an equatorium,[1] and wrote an Arabic treatise on its construction and use. The treatise had the title *Nuzhat al-Ḥadā'iq* (Excursion to the Gardens) and the equatorium was called *Ṭabaq al-Manāṭiq* (Plate of the Deferents). In the *Nuzha,* he also described another instrument, named *Lawḥ al-Ittiṣālāt* (Plate of Conjunctions). The main purpose of both instruments was to reduce the amount of calculation astrologers needed to do. About ten years later, while Kāshānī was working at Samarkand Observatory, he wrote an Arabic supplementary tract to the *Nuzha*, in which he added a set of ten appendices. Most of them describe new methods of construction and give simple and more precise instructions for using the equatorium.

During the reign of Sultan Bayezid II (from 1481 to 1512), most probably in Istanbul, an anonymous astronomer composed an untitled Persian treatise about Kāshānī's equatorium, and dedicated it to the Sultan.[2] Having found the only manuscript of this Persian treatise, the late Edward S. Kennedy surmised that the original work of Kāshānī was lost.[3] He started his investigation of the equatorium using the Persian text and published two papers, one on the Plate of Conjunctions in 1947,[4] and the other on the equatorium in 1950.[5] In 1951

1 Equatoria were similar to standard astrolabes in appearance and in the Middle Ages were generally used for determining planetary longitudes without numerical calculations.

2 The only extant manuscript of this treatise is now preserved at Princeton University Library, Garret collection, no. 75. See Mohamad E. Moghadam, Yahya Armajani, *Descriptive Catalog of the Garret Collection of Persian, Turkish and Indic Manuscripts Including some Miniatures in the Princeton University Library* (under the supervision of Philip K. Hitti), Princeton University Press, Princeton, 1939, pp. 35–36.

3 Edward S. Kennedy, "Al-Kāshī's 'Plate of Conjunctions'", *Isis* 38 (1947), pp. 56–59: 57.

4 *Ibid.*

5 E.S. Kennedy, "A Fifteenth-Century Planetary Computer: Al-Kāshī's 'Ṭabaq Al-Manāṭeq' I. Motion of the Sun and Moon in Longitude", *Isis* 41 (1950), pp. 180–183.

 | DOI:10.1163/9789004412842_008

he discovered that a manuscript of Kāshānī's *Nuzha* was preserved in the India Office Library in London, but he carried on his research without paying much attention to this original work of Kāshānī. The result was two more papers in the same year, 1951.[6] It was only in his last paper in 1952 that he included some information about the original Arabic text.[7] Eventually, in 1960 he published a book on the instrument entitled *The Planetary Equatorium of Jamshīd Ghīyāth al-Dīn al-Kāshī*,[8] which included a facsimile of the Persian text, an English translation and a commentary. Kennedy also discussed briefly some of the appendices Kāshānī published in the supplementary tract.

As nearly all of Kennedy's contribution is based on the Persian treatise of the anonymous Ottoman astronomer, it is worth studying Kāshānī's own text to learn more about his equatorium and his scientific career. In the present chapter I describe some of the features of the original text and explain how the three additional plates described in the *Nuzha* can be used to find the true longitudes of the superior planets (Mars, Jupiter and Saturn) and of Venus. I also show how the equatorium can be used for different geographical longitudes.

1 The *Nuzhat al-Ḥadāʾiq*

Nuzhat al-Ḥadāʾiq, which has no dedication, consists of an introduction, two articles on the equatorium, and a conclusion on the *Lawḥ al-Ittiṣālāt*, an ingenious device for performing linear interpolation to pinpoint graphically the time of day at which a planetary conjunction will be observed at a certain terrestrial longitude. The list of extant manuscripts of *Nuzhat al-Ḥadāʾiq*, sorted into chronological order, is as follows:

1. Tehran, Malik National Library, MS 3180.5, copied in 830 H (1427 CE), pp. 319–357 plus the supplementary tract pp. 358–382.
2. Dublin, Chester Beatty Library, MS 3640.2, (probably) copied in the fifteenth century, fols. 112v–126v.
3. Tehran, Central Library of Tehran University, MS 2508, copied in 964 H (1557 CE), pp. 1–40 plus the supplementary tract pp. 41–58.

---

6 E.S. Kennedy, "A Fifteenth-Century Lunar Eclipse Computer", *Scripta Mathematica* 17 (1951), pp. 91–97; E.S. Kennedy, "An Islamic Computer for Planetary Latitudes", *Journal of the American Oriental Society* 71 (1951), pp. 13–21.

7 E.S. Kennedy, "A Fifteenth-Century Planetary Computer: al-Kāshī's Ṭabaq al-Manāṭeq", *Isis* 43 (1952), pp. 42–50: 45.

8 E.S. Kennedy, *The Planetary Equatorium of Jamshīd Ghīyāth al-Dīn al-Kāshī*, Princeton University Press, Princeton, 1960.

4. Mumbai, Mulla Firuz Library, MS 20.2, (probably) copied in the sixteenth century.
5. Tehran, National Library of Iran, MS 6984.2, copied in 1277 H (1860 CE), fols. 16v–33v plus the supplementary tract fols. 33v to 43r.
6. London, British Library (formerly in the India Office Library), copied in 1280 H (1863 CE), pp. 1–92.[9]

Kāshānī's treatise also appeared in a lithograph edition (pp. 250–288) copied from the first manuscript.[10] Copies of this are now preserved in various Iranian libraries including the Majlis Library (MS 228) and the National Library of Iran (MS 2099). The supplementary tract is attached to the manuscripts 1, 3 and 5 and to the lithograph edition (pp. 289–313).[11]

The first manuscript was transcribed in Samarkand while Kāshānī was living there by ʿAbd al-Razzāq ibn Muḥammad, who is also known as Muʿīn al-Munajjim Kāshānī, an astronomer who might have come to Samarkand at Kāshānī's request or on his recommendation. The present chapter is based on this first manuscript, which is the oldest, the most intact, and the most authentic extant manuscript. Page numbers in the text refer to this manuscript.

The first article is devoted to the construction of the equatorium (pp. 319–336), which has four main components, a ring, a plate, an alidade, and a ruler (*Fig. 7.1*). Kāshānī describes the standard construction of each part and then suggests fourteen different plates (pp. 329–333) each of which differs slightly from the standard plate. He claims that each plate has some advantages over

9 Details of most of the Arabic manuscripts held by the British Library are available only in printed catalogues and handlists, and are not included in the main library catalogue. This manuscript is listed in E. Denison Ross, Edward G. Browne, *Catalogue of two collections of Persian and Arabic manuscripts preserved in the India Office Library*, Eyre and Spottiswoode, London, 1902, entry no. CCX. The India Office Library in London was transferred to the British Library in 1982. This copy of the *Nuzha* contains none of the appendices.

10 This volume consists of three treatises by Kāshānī: *Miftāḥ al-Ḥisāb*, *Nuzhat al-Ḥadāʾiq* and its supplements. Muhammad Ṣādiq ibn Abulqāsim Ḥussaynī Khwānsārī published the lithograph edition in Tehran in 1306 H (1889 CE).

11 I have produced a critical edition of *Nuzhat al- Ḥadāʾiq* and the appendices as a part of my PhD thesis based on the lithograph edition and the listed manuscripts, except the fourth one. Hamid Bohloul, "تصحیح، ترجمه وشرح رسالهٔ نزهة الحدائق والحاقات آن اثر غیا تصحیح، ترجمه وشرح رسالهٔ نزهة الحدائق والحاقات آن اثر غیاث الدین جمشید کاشانی همراه با بررسی منشأ، ساختار وکاربردهای ابزار موسوم به «صفیحه» در نجوم دورهٔ اسلامی (A Critical Edition, Translation, and Commentary of Ghiyāth al-Dīn Jamshīd Kāshānī's *Nuzhat al-Ḥadāʾiq* and its Appendices along with a Survey of Origin, Construction, and Use of the Equatorium [Ṣafīḥa] in the Islamic Astronomy)", unpublished PhD thesis, Institute for Humanities and Cultural Studies, Tehran, 2017.

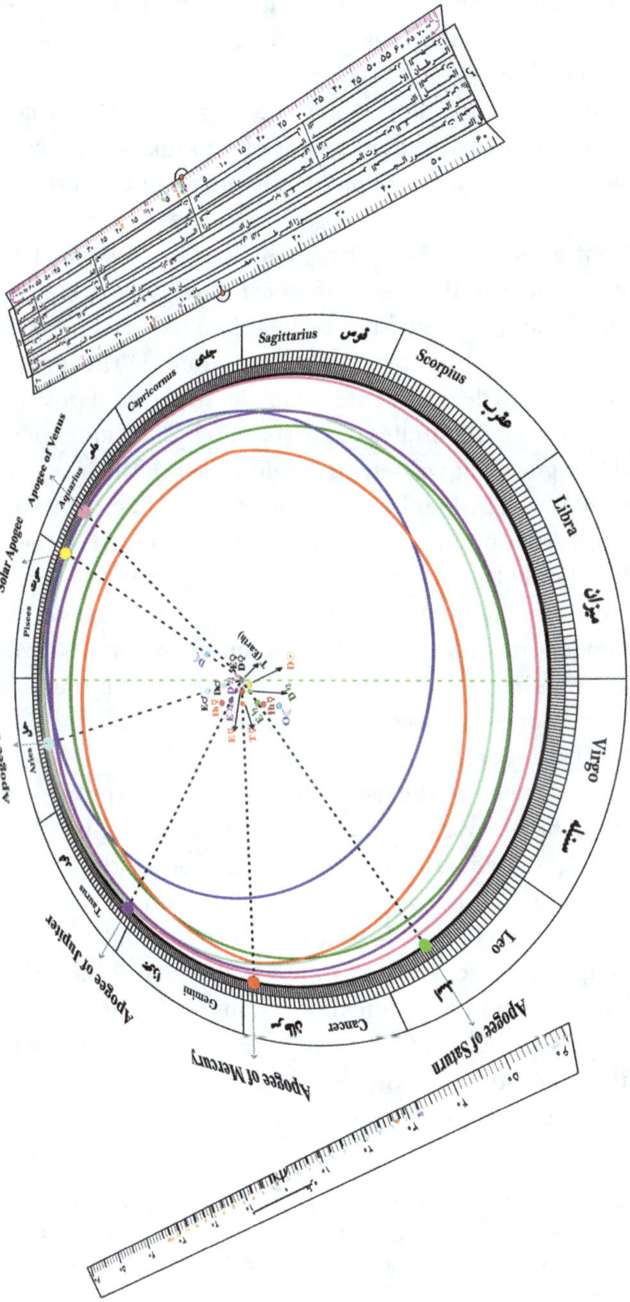

FIGURE 7.1 The ruler, the equatorium (the standard plate in the ring), and the alidade (left to right), drawn to scale

the others. Some of them are easier to make and draw, and some are easier to use. Finally, he picks a combination of the standard and parallel plates as the most favourable design without stating his reasons.

The first plate, called the standard plate, and the ring have been discussed fully in Kennedy's contributions. It is, however, useful to take a fresh look at their design, based on Kāshānī's own explanations, to puzzle out how the other plates originated.

The first component to be made is the ring, which is not described clearly but must have been similar to the mater of an astrolabe. According to the *Nuzha*, "the diameter of the ring should not be less than half of a great cubit but it would be better to make it two or three *Hāshimī* cubits".[12] The inner part of the ring is to be graduated in degrees of arc or smaller divisions if possible. It is recommended to graduate it in minutes of arc (p. 348).[13] The names of signs of the zodiac are to be engraved or written on the outer part of the ring.

The plate is to be mounted inside the ring in such a way that it can rotate easily (p. 319). Kāshānī's description is incomplete in this section. He does not state whether the ring should be attached to another plate (as its base) or not. It is also not clear how to install the moveable plate inside the ring. Having become aware of these shortcomings, Kāshānī attempted to correct them by giving detailed instructions in the fourth appendix. He carries on by describing how to draw permanently or to engrave the orbs of all planets, but not that of the Sun, on the plate. The rim of the plate plays the role of the Sun's deferent.

The second article is devoted to the use of the equatorium (pp. 336–354). Most equatoria, both extant physical examples and those described in medieval literature, were designed for finding planetary longitudes only, but Kāshānī's equatorium could also be employed for determining planetary latitudes, planetary distances from the Earth, time of occurrence and magnitudes of lunar and solar eclipses, the equation of time, etc. He claims implicitly in the introduction of *Nuzha* that in common applications his equatorium could take the place of a *Zīj* (astronomical tables) when high accuracy was not expected because with this instrument the desired results could be achieved in the easiest and fastest possible way (p. 319). Although the equatorium had

12 Unfortunately it is not possible to precisely convert these units to modern measurement units. For more information see: Kennedy, *The Planetary Equatorium …*, cit. (n. 8), p. 165. In the seventh appendix of the *Nuzha*, Kāshānī stated that if we want to graduate the rim of the ring to a minute of arc the radius of the circle must be three cubits (pp. 376–377). If we consider that there should be at least 0.5 mm between the divisions, the circumference of the rim must be 10.8 m and its radius is about 1.72 m. Therefore one cubit must be not less than about 57 centimeters.

13 In the figures of Kāshānī's equatorium, the ring's subdivisions are equal to 30 minutes of arc.

fewer uses than a *Zīj*, the equatorium could be used to solve the most common problems for which a *Zīj* was normally used.[14] All the different plates could be used in the same way for these applications except for the calculation of planetary longitudes for which there are different sets of instructions in the *Nuzha* (pp. 339–343).

Kāshānī seems not to have been completely satisfied with his invention and some of its applications. Accordingly, he offers a number of modifications in the appendices, and gives a detailed description of a new device.

The first appendix provides the instructions for finding lunar longitude with a plate on which a curve approximately similar to an *ihlīlajī*[15] has been drawn to represent the lunar orbit (p. 358).

In the second appendix Kāshānī describes how to draw a curve similar to an *ihlīlajī* to represent the orbits of the Moon and Mercury (pp. 358–360). In the *Nuzha*, the instructions are to draw a circular lunar orbit, but Mercury's orbit can be drawn either circular or as an *ihlīlajī* (pp. 335–336).

In the third appendix, Kāshānī gives a detailed account of the procedures that a user should follow for finding planetary latitudes (pp. 360–364). Several different instructions are described for this in the *Nuzha* (pp. 346–349). A simpler version of alidade is also described in this appendix.

In the fourth appendix, Kāshānī describes how to make the plate and the ring and then explains the procedure for installing the alidade on the plate and ring (pp. 364–367). The suggested procedure, along with other evidence, suggests that Kāshānī may not have been involved in the making of an equatorium before writing the *Nuzha* because in it he just said that the alidade could be installed on the plate like the alidade of an astrolabe (p. 339). This method would prevent the instrument from working correctly and some of its applications, for example the measuring of lunar latitude and the longitude of Venus, would have been impossible.

The fifth appendix has two main sections (pp. 367–375). In the first, Kāshānī gives a full account of how to draw 24 concentric circles on the back of the plate, how to graduate them, how to make a narrow metal ring and how to make two extra rulers. In the second section he explains how to use all of these components to determine graphically the mean longitudes of the planets, the apogee of the Sun at any desired date,[16] and other data that can be derived from a table of mean motions in a *Zīj*.

---

14 Kennedy, *The Planetary Equatorium* ..., cit. (n. 8), p. 22.

15 This is a closed shape created from two equal arcs of two equal circles. Both of the arcs are smaller than a semi-circle. The shape approximates to an ellipse.

16 In the Ptolemaic or geocentric system, the Sun is assumed to rotate around the Earth which is stationary at the centre of the universe. The centre of the Sun's circular orbit is away from the centre of the Earth. Therefore, the distance from the Earth to the Sun will

The sixth appendix is devoted to the description of the most versatile plate (pp. 375–376). This is not a completely new plate but a combination of other plates.

In the seventh appendix, Kāshānī deals with practical problems of constructing an equatorium with a radius of about 1.7 m the rim of which can be graduated to minutes of arc (pp. 376–380). The graduation of a circle in minutes of arc requires 21,600 divisions on the circumference of the circle giving an interval of about 0.5 mm between any two graduations, which is about the lower limit for a practical instrument. Considering the difficulty of constructing such a huge instrument in the fifteenth century, and that instruments for casting horoscopes or similar purposes did not need great accuracy, Kāshānī seems to have attempted to describe an ideal instrument.

The eighth appendix describes two new applications for the instrument, determining the equation of time and the co-ascensions of the ecliptic, and gives instructions for these operations (pp. 380–381).

In the ninth appendix, Kāshānī gives instructions for using the plates described in the sixth appendix and those of the large equatorium described in the seventh appendix solely for finding planetary longitudes and distances of the planets from the Earth (pp. 381–382).

In the last appendix, Kāshānī gives the reason for naming the instrument *Ṭabaq al-Manāṭiq*. In the sequel, he claims that some of his friends persistently asked him to change the name of the equatorium from *Ṭabaq al-Manāṭiq* to *Jām-i Jamshīd* (Jamshīd's Cup), which is the name of a magic instrument belonging to a legendary King of Persia (p. 382).

2 The Anonymous Persian Treatise

Like the *Nuzha* the anonymous treatise consists of an introduction, two articles and a conclusion. In his introduction the anonymous author claims to have constructed an equatorium based on what Kāshānī has written and then compiled this treatise on its construction and use.[17] He also claims to have tested the equatorium and obtained results entirely consistent with the results obtained using *Zījes*.[18] This is the only document that suggests that at

vary during its rotation around the Earth. The farthest and least distances of the Sun from the Earth are defined as the Sun's apogee and perigee.

17 Kennedy, *The Planetary Equatorium ...*, cit. (n. 8), pp. 24–25; p. 24 is the facsimile of fol. 4r of the manuscript and p. 25 gives its English translation.

18 *Ibid.*

least one astronomer constructed Kāshānī's equatorium in the late fifteenth or early sixteenth centuries.

In the first article, as well as making a few comments of little significance, the author describes a simple improvement to the instrument. In the *Nuzha*, Kāshānī writes that the standard plate should be able to rotate smoothly inside the ring and that it can be fixed in any desired positions by using melted wax. If it is necessary to turn the plate again the user should leave the instrument in the Sun's rays so that the wax will melt, allowing the plate to be moved (p. 339). Instead of this rather unusual instruction the anonymous author suggests that a number of small holes should be made in the circumference of the plate. A projecting piece with a hole in it should be attached to the ring where the solar apogee is assumed to be. Then the plate can be fixed in any position with a small peg inserted through the holes.[19] The rest of the instructions in the Persian text remain faithful to the *Nuzha*, except for the alidade where the description from the third appendix is substituted.[20]

The second article of the Persian text is devoted to the use of the equatorium. The author selects just some of the uses explained in the *Nuzha* and the appendices and omits the others. He translates the relevant sections from Arabic into Persian but the translation is not verbatim: he adds some new information,[21] tabulates some of the parameters,[22] and summarizes some of the original instructions.[23] The author completely omits the variant plates described in the *Nuzha*. He also discusses what is written in the third appendix about planetary latitudes instead of the instructions in the *Nuzha* and he translates the eighth appendix completely from Arabic into Persian.

Because Kennedy's study is primarily based on the Persian text he has not studied the second, sixth, seventh, and ninth appendices at all. Furthermore, his concise description of the variant plates,[24] based on Kāshānī's *Nuzha* and the fifth appendix,[25] and his references to the first and fourth appendices,[26] are quite insufficient and need further scholarly studies.

19 Kennedy, *The Planetary Equatorium* ..., cit. (n. 8), pp. 32–35 and 165.

20 *Ibid.*, pp. 54–57 and 180–181.

21 *Ibid.*, pp. 120–121; Ghīyāth al-Dīn Jamshīd Kāshānī, "*Nuzhat al-Ḥadā'iq*", copied in 830 H (1427 CE), Tehran, Malik National Library, MS 3180.5, p. 334.

22 Kennedy, *The Planetary Equatorium* ..., cit. (n. 8), pp. 98–99; Kashani, op. cit. (n. 21), p. 362.

23 Kennedy, *The Planetary Equatorium* ..., cit. (n. 8), pp. 180–183.

24 Kennedy, "An Islamic Computer ...", cit. (n. 6), p. 45; Kennedy, *The Planetary Equatorium* ..., cit. (n. 8), pp. 175–176.

25 *Ibid.*, pp. 189–190.

26 *Ibid.*, pp. 171, 173 and 183.

3 Ptolemaic Planetary Theory and Equatoria

In order to understand how Kāshānī's equatorium represents the motion of the superior planets, Mars, Jupiter and Saturn, in longitude it will be helpful to review the outlines of Ptolemaic planetary theory. In the Ptolemaic system, each planet (P in *Fig. 7.2*) is attached to an epicycle which rotates uniformly around its centre O in a counter-clockwise direction. The centre of the epicycle O moves in the same direction on an eccentric circle called the deferent, the centre of which (D) is at a constant distance (the eccentricity, e) from the centre of the Earth T in a direction towards the planet's apogee A. The line that connects the centre of the Earth to the apogee (TA) is called apsidal line. The motion of the centre of the epicycle is not uniform around the centre of the deferent but about a point, called the equant (E), located on the apsidal line at a constant distance, equal to the eccentricity of the deferent, from the centre of the deferent towards the planet's apogee. Therefore the angle ϑ in *Fig. 7.2* (called the mean centrum) is a linear function of time. A point $\overline{\lambda}_P$ can be defined that moves on the ecliptic (represented by the outer circle in *Fig. 7.2*) around the centre of the Earth at a constant rate equal to the velocity of O as measured from E, such that $T\overline{\lambda}_P$ is always parallel to EO. The angle $VT\overline{\lambda}_P$ is known as the mean longitude of the planet. The mean longitude as a function of time, and other parameters shown in *Fig. 7.2*, including the angle α and the angle β, are tabulated in tables of mean motions in Ptolemy's *Almagest* and in Islamic *Zījes*.

After determining the mean longitude at a given time from a table of mean motions, two corrections must be applied to it to find the true longitude $VT\lambda_P$, which gives the position λ_P of the planet as seen from the Earth T. The first correction is for the angle α, which must be subtracted from the mean longitude if ϑ is less than 180° and added to it if ϑ is greater than 180°. The second correction is for the angle β, which must be added if the angle a_v shown in *Fig. 7.2* is less than 180° and subtracted if a_v is greater than 180°. The motion of the Sun can be modeled much more simply, but for Mercury and the Moon, the motions become more complex and they are not studied in this paper.

Working with astronomical tables to find planetary longitudes is exceedingly complex for a mediocre astronomer and time-consuming for even a skilled astronomer. The calculations for determining planetary latitudes or predicting eclipses are even more difficult. It takes, for instance, about half an hour for a professional astronomer to compute a planet's true longitude.[27] It was

27 Julio Samsó, "Astrology", in *The Different Aspects of Islamic Culture: Science and Technology in Islam*, UNESCO Publishing, Paris, 1998–2016, 6 vols.: v. 4 (ed. by A.Y. Al-Hassan, 2001), pp. 267–296: 287.

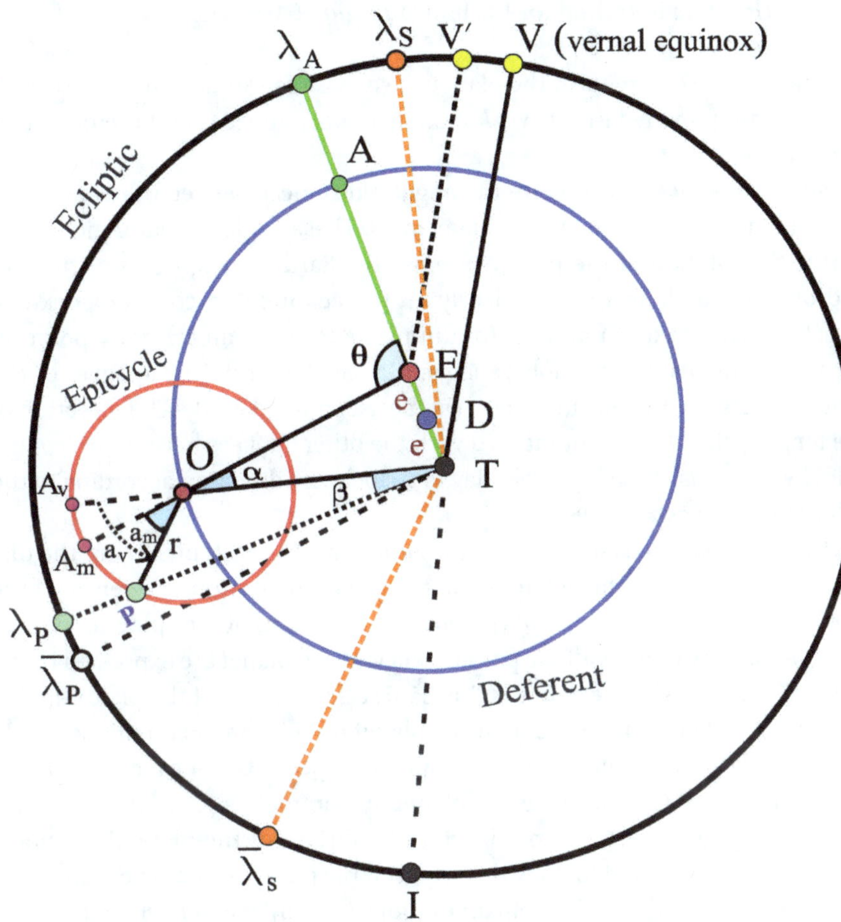

FIGURE 7.2 Ptolemaic planetary theory for superior planets and Venus

customary for medieval astronomers to try to solve the astronomical problems without numerical calculations. They invented ingenious instruments like astrolabes, quadrants, *qibla* indicators and sundials. Even though the first two of these had observational functions they were devised mainly to decrease the number of calculations or even to eliminate them. Another set of instruments, equatoria, based on the Ptolemaic system, was invented in order to get rid of long and painstaking computations of planetary longitudes. Using this instrument even a novice astronomer could easily compute a planet's longitude in less than five minutes. The accuracy of the results was completely acceptable for astrologers intending to cast horoscopes for their ordinary clients.

4 The Data Required for Kāshānī's *Ṭabaq al-Manāṭiq*

For drawing the deferents of the planets, two main sets of data are necessary,[28] and they are provided in the *Nuzha*, the planetary apogees and the planetary parameters.

Planetary Apogees: These are the angular distances between the apogee of the Sun and those of the five visible planets. These angles remain unchanged during the rotation of the planets around the Earth. The apogees of the Sun and of the five planets move linearly on the ecliptic in a counter-clockwise direction. The maker of an equatorium needs to select an arbitrary point on the rim of the plate as the solar apogee. Having done this, the apogees of the other planets, or their apsidal lines, can be ascertained easily. The direction of the lunar apsidal line is not marked with the other apogees because of its high velocity, which is about 11° 13′ per day in a clockwise direction according to the values in the *Nuzha* (p. 328).[29]

Planetary parameters: The relevant parameters for each planet are the distance of the centre of the deferent and the equant point from the centre of the Earth, the radius of the deferent (p. 320), and the radius of the epicycle (p. 324). In Ptolemaic astronomy, all the parameters for each planet are expressed in sixtieths of the radius of the deferent. Kāshānī sets the radius of the plate equal to sixty units and gives all of the required planetary parameters on this scale. He shortens all the parameters for each planet by a special factor called the 'norming coefficient'.[30] Compressing the planetary parameters prevents the circles representing the deferents from leaving the surface of the plate. This data is the information needed to draw the orbits on the plate. The data needed to use the equatorium is the table of mean motions (p. 328) which Kāshānī provides from his *Zīj-i Khāqānī* (p. 327).[31] The planetary apogees, mean motions, and anomalies[32] can be ascertained from this table for any required time. According

28 Sebastian Leonard Dundas Falk, "Improving Instruments: Equatoria, Astrolabes, and the Practices of Monastic Astronomy in Late Medieval England", unpublished PhD thesis, University of Cambridge, Cambridge UK, 2016, p. 125.

29 In his *Lubāb Iskandarī* (fol. 7r), Kāshānī gives 11° 9′ 7″ 14‴ per day for the motion of lunar apogee, which is more precise than the rounded value of the *Nuzha*. See Kāshānī, "*Lubāb Iskandarī*" ["*Iskandarī Kernels*"], c. 812 H (1409–1410 CE), Tehran, Majlis Library, MS 6312.

30 Kennedy, *The Planetary Equatorium* ..., cit. (n. 8), p. 169.

31 Kāshānī compiled this *Zīj* about two years before *Nuzha*. For more information see: E.S. Kennedy, *On the Contents and Significance of the Khāqānī Zīj by Jamshīd Ghīyāth Al-Dīn Al-Kāshī*, « *Islamic Mathematics and Astronomy* 84 », Institute for the History of Arabic-Islamic Science, Frankfurt, 1998, p. 26.

32 In Ptolemaic system, the angle measured at centre of epicycle between epicyclic apogee A_m and the planet P is called the anomaly. This is angle $A_m\,O\,P$ in *Fig. 7.2*.

to the *Nuzha*, the table of mean motions should be drawn on the face or back of the equatorium (p. 327).

Then the instrument must be set for the time at which we want to find the planetary longitudes. For this the longitude of the solar apogee is needed for that time (p. 339), which can be found from the table of mean motions or by using the device described in the fifth appendix. Lastly the solar apogee mark must be put in the correct position on the graduated ring. Then all planetary apogees, except the lunar, are in the correct adjustment for the given time.

5 The Use of the Standard Plate

According to Ptolemaic planetary theory, two corrections must be determined and applied to the mean longitude of a superior planet—Mars, Jupiter or Saturn—to find its true longitude.[33] First, the equation of the centre has to be found. The alidade pivoted at the centre of the plate (T in *Fig. 7.3*, which shows Mars, but the principle is the same for other superior planets) is turned so that it points to the mean longitude of the planet, $\bar{\lambda}_P$, on the graduations on the ring. The ruler is put on the plate so that its edge lies on the equant point E and it is made parallel to the alidade. The intersection of the ruler with the deferent O gives the position of the centre of the epicycle of the planet. The arc between the pointer of the alidade and the point where the ruler crosses the ring, $\bar{\lambda}_P \lambda_O$, gives the magnitude of the first correction (p. 339).

To find the second correction, called the 'equation of anomaly' for the superior planets, Kāshānī takes advantage of a principle in Ptolemaic astronomy, according to which the epicyclic radius passing through the planet (OP) always remains parallel to the mean Sun ($T\bar{\lambda}_S$) (p. 338). In the description of the alidade Kāshānī instructs the instrument maker either to put marks on the edge of the alidade at the radius of each planet's epicycle or to mark on the plate six circles of those radii concentric with the centre of the plate (p. 324). The pointer of the alidade is rotated to the mean solar longitude $\bar{\lambda}_S$. A mark (R) is to be put on the plate either at the edge of the alidade against the mark representing the radius of the epicycle or at the intersection of the alidade

33 In *Fig. 7.2*, O uniformly rotates around E. As a result it does not uniformly rotate around T. But we can define a point on the ecliptic with the same velocity of O which uniformly rotates around T. This point $\bar{\lambda}_P$ is called mean longitude of the planet and its amount is equal to the angle $VT\bar{\lambda}_P$. The arc between the vernal equinox (or the first point of Aries) and the position of the planet on the ecliptic as viewed from the Earth is considered the true longitude. In *Fig. 7.2* it is $V\bar{\lambda}_P$ or $VT\bar{\lambda}_P$. The mean longitudes are given in mean motion tables. But we have to compute true longitudes.

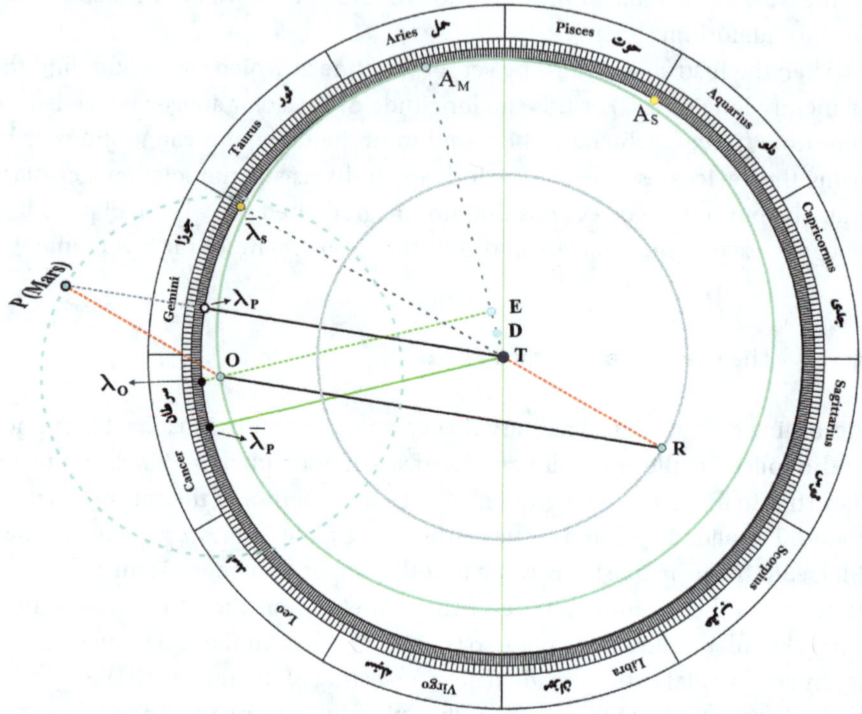

FIGURE 7.3 Mars in the standard plate (the blue circle represents the deferent of Mars and the grey circle represents the epicycle of Mars)

with the circle of the epicycle in the direction opposite to the solar apogee (p. 342). Next the edge of the ruler is laid across this point *R* and the centre of the planet's epicycle *O*. Then the alidade is turned to be parallel to the ruler. The position of the point of the alidade on the graduated ring pinpoints the true longitude λ_P of the planet at the given time (p. 342). The arc $\lambda_O\,\bar{\lambda}_P$ gives the magnitude of the second correction to be applied to the mean longitude.

6 The Single Apogee Plate

The two primary characteristics of the single apogee plate (*Fig. 7.4*) are that it has a single apogee direction and that it is a fixed plate (p. 329). There is a single point on the plate that is the apogee for all the planets. Therefore all the eccentric centres and equant points are situated on the line, called the equatorial diameter, that connects the first point of Aries to the first point of

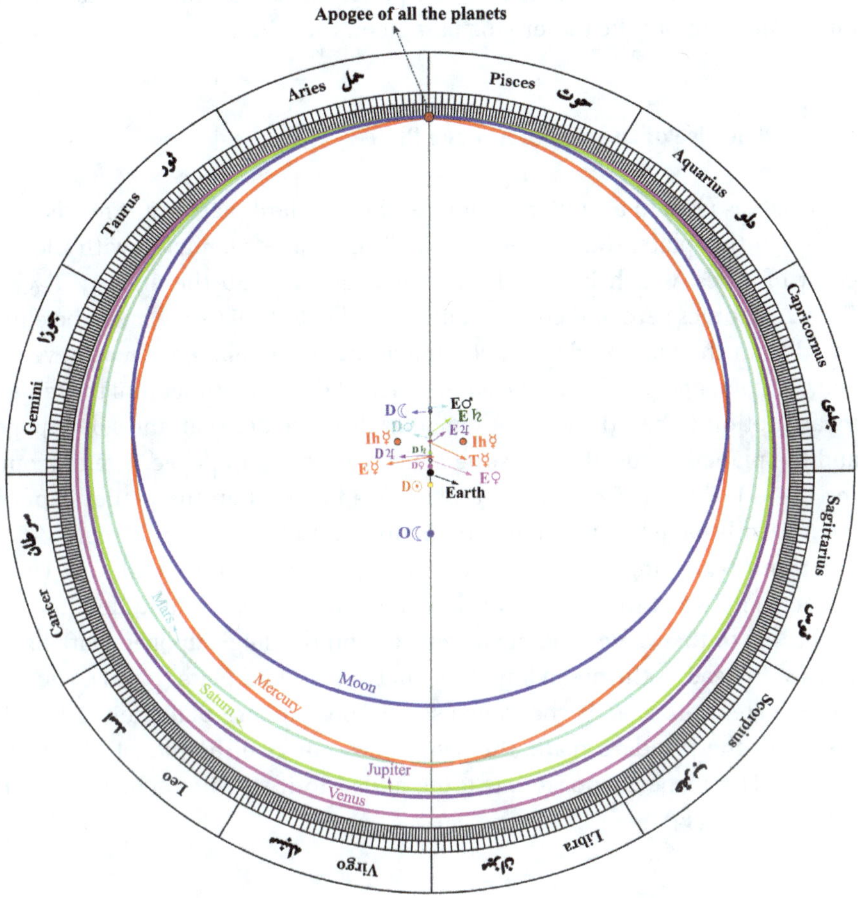

FIGURE 7.4 The single apogee plate (drawn to scale)

Libra, directly opposite it in the zodiac (p. 321). These features considerably simplify the construction of the equatorium because there is no need to have a separate ring and plate or to include a means for attaching the plate to the ring. The instrument maker only has to make the plate, and then draw the ring and its graduations on it. All the deferents and important points are drawn on it (p. 329). The user does not have to set the instrument to the required time. Although these modifications make the construction of the equatorium easier, Kāshānī was aware that it was rather more complicated to use. As with the standard plate, Kāshānī shortens all the planetary parameters and expresses them in sixtieths of the radius of the plate. Kennedy calculated a parameter called the 'norming coefficient' for each planet and clarified Kāshānī's underlying parameters.[34] For the other plates, these norming coefficients are rather different but the underlying parameters are similar.

7 The Use of the Single Apogee Plate

This plate is used in a similar manner to the standard plate but it needs more steps. On this plate, the apogees λ_A of all the planets coincide with the first point of Aries, which is also the vernal equinox V. So the angle $V\,T\,\bar{\lambda}_A$ in *Fig. 7.2* becomes zero and consequently the effect of this angle must be eliminated from the magnitude of mean longitude of the planets. To achieve this end, the solar apogee λ_S must be determined at the given time from the table of mean motions. Then the constant angular distance between the solar apogee and the planet in question must be added to the solar apogee. The outcome, the angle $V\,T\,\lambda_A$ on *Fig. 7.2*, is the planet's apogee. Then the planet's apogee must be subtracted from the mean longitude of the planet, the angle $V\,T\,\bar{\lambda}_P$. The result is the angle $\lambda_A\;T\,\bar{\lambda}_P$ (the mean centrum) and is equal to ϑ. This is the angle between the planet's apsidal line and its mean longitude direction. According to the *Nuzha*, the mean centrum on the single apogee plate has to be used instead of the mean longitude on the standard plate (p. 341). The subsequent steps are exactly the same as with the standard plate but, unlike the case with the standard plate, the result does not give the planet's longitude directly. The planet's apogee has to be added to it so as to achieve the true longitude (p. 343).

8 The Fixed Plate

The fixed plate is a type of plate that retains all the features of the standard plate but it is fixed instead of being able to rotate in the ring. It is very similar to the previous plate. It is designed for a particular time and all the apsidal lines are fixed for that time. It operates like the standard plate but two adjustments have to be made. First, the solar longitude has to be calculated for the time in question. To do this, if this time is before the time that the plate has been designed for, the difference between the solar apogee on the plate and the solar apogee for the time in question must be subtracted from the mean

34 Kennedy, *The Planetary Equatorium …*, cit. (n. 8), p. 169.

motions of the planet. If the required time is after the time for which the plate is designed, the difference must be added to the mean motions (p. 333). The result is used where the mean motion would be used with the standard plate. The other steps in the calculation are the same. In the first case, where the required time is before the time for which the plate was designed, the difference of the two solar apogees must be added back to the result to achieve the true longitude and in the second case it must be subtracted (p. 333). This is the second adjustment. Looking at this another way, the first adjustment sets the mean motion at the proper angular distance with respect to the true apsidal line and the second adjustment sets the result at the proper angular distance with respect to the vernal equinox or first point of Aries, which is the reference point for measuring the longitude.

9 The Parallel Deferent Plate

The centre of the parallel deferent plate is assumed to be the centre of all the deferents (except that of Mercury). Consequently the deferents are parallel to each other and the orbits do not intersect. The striking effect of this shift is that there is no single point for the centre of the Earth. It varies for each planet. This centre and the equant point lie on the apsidal line of each planet at a distance apart equal to their eccentricities, symmetrically on either side of the centre of the plate (pp. 330–332). Kāshānī writes that the radii of the epicycles of the planets should be marked along the edge of the alidade, or six circles with the radii of the epicycles of the planets should be drawn concentric with the centre of the plate. The plate can be rotated, and has to be set for the required time.

After setting the plate, the mean longitude of the planet in question—in this example a superior planet—must be extracted from the table of mean motions. Then the true longitude must be found by following Kāshānī's instructions in nine steps:

1. The alidade, which is fixed at the centre of the plate, is rotated to point to the planet's mean longitude ($\bar{\lambda}_P$, in *Fig. 7.5*) on the graduated ring.
2. Next the ruler must be placed parallel to the alidade with its edge passing through the equant point E. The intersection of the planet's deferent and the edge of the ruler (F) is called the fictitious centre.
3. The ruler is then laid across the fictitious centre F and the centre of the Earth T.
4. The alidade is turned to be parallel to the ruler.

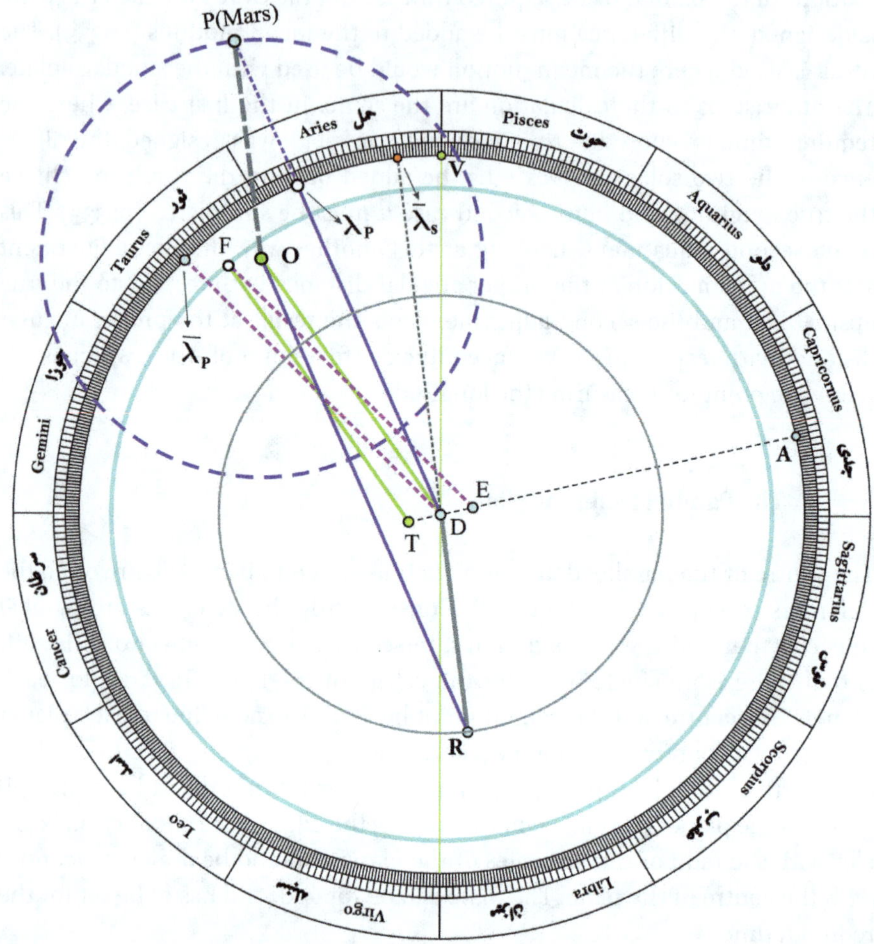

FIGURE 7.5 Mars on the parallel plate

5. A line is drawn along the edge of the alidade parallel and equal in length to the line *FT*, to a certain point *O*. Such a point is called the planet's epicyclic centre (pp. 340–341).
6. The pointer of the alidade is set on the mean Sun $\bar{\lambda}_S$.
7. A point *R* is marked on the plate in the direction opposite to the solar apogee either at the edge of the alidade where the radius of the epicycle has been marked or at the intersection of the alidade with the circle of the epicycle marked on the plate (p. 342).
8. The edge of the ruler is put across the centre of the epicycle *O* and point *R* and then the alidade is placed parallel to the ruler.

9. The pointer of the alidade (which is near to the epicyclic centre) indicates the true longitude of the planet (p. 342).

For determining the longitude of Venus the sixth step must be replaced by the following:

6b. The mean longitude of the Sun must be added to the mean anomaly of Venus which can be done using the table of mean motions.

The result is called the compound anomaly (p. 327). The pointer of the alidade must be turned to the compound anomaly on the graduated ring (p. 342).

This method is valid and equivalent to using the standard plate. Although the rim of this plate represents the ecliptic as on the other plates, its centre is not the centre of the Earth *T*. Thus the deferent is not drawn in its proper place in relation to the ecliptic and the Earth. As a result, an extra correction needs to be applied so that the position of the centre of the epicycle can be found. In *Fig. 7.6*, let the dashed circle be the ecliptic on the standard plate, and the red circle the ecliptic and the blue circle the deferent on the parallel plate. The angle $V' T \bar{\lambda}'_P$ is the mean longitude of the planet and equal to the angle $V D \bar{\lambda}_P$. On the standard plate, a line parallel to the $T \bar{\lambda}'_P$ must be drawn from the equant point whose intersection with the deferent at point *F* will be the centre of the epicycle for a person observing it from *T*. If a line parallel and equal to *TF* is drawn from *D*, then the other end of this line, point *O*, will represent the centre of the epicycle for an observer at *D*. So the angles $F T \bar{\lambda}'_P$ and $O D \bar{\lambda}_P$ are equal to the equation of the centre and must be added to the mean longitude or subtracted from it. The other steps are the same as for the previous plate.

10 Geographical Longitude and the Equatorium

In his *Nuzha*, Kāshānī copies all the planetary parameters and the table of mean motions from his *Zīj-i Khāqanī*. Because of this the results are accurate for the terrestrial longitude of the city of Shiraz, for which the *Zīj* was compiled. According to Kāshānī, Shiraz was 88° east of the Fortunate Isles.[35] He proposed the following method for those who would like to employ the equatorium for other longitudes:

1. The difference between two longitudes is to be calculated.
2. The result is divided by 15.

35 E.S. Kennedy, M.-H. Kennedy, "Al-Kāshī's Geographical Table", *Transactions of the American Philosophical Society* 77, 7 (1987), pp. 1–45: 29.

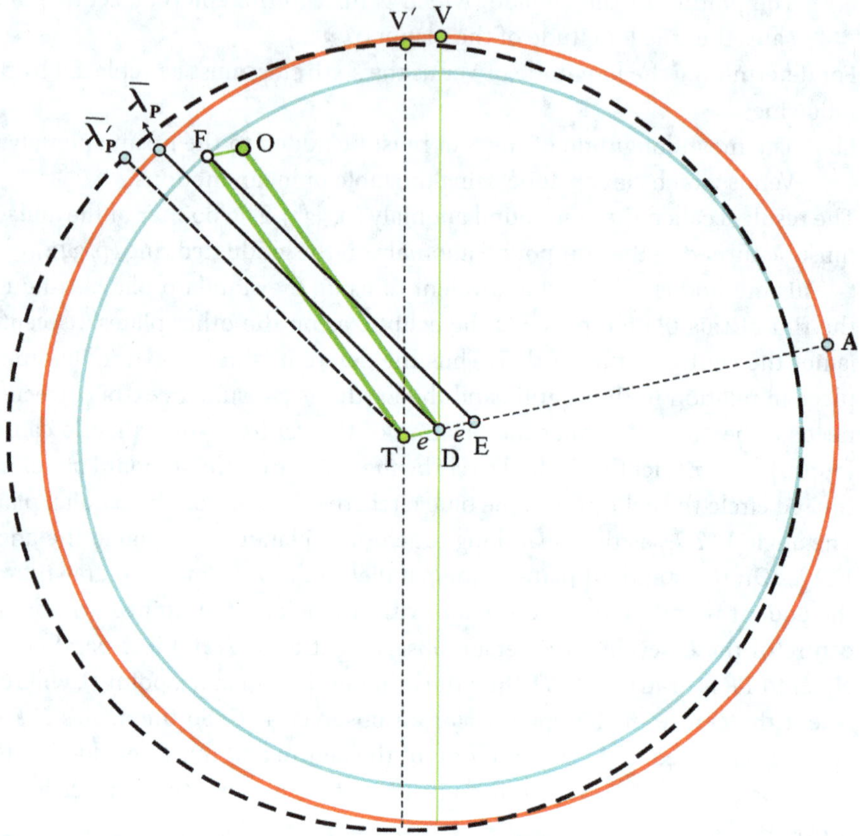

FIGURE 7.6 Equality of the parallel plate with the standard plate

3. The quotient is added to the given longitude if the place in question is west of Shiraz and subtracted if it is east of Shiraz (p. 327).

This method is easily shown to be correct given that all the celestial bodies appear to move at a rate of 15 degrees per hour.

11 Conclusion

Kāshānī follows the Ptolemaic planetary theory of longitude closely in the *Nuzha* except that he employs new parameters for the epicyclic radius of Mars and the eccentricity of Venus. All of the values of the parameters have been taken from his *Zīj-i Khāqānī* but they have been rounded up or down. In his *Zīj,* Kāshānī discussed a method for calculating planetary longitudes which was faithfully Ptolemaic.[36] Consequently if an astronomer employs the *Zīj* and the ideal equatorium with a very large radius to calculate a planet's true longitude, he would obtain nearly identical results. Small differences would occur because of the rounding of the parameters. As stated above, to achieve results accurate to minutes of arc with an equatorium, the plate would have to have a radius of at least 1.7 metres. For casting a horoscope much smaller instrument would have been sufficiently accurate.

36 Kennedy, *On the Contents and Significance ...*, cit. (n. 31), p. 17.

CHAPTER 8

The Introduction of the Telescope into Iran before the Nineteenth Century

Seyyed Hadi Tabatabaei

1 Introduction

The birth of the new astronomical knowledge in the West occurred in the sixteenth century, and the first use of telescopes for astronomical applications in Iran was at the beginning of the seventeenth century during the rule of the Safavid dynasty. Before the establishment of the *Dār al Funūn*, the first modern institution of higher education in Iran, in the year 1851 various dynasties had ruled Iran and Iranian science had undergone various changes. The introduction of modern astronomical instruments such as telescopes into Iran before the establishment of modern schools and colleges is an important neglected issue in the history of science in Iran and deserves more research.

To track the introduction of the telescope in Iran it is necessary to study the first scientific instruments which arrived in the country between the sixteenth and the nineteenth centuries by using contemporary sources—not only books about Iranian history, but also the notes and itineraries of foreign travellers in Iran and Iranian books about modern astronomy written in that period. A brief look at Iran's history reveals that modern science arrived from Europe, India, the Ottoman Empire and Russia. In this process ambassadors, scientists, students and travellers played very important roles.

One of the first transfers of modern instruments to Iran occurred in the pre-Safavid period, i.e. before 1501. During the reign of Uzun Hassan (1423–1478), the ruler of the Aq Qoyunlu Turkomans, ambassadors from Venice brought guns, cannon and eyeglasses to Iran, so Iranians became familiar with eyeglasses and lenses.[1] At this time eyeglasses were seen as a medical device by one of the great Iranian poets, Abdolrahman al-Jami (1414–1492) in a poem in which he calls them "European glasses" and refers to their use by him:

1 Hossein Mahboobi Ardakani, *Tarikh-e Moassesate Tammadoni e Jadid dar Iran* (*The History of Modern Civilization Institutes in Iran*), University of Tehran, Tehran, 1370 H/1991 CE, 3 vols.: v. 1, p. 5.

 | DOI:10.1163/9789004412842_009

My two eyes have no efficiency,
Unless I make them four [eyes] by [using] eyeglasses.[2]

Paintings of the Safavid period, especially paintings by Reza Abbasi (1565–1635) and his students suggest that eyeglasses were used widely (*Fig. 8.1*).

There is interesting information about the state of science and about scientific instruments in Iran in Western travel books such as the travelogue by John Chardin (1643–1713).[3] In it he considers the most important astronomical instrument of Iran to be the astrolabe, and he points out that Iranian astrolabes are better and more accurate than European ones.[4] As Chardin said, astronomy in the Safavid dynasty was based on the works of Ptolemy, and therefore the common challenges of European science were not those of the scientific communities of Iran. Nevertheless, some of the new Western ideas and inventions came to Iran, and the telescope was one of them.

2 Evidence of the First Telescopes in Iran from Persian Texts

A letter about Tychonic cosmology dated 1624 from Pietro della Valle to his astronomer friend in Iran Zeynal-Din Lari is the oldest known Persian text about modern astronomy. This letter and its importance were described for the first time by Aydın Sayılı,[5] and Sonja Brentjes has also discussed some parts of it.[6] There are two manuscript copies of this letter in the Vatican Library, and

2 "Az Do Chashme Man Nayayad Hich Kar/Az Ferangi Shishe Nagashteh Chahar" (Persian: از دو چشم من نیاید هیچ کار / از فرنگی شیشه ناگشته چهار). From: Farzam Haghighi, "Tafanonni Dar Adab-e Farsi: Bazyabi-e Eynak Az Motoun-e Sadeha-ye Nohom Ta Davazdahom [A survey in Persian literature: eyeglasses in fifteenth to nineteenth century texts]", *Majale-ye Sher Pajhohi (Boostan-e Ada)* [*Poem Researches magazine*] 7, 23 (1394 H /2015 CE), p. 101.

3 Sir John Chardin travelled twice to Iran between 1665 and 1677. He was familiar with the Court. Because of his facility with the Persian and Turkish languages, his travel book is very important for understanding the society and everyday life of Iran. See John Emerson, "Chardin, Sir John", in *Encyclopaedia Iranica*, Encyclopaedia Iranica Foundation Inc., New York, 1982-: v. 5, 4 (1992), pp. 369–377.

4 John Chardin, *Safarnameh: Tarikh-e Olum-e Iranian* [*Chardin's Travel book: History of Iranian Sciences*] (tr. by Mohhemed Abbasi), Amir Kabir, Tehran, 1338 H/1959 CE, pp. 145–148.

5 Aydın Sayılı, "An Early 17th Century Persian Manuscript on the Tychonic System", *Anadolu* (*Anatolia*) 3 (1958), pp. 84–87.

6 Sonja Brentjes, "Early Modern European Travellers in the Middle East and their Reports about the Sciences", in N. Pourjavady, Ž. Vesel (eds.), *Sciences, techniques et instruments dans le monde iranien (Xe–XIXe siècle)*, Institut Français de Recherche en Iran et Presses Universitaires d'Iran, Tehran, 2004, pp. 379–420.

FIGURE 8.1 Portrait of Reza Abbasi with spectacles, painted in 1673, after his death, by Moiin Mosavvar
PRINCETON UNIVERSITY, USA

Kamran Amir-Arjomand has published an edited version of it with a very complete introduction (*Fig. 8.2*).[7] This text had been written by Christoforo Borri (1583–1632) in 1613. He was an Italian Jesuit and astronomer and in 1623–1624

7 Kamran Amir-Arjomand, "Enteqal-e Elm Dar Ahd-e Safavi: Resaleii Dar Tashrih-e Heyat-e Jadid Bar Asas-e Nazar-e Tycho Brahee [Transfer of Knowledge to Iran in the Safavid Era: A Letter on Modern Astronomy Explaining the Tychonic Model]", *Tarikh-e Elm: Iranian Journal*

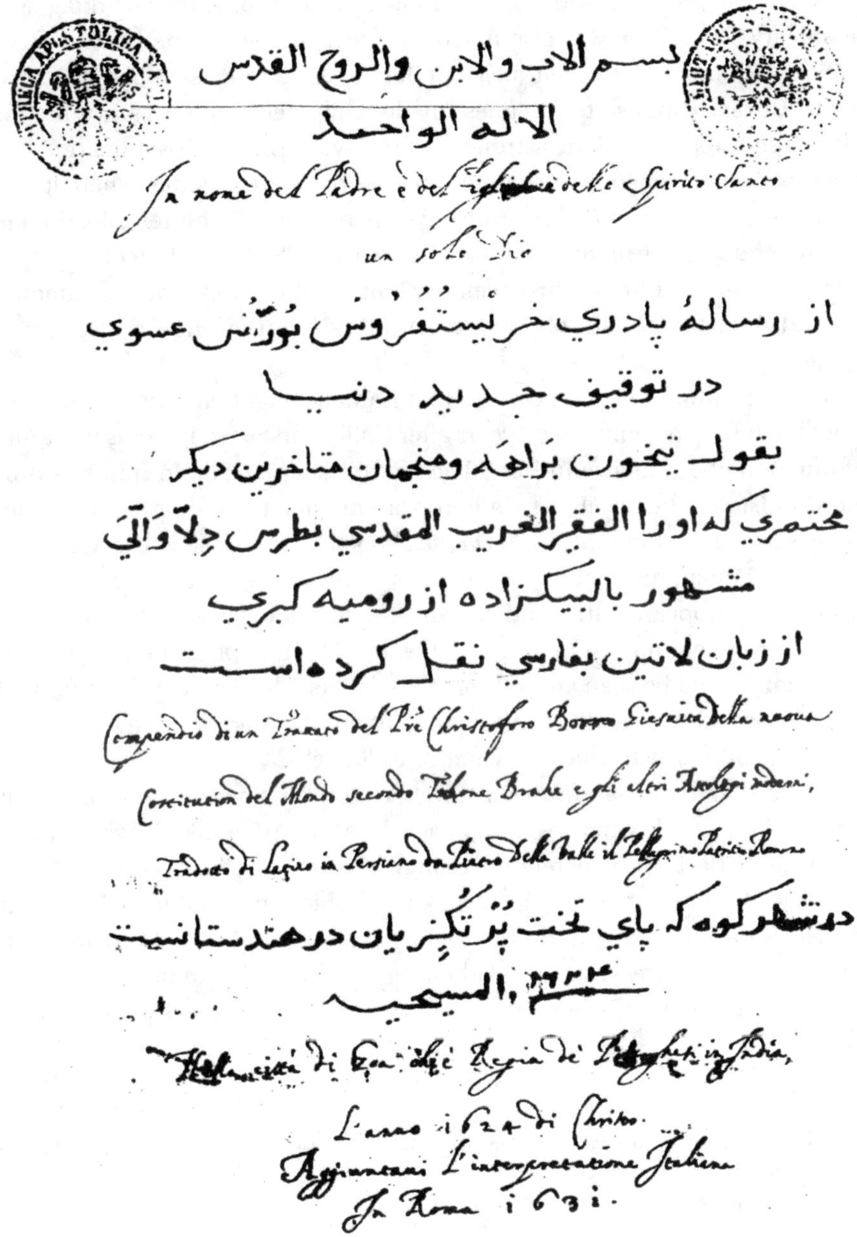
بسم الاب والابن والروح القدس
الاله الواحد
In nome del Padre e del figliuolo e dello Spirito Santo
un solo Dio
از رسالهٔ پادري خرستفروس بورنس عسوي
در توقيف جديد دنيا
بقول تيخون براهه ومنجمان متاخرين ديگر
مختصري كه اورا الفقير الغريب المقدسي بطرس دلا والي
مشهور بالبيكزاده از روميه كبري
از زبان لاتين بفارسي نقل كرده است
Compendio di un Trattato del P.re Christoforo Borro Giesuita della nuova
Costitutione del Mondo secondo Tichone Brahe e gli altri Astrologi moderni,
Tradotto di Latino in Persiano da Pietro della Valle il Pellegrino Patritio Romano
در شهر گوه كه پاي تخت پرتگيزيان در هندستانست
المسيح
Nella città di Goa ... Regia de' Portoghesi in India,
L'anno 1624 di Christo.
Aggiuntavi l'interpretatione Italiana
In Roma 1631.

FIGURE 8.2 The first page of the Persian manuscript by della Valle from Kamran Amir-Arjomand, "Enteqal-e Elm Dar Ahd-e Safavi …", *Tarikh-e Elm* 9, 1 (2011)

he was in Goa, in India, on his way from Asia to Europe after working as a missionary. Della Valle was also in Goa in the same period, on his way from Isfahan to Rome. Della Valle translated Borri's text into Persian and sent it to Zeynal-Din in response to a request made while della Valle was in Iran, for information about modern astronomy for Zeynal-Din to disseminate.[8] The telescope is mentioned in Persian for the first time in this letter, where it was called the *alat-e cheshmak deraz* (long-eye instrument). In this text, Borri/della Valle wrote about the moons of Jupiter and said that "there are four or five [sic.] little things around Jupiter first seen by Galileo using a long-eye instrument". This text also described the observation of sunspots by the long-eye instrument or telescope.[9]

From this information we can say that Iranians might have been informed about the telescope only fifteen years after Galileo first used it, but is there any information about its use in Iran? The first telescope made in Iran was produced in Isfahan before 1659 by a European missionary and priest. This telescope is mentioned in *Navadir al-Olum,* a book about new sciences written in Persian by Mohammad Salih Gazvini in 1659. The book shows that by this date a well-read European mathematician and engineer had constructed an instrument with which more stars were visible, the Moon appeared enlarged, and small 'stars' could be seen around some of planets. This telescope was formed of two lenses at either end of a rectangular tube and it was about a metre long.[10] The maker of this instrument was named as Rafael.

From a search of the European travellers and missionaries to Iran at that time we can say this man was Raphaël du Mans (1613–1696), who was in Isfahan as a priest.[11] A book referred to in the *Navadir al-Olum* is *Riaz ol-Olama* by Mirza Abdollah Isfahani (1657–1722), who added more information about the telescope with the remark "this is not such an amazing and unknown instrument because we saw a lot of them in Isfahan and Istanbul and it is an instrument for looking at objects far from us".[12] Thus there was a telescope in

---

for the History of Science 9, 1 (Persian section; 1390 H/2011 CE), pp. 1–26. See also Brentjes, op. cit. (n. 6), p. 410.

8 Amir-Arjomand, op. cit. (n. 7), p. 5.

9 *Ibid.*, p. 19.

10 The length of the telescope was given as two *Zar'e*. A *Zar'e* is an old unit of length of about 45 to 50 cm.

11 On Father Raphaël du Mans, born Jacques Dutertre, see Richard Francis, "Du Mans, Raphaël", in *Encyclopaedia Iranica,* cit. (n. 3), v. 7, 6 (1996), pp. 571–572.

12 Abbas Eqbal-e Ashtiani, "Avalin Doorbin Nojumi Dar Iran [The first Telescope in Iran]", *Yadegar* 20 (1325 H/1946 CE), pp. 33–36. *Doorbin* is Persian for binocular.

Iran in Isfahan in 1659 and there were more telescopes in Iran by 1722, by which time it was a far from unknown instrument.

Iranians who were living in India and Indians familiar with the Persian language played a very important role in the transfer of modern science and technology to Iran, and there are Indian texts and documents about telescopes written in Persian that need to be examined. The oldest Persian treatise about modern astronomy that is not translated from Western sources was written in India by an Iranian between 1769 and 1772, and this treatise, *Resale-I dar Esbat-e Hey'at-e Jadid* (A Treatise on the Proof of Modern Astronomy), is one of the first texts written by an Iranian about Western science. The author of the book was Abu Taleb ibn Hassan Hosseini Safavi who may have been associated with the East India Company because he described himself as being in contact with "philosophers and mathematicians of the Christians". This text is in two sections and each has several chapters. The first section is about geography. The second is on astronomy, and has some schematic drawings of planets, such as Saturn with its moons and rings and Jupiter with its moons, and it also describes mountains and craters on the Moon's surface.[13] These drawings and other details show that this man was familiar with telescope observations and may himself have observed the sky with a telescope.

One of the lesser known Persian treatises about modern astronomy that has information about the telescope is *Hey'at-e Jadid-e Engrizi* (New English Astronomy), written by Mir Mohammad Hossein Isfahani in 1797 in India in Persian. The name shows that the author was originally from Iran. The text was written after his two-year visit to England and is in two parts, one about Western geography and newly-discovered lands and the other about modern astronomy and especially about the telescope invented by Newton, with which new instrument Europeans had discovered new comets and moons, sunspots, the rings of Saturn and the crescents of Mercury and Venus. The only known manuscript of this treatise is in the Maulana Azad Library of the Aligarh Muslim University of India.[14]

Another Indian manuscript about modern astronomy is *Maimua Shamsi* (Solar System) written in Persian by Maulavi Abol-Kheyr in 1808.[15] The information in the manuscript came from English sources and the translation was

13 For an edition of Abu Taleb ibn Hassan Hosseini Safavi's book with an introduction and a commentary, see Hossein Masoumi Hamedani, "Resale-I dar Esbat-e Hey`at-e Jadid", *Ma'aref* 2, 2 (1363 H/1984 CE), pp. 112–120.

14 Mir Mohammad Hossein Isfahani, *Hey'at-e Jadid-e Engrizi* [*New English Astronomy*], 1212 H/1797 CE, Aligarh (India), Aligarh Muslim University, Maulana Azad Library, MS 18.1. My special thanks to Dr. Reza Dehgahi who sent me his writings about this manuscript.

15 Maulavi Abol-Kheyr, *Majmua Shamsi*, undated, Tehran, Majlis Library, MS 8897, fol. 20v.

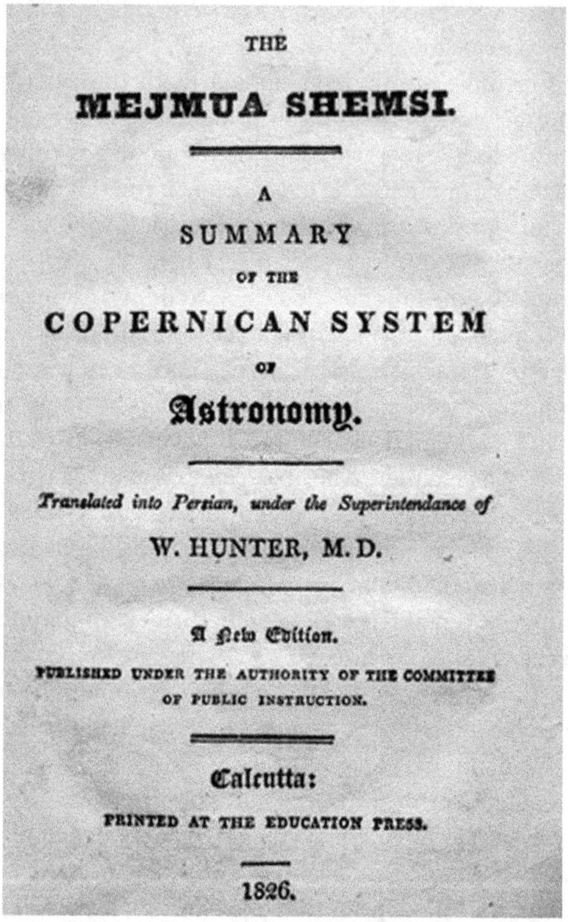
THE

MEJMUA SHEMSI.

A

SUMMARY

OF THE

COPERNICAN SYSTEM

OF

Astronomy.

Translated into Persian, under the Superintendance of

W. HUNTER, M. D.

A New Edition.

PUBLISHED UNDER THE AUTHORITY OF THE COMMITTEE OF PUBLIC INSTRUCTION.

Calcutta:

PRINTED AT THE EDUCATION PRESS.

1826.

FIGURE 8.3 Title page of the printed edition of *The Mejmua Shemsi*, Calcutta [Kolkata], 1826

supervised by the British orientalist William Hunter (1755–1812). Some sources claim that it was written by Hunter.[16] It was published in Kolkata in 1826 (*Fig. 8.3*),[17] and it describes telescope observations of the planets. Chapter five, for example, is about the moons of planets. After these books there were others

16 Kamran Amir-Arjomand, "The Emergence of Scientific Modernity in Iran: Controversies Surrounding Astrology and Modern Astronomy in the Mid-Nineteenth Century", *Iranian Studies* 30 (1997), pp. 5–24: 9

17 [Maulavi Abol-Kheyr], *The* Mejmua Shemsi. *A Summary of the Copernican System of Astronomy, Translated into Persian under the Superintendence of W. Hunter, M.D.*, Printed at the Education Press, Calcutta [Kolkata], 1826.

written in India with information about the telescope and most of them were known in Iran.

During this period some travellers to Europe visited European scientific centres. One of them was Mirza Abu Taleb Khan Isfahani (1752–1806), who travelled to England in 1800 where he visited observatories, notably the one attached to Oxford University, and saw a large orrery exhibited in a London play-house. He wrote about hundreds of telescopes and astronomical instruments in his travel book.[18] Three years before Abu Taleb Khan's book, there had been another book in Persian about modern astronomy and the telescope published in India, called *Tohfat al-Alam* and written by Mir Abd al-Latif Khan Shoushtari, who stated that the telescope had been invented by Copernicus.[19] However, this book is so full of mistakes about astronomy that it cannot be considered a real scientific book and should be ignored.

After the defeat of Iran in the war with Russia (1804–1813) the Iranian government realized that one of the reasons for the country's defeat was the use of modern science and technology by Russia, so it decided to introduce new science and technology, and sent groups of students to Western countries and set up boards of foreign military and technical experts to improve the military capability of the army. Students in different branches of science and technology travelled to study in Britain, France, Russia and other countries and returned to high positions and important roles in Iran. One of these men was Mirza Saleh Shirazi. He studied natural philosophy in England and mentioned the Herschel telescope in a part of his travelogue. On 3 October 1818 he visited the famous astronomer William Herschel, by then an old man, and talked with his sister and his son. He saw various parts of his observatory and the telescopes. He noted the dimensions and other details of Herschel's great telescope. Mirza Saleh asked about the cost of the instrument, and was told that "the expenses

18 Mirza Abu Taleb Khan Isfahani, *Masir-e Talebi ya Safarnamey-e Mirza Abu Talib Khan* [*Talebi Route or Mirza Abu Talib Khan's Travel Book*] (ed. by Hossein Khadiv-Jam), Elmi va Farhangi, Tehran, 1383 H/2004 CE, p. 116. The book was published in India and Iran several times because it was an excellent source of information about Western countries as seen through the eyes of an Eastern man. For an English translation see Mirza Abu Taleb Khan, *The Travels of Mirza Abu Taleb Khan in Asia, Africa and Europe, during the years 1799, 1800, 1801, 1802, and 1803* ... (tr. by Charles Stewart), Longman, Hurst, Rees, Orme and Brown, London, 1814, 2 vols. In particular, see v. 1, p. 168 for the reference to the observatory in Oxford (which would have been the Radcliffe Observatory) and v. 1, p. 211 for the exhibition of the orrery.

19 Mir Abd al-Latif Khan Shoushtari, *Tohfat al-Alam va Zeyl-e al-Tohfe* (ed. by Samad Movahhed), Ketabkhane-ye Tahouri, Tehran, 1363 H/1984 CE, p. 300.

of the big and small stands [telescope fixings] and the three telescopes has been 20,000 pounds".[20]

During this period Iranians became more familiar with scientific books and more able to translate them, so the new generation became more aware of modern science. At the same time, some diplomatic groups, after visiting scientific centres in Western countries suggested to the government that similar institutes and centres should be established in Iran.[21]

One of the very important books that was translated into Persian in Iran was *Tarjome-ye Heya't-e Jadid* (Translation of [a Treatise on] Modern Astronomy), translated by Mirza Masoud-e Ansari who was an Iranian diplomat. After he learned French, he translated it from a French book by the order of the Royal Prince (Abbas Mirza).[22] It is a very technical book, suitable for well-educated astronomers. It records some telescope observations including sunspots seen with a large telescope,[23] the mountains of the Moon,[24] the moons around the planets,[25] and the rings of Saturn.[26]

The process of bringing modern astronomy and science to Iran was accelerated by the work of missionary boards that established modern schools, and the educational books translated and used by these missionary groups are another source of information about the telescope in Iran. Notable among these is a manuscript composed in Persian by an American missionary, James Lyman Merrick under the title *Risale-ye Elm ol-Nojum* (A Treatise on Astronomical Science). Written in 1841 in Iran, the treatise includes some detailed drawings of planets as seen through a telescope (*Figs. 8.4–8.7*).[27]

---

20 The travelogue has been published twice in Iran in 1968 and 2009 (from a manuscript in the British Library, London, Persian collection, MS Add 24034), but has some lacunae, including the costs of the telescopes and related equipment. This information was found by Abbas Eqbal Ashtiani in a manuscript in a private collection in Iran. Ashtiani says that there is another manuscript of the travelogue in the British Library, but he does not give its number. See: Abbas Eqbal Ashtiani, "Molaghat-e Do Tan Irani ba Hershel [A Meeting of Two Iranians with Herschel]", *Yadegar* 2 (1323 H/1944 CE), pp. 76–78.

21 Amir Sadat-Mousavi, "Nojum-e Jadid Pish az Dar ol-Fonun [Modern Astronomy before the Dar al-Fonun]", *Tarikh-e Elm: Iranian Journal for the History of Science* 12 (1393 H/2014 CE), pp. 223 and 227.

22 Masoud ibn Abd ol-Rahim Ansari Garmrudi, *Tarjome-ye Heya't-e Jadid*, translated in 1250 H/1834 CE, Tehran, Madrese-ye Shahid Motahhari Library, MS 616, fols. 7r–7v.

23 *Ibid.*, fol. 26r.

24 *Ibid.*, fol. 50r.

25 *Ibid.*, fol. 60r.

26 *Ibid.*, fol. 61r.

27 James Lyman Merrick, "*Elm ol-Nojum*", copied 1277 H/1860 CE: Tehran, Majlis Library of Iran, MS 14503. See Seyyed Hadi Tabatabaei, "The advent of modern astronomy into Iran before 'Dār al Funūn' and a critical edition of the *Elm ol-Nojūm* (by an unknown author and translator, translated in 1257 Hijrah)", MSc degree thesis in History of Science, University of Tehran, Institute for the History of Science, September 2014.

FIGURE 8.4 Drawings of Venus as seen through a telescope, from Merrick, "*Elm ol-Nojum*"
TEHRAN, MAJLIS LIBRARY OF IRAN, MS 14503, FOL. 10R

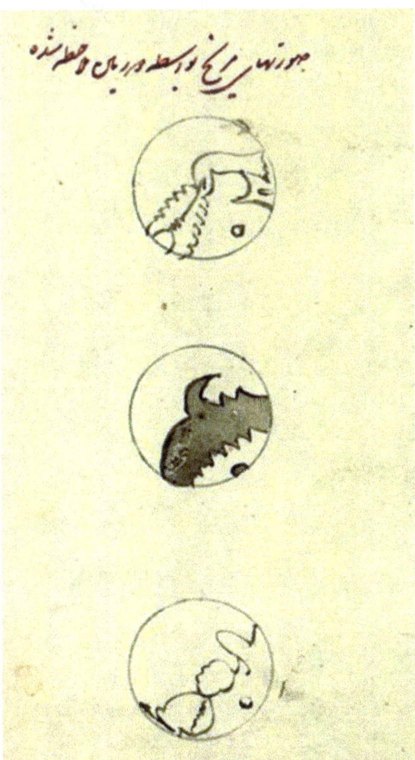

FIGURE 8.5
Drawings of Mars as seen through a telescope, from Merrick, "*Elm ol-Nojum*"
TEHRAN, MAJLIS LIBRARY OF IRAN, MS 14503, FOL. 44V

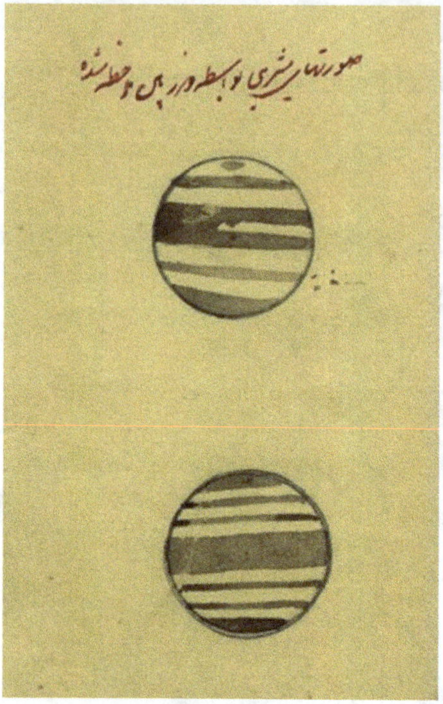

FIGURE 8.6
Drawings of Jupiter as seen through a telescope, from Merrick, "*Elm ol-Nojum*"
TEHRAN, MAJLIS LIBRARY OF IRAN, MS 14503, FOL. 51R

FIGURE 8.7
Drawing of Saturn as seen through a telescope, from Merrick, "*Elm ol-Nojum*"
TEHRAN, MAJLIS LIBRARY OF IRAN, MS 14503, FOL. 55V

Exactly in the middle of the nineteenth century (1850–1851) the first Iranian higher educational institution, the *Dār al Funūn*, opened. This date can be regarded as the beginning of higher education in Iran, after which modern instruments and sciences became familiar to everyone in Iran.

CHAPTER 9

Hugo Masing's Golitsyn-Vilip Seismographs

From Tartu to Five Continents

Janet Laidla

1 Introduction

At the end of the nineteenth century seismology appeared as a research field at the University of Tartu Observatory. It culminated in the 1920s with the improved Golitsyn-Vilip seismographs constructed at Hugo Masing's fine mechanics workshop (*Figs. 9.1* and *9.2*) and sold from the small university town of Tartu to locations all around the globe.

The purpose of this chapter is to understand how seismographs from Tartu were marketed to be used all over the world. It is based on papers from Hugo Masing's seismograph-making business. The majority of these papers are the correspondence between him and the individuals or institutions that ordered his instruments (especially in the years 1925–1936, during which period Masing had to establish himself as the maker of world-class seismographs). The content of the available sources dictated that the emphasis should be on his marketing strategies that can be determined from his correspondence. The correspondence is in three files containing more than a thousand pages of letters and often the draft versions of Masing's replies.[1] As the innovations in the seismographs in question have been discussed previously elsewhere[2] and the marketing strategies of instrument makers have not been studied in Estonia,[3]

1 National Archives of Estonia (*Rahvusarhiiv*), H. Masingi Seismograafiline Tööstus [H. Masing's Seismographical Industry], 1925–1940, RA, ERA.3103.1.2 (Tallinn), RA, ERA.3103.1.3 (Tallinn) and RA, ERA.3103.1.4 (Tallinn). The National Archives have branches in Tallinn, Tartu, Rakvere and Valga. The bulk of the collection is in Tallinn and Tartu where the main reading rooms are situated. In this article locations are indicated at the time of writing but they may change if the collections are moved to modern storage facilities in Tartu.

2 See for example M. Elango, "Johan Vilip—väljapaistvaim Eesti füüsik käesoleva sajandi esimesel poolel" [Johan Vilip—a leading Estonian physicist from the first part of the twentieth century], *Teaduse ajaloo lehekülgi Eestis* 1 (1968), Eesti Loodusuurijate Selts, Tartu, pp. 104–111; and Enn Hendre, "Prof. Johan Vilip's Contribution to Scientific Instrument Engineering", *Museum of Tartu University History: Annual 1996*, Museum of Tartu University History, Tartu, 1997, pp. 50–59.

3 See for example A.D. Morrison-Low, *Making Scientific Instruments in the Industrial Revolution*, Ashgate, Aldershot, 2007; Peter de Clercq, "Exporting Scientific Instruments Around 1700:

 | DOI:10.1163/9789004412842_010

FIGURE 9.1 Hugo Masing's Golitsyn-Vilip seismograph from the University of Tartu Museum collection
IMAGES: ANDRES TENNUS/UNIVERSITY OF TARTU

investigating this correspondence seemed as an interesting approach to the study of the history of scientific instruments.

2 History of Seismology in Tartu

The history of seismology at Tartu could have begun with Prince Boris Golitsyn (1862–1916) but during his short stay at Tartu University he was not yet interested in seismology. Nevertheless, in Tartu he taught a young physics student, Johan Vilip, who later became known for improving the seismographs this chapter is about. Golitsyn studied at the Naval Academy in Nikolajevsk and later at the University of Strasbourg. In 1891 he became a *privatdozent* at the

The Musschenbroek Documents in Marburg", *Tractrix: Yearbook for the History of Science, Medicine, Technology and Mathematics* 3 (1991), pp. 79–120.

FIGURE 9.2 Hugo Masing's Golitsyn-Vilip seismograph from the University of Tartu Museum collection
IMAGES: ANDRES TENNUS/UNIVERSITY OF TARTU

University of Moscow from where he took a temporary post at the University of Tartu.[4]

4 Tõnu Reinet, *B.B. Golitsõni ja A.I. Sadovski elu ja tegevus ning nende mõju rahvusvaheliselt tunnustatud Eesti teadlase J. Vilipi kujunemisele* [The life and works of B.B. Golitsyn and

Golitsyn was a member of the seismographic commission of the Russian Academy of Sciences.[5] The first vertical and horizontal seismographs made according to his specifications became available in 1906–1907. They were based on his theory of galvanometric registration developed between 1903 and 1908. Golitsyn's ambition was to build seismographs that had longer periods of oscillation, and that as a consequence had vibration frequencies that could be used for the detection and registration of distant earthquakes. Golitsyn was also hoping to determine the centres of the earthquakes and the angle of their emergence.[6] Under Golitsyn several new seismological stations were founded and equipped with the newest available instruments. The new Golitsyn seismographs were used and tested at the central Pulkovo (Saint Petersburg) station and also at the Tartu station, and were subsequently exported to stations in the United States, England and France.[7]

The beginnings of seismological research at Tartu Observatory can be traced back to Grigori Levitski (1852–1917) who came to Tartu in 1894, a few years after Golitsyn. The once-famous observatory of Tartu was typical of late nineteenth-century observatories in looking for new fields to study, and both astrophysics and geophysics (including seismology) emerged as new areas of interest.

Levitski showed signs of interest in seismology while he was still working at Kharkiv Observatory (now in Ukraine), where he also founded a seismological station. In 1896 he ordered four different horizontal pendulums to be used at the new seismological station in Tartu (a Rebeur-Paschewitz pendulum with photographic registration, two Zöllner pendulums and a Zöllner heavy pendulum with mechanical registration). The Tartu station was also responsible for testing new seismological instruments that were sent to Russia (for example the Stückrath heavy pendulum). For a period of time the Tartu station served as one of the central points where data from other stations in Russia was gathered and prepared for publication. Levitski improved the Rebeur-Paschewitz instrument by constructing a better pendulum, and the Zöllner pendulum by increasing its weight, helped by the instrument maker Repsold (of Hamburg). Levitski was later assisted by a local instrument maker, Bernhard Schulze. The instruments were installed in the cellar of the observatory, but also in the old

A.I. Sadovski and their influence on the internationally recognized Estonian scientist J. Vilip], University of Tartu, Tartu, 1965, pp. 9–12.

5 *Ibid.*, p. 13.

6 Hendre, op. cit. (n. 2), p. 52.

7 Reinet, op. cit. (n. 4), pp. 13 and 27.

gunpowder cellar situated nearby. Both provided good conditions with only minimal variations in temperature.[8]

Levitski's interest in seismology was shared by a colleague at the observatory, Aleksandr Orlov (1880–1954),[9] who worked in Tartu between 1905 and 1906 and from 1907 until 1912. He became interested in seismology when he was studying in Göttingen during his European tour, between 1902 and 1905. After 1912 Orlov returned to Ukraine and worked at Odessa and Poltava. His main research interest was in the influence tides had on the movement of the Earth's core. Mostly he used the horizontal pendulums that were situated in the gunpowder cellar. He was a member of the central seismological committee of the Russian Imperial Academy of Sciences and participated at a seismological conference in Manchester in 1911.[10]

3 The Golitsyn-Vilip Seismograph

3.1 *Johan Vilip*

As previously mentioned, one of Boris Golitsyn's students was a young Estonian, Johan Vilip (1870–1942). Golitsyn invited Vilip to work as an assistant at the Physics Laboratory at the Naval Academy, and at the Imperial Academy of Sciences at St Petersburg in 1896. Vilip became interested in seismology in 1907–1908 just at the time Golitsyn had improved his seismographs. In 1911 Vilip was promoted to the position of Director of the Pulkovo seismological station. The collaboration between Vilip and Golitsyn lasted until the latter died in 1916.[11] Vilip was elected head of the Physics Department and professor of Physics at Tartu University in 1920.

Johan Vilip improved both the horizontal and vertical seismographs constructed by Golitsyn.[12] He made several small improvements to the horizontal seismograph, the most important of which was to eliminate the problem

8 Grigori Želnin, "Tartu tähetorni tegevus aastail 1894–1908" [Activities of Tartu Observatory 1894–1908], *Tartu Tähetorni kalender* 44 (1968), pp. 51–61.

9 Želnin, "Tartu tähetorni tegevus aastail 1908–1918" [Activities of Tartu Observatory between 1908–1918]", *Tartu Tähetorni kalender* 45 (1969), pp. 48–61 and 51–52.

10 Želnin, *Tartu Tähetorni kalender* 44 (1968), cit. (n. 8), pp. 57 and 61; Želnin, *Tartu Tähetorni kalender* 45 (1969), cit. (n. 9), p. 56.

11 Elango, op. cit. (n. 2), p. 104–106; Hendre, op. cit. (n. 2), p. 51.

12 For the horizontal seismograph see J. Vilip, "On New Precision-Seismographs", *Acta et Commentationes Universitatis Tartuensis* (*Dorpatensis*) A X.7 (1926). For the improved vertical seismograph see J. Vilip, "A Galvanometrically Registering Vertical Seismograph with Temperature Compensation", *Acta et Commentationes Universitatis Tartuensis* (*Dorpatensis*) A XX.6 (1930).

caused by diamagnetism of the damper of the pendulum that changed the period of oscillation. The vertical seismograph was redesigned into an almost new instrument. By using special weights that were moved further away from the axis of oscillation and two very precise bimetallic (invar and brass) coils, Vilip was able to reduce the effect of temperature on the length of pendulum and therefore on the zero-point. This meant that the seismographs no longer had to be stationed in special multi-walled cellars that protected them from changes in temperature. The new and improved vertical seismograph was the topic of Vilip's doctoral dissertation.[13] He also improved the suspension system of the pendulum by adding a helical spring which compensated for the changes in the period of oscillation. Instead of a Deprez-d'Arsnoval aperiodic galvanometer, a Hartmann & Braun galvanometer was used in the improved seismographs.[14]

3.2 *Hugo Masing and His Workshop*

The reinstating of Tartu University in 1802 soon created a market for precision instruments. Although many instruments were ordered from abroad, the limited budget encouraged more local solutions. After Estonia became an independent republic in 1918 and after the language of instruction at the university became Estonian in 1919 the university continued to need locally-made scientific instruments. These were provided by the university workshop headed by Jaan Muuga (1891–1968),[15] and also by several small companies and private mechanics in Tartu. Paul Siilbaum, for example, was the favoured mechanic of the Tartu Observatory.[16]

One of these smaller workshops was founded by Eduard Pikkand (1892–1984) who started his fine mechanics workshop in 1927. It produced scientific instruments and models for schools, and more complex devices for university clinics and laboratories.[17] Another shop was established by Evald Neugard and called Experimentator (1914–1940). At first this produced physics apparatus.

13 Elango, op. cit. (n. 2), p. 107.

14 Hendre, op. cit. (n. 2), pp. 55–57.

15 On Jaan Muuga see Hillar Palamets, "Ülikooli teenistuses: meenutusi ja materjale ülikooli tehnilisest teenistusest viimase poolsajandi jooksul" [At the service of the university: memoires and sources from the university technical services for the past half century], Manuscript at the University of Tartu Library, 141.89.

16 "Aruanded teadusliku töö ja õppetegevuse alal" [An account of research and teaching], National Archives, Tartu Observatory, RA, EAA.5374.1.351 (Tartu).

17 "Pikkand, Eduard; peenriistade tööala" [Pikkand, Eduard, field of fine mechanics], National Archives, Tartu Aptitude Board, RA, ERA.3167.1.904 (Tallinn).

After the First World War the workshop produced educational devices in the fields of mechanics, electronics and physics.[18]

Of all the small workshops, it is Hugo Masing's that had the most international recognition.[19] In 1901 Boris Golitsyn invited Masing to St Petersburg to work at the Institute of Physics of the Imperial Academy of Sciences, where he eventually began constructing seismographs for Golitsyn.[20] After the Russian revolution and Estonian declaration of independence he came to Tartu to set up his workshop there. We know much less about Masing's other products, but according to an advertisement from 1924 the workshop produced meteorological, geodetic, mathematical and chemical instruments, and balances.[21]

The partnership with Johan Vilip was only one example of Masing's successful collaboration with scientists at the University of Tartu. Alfred Fleisch worked as the Professor of Physiology and Chemistry at the university between 1926 and 1932. In addition to teaching and research he also designed original devices, the pneumotachometer for studying respiration and the pulse time recorder or interval recorder for studying blood circulation.[22] The pulse time recorders were also constructed by Masing's company but the correspondence implies that they were sold through Fleisch as the instruments rarely occur in correspondence with other clients.

In addition to full instrument sets Masing also sold replacement parts (light bulbs most often) and provided lengthy instructions for setting up the

18 "Mehaanikatöökoda Experimentator—E. Neugard" [Experimentator Mechanical Workshop—E. Neugard], National Archives, Chamber of Commerce and Industry, RA, ERA.891.2.4628 (Tallinn); "Experimentator—mehaanikatöökoda Tartus" [Experimentator—a mechanics workshop in Tartu], National Archives, Estonian Workers Insurance Association, RA, ERA.842.1.952 (Tallinn).

19 In addition to the sources mentioned in the introduction see also "Teaduslike aparaatide tööstus Hugo Masing" [Hugo Masing's scientific equipment business], National Archives, Chamber of Commerce and Industry, RA, ERA.891.2.4918 (Tallinn); "Masing, Hugo Tartus, Lai nr. 17 teaduslike aparaatide tehas" [Masing, Hugo, 17 Lai St., Tartu, scientific equipment factory], National Archives, Ministry of Economy, RA, ERA.969.4.1020 (Tallinn).

20 Leili Kriis, "Physiology Professor Alfred Fleisch (1892–1973) and His Legacy at the University of Tartu", *Acta Baltica Historiae et Philosophiae Scientiarum* 4 (2016), pp. 70–96: 77.

21 *Postimees*, 1 April 1924, p. 11. *Postimees* is an Estonian daily newspaper. This advertisement was part of a larger feature introducing the industry in Tartu. There is a similar list in a document dating from 1938; see "Masing, Hugo Tartus, Lai nr. 17 teaduslike aparaatide tehas" [Masing, Hugo, 17 Lai St., Tartu, scientific equipment factory], National Archives, Ministry of Economy, RA, ERA.969.4.1020 (Tallinn).

22 Kriis, op. cit. (n. 20).

instruments.[23] In 1928 Masing reported that he had written to the Edison Lamp Works of the General Electric Company to inquire if they could start to manufacture light bulbs according to Masing's drawings so that seismological stations in the United States would not have to order from him.[24]

4 Other Seismographs in the Market at the Time

Estonian historiography regards both Golitsyn's early seismograph and the later Golitsyn-Vilip seismograph as the best available at the time. It cannot be claimed conclusively that the seismographs Masing made to these designs were the only ones available or considerably better than the others. However, they were remarkable enough to be of interest to several museums. In 1930 The Science Museum in London asked if they could include a seismograph in their exhibition. Masing responded that he was not able to make one for free just to be exhibited and sent good-quality images instead.[25] Another museum interested in Masing's seismographs was the Museum of Science and Industry in Chicago.[26]

Naturally both Vilip and Masing were interested in other similar instruments on the market. There is a letter from the Cambridge and Paul Instrument Company sent to the Institute of Physics that Vilip very probably forwarded to Masing because the company was also making Golitsyn type seismographs.[27] The Cambridge and Paul seismograph was shown to be a serious competitor in a letter about Masing's requirement to pay half the cost of a seismograph in advance. The Reverend Tondorf of the Seismological Observatory of Georgetown University, Washington D.C., considered this an insult and threatened to order a Cambridge and Paul seismograph instead.[28]

Another alternative was a torsion seismograph by Anderson and Wood that was mentioned by Nörlund.[29] We can also see from the correspondence which other seismographs were in use at stations that were interested in the

23 For example, Masing to Mr. Hubert J. Brockman, 25 October 1928, National Archives, cit. (n. 1), RA, ERA.3103.1.2, fol. 16.

24 Masing to James B. Macelwine, 3 March 1928, *ibid.*, fol. 253.

25 Correspondence with the Science Museum, 1930, *ibid.*, fols. 188–199.

26 L. Otermba to Masing, 12 May 1932, *ibid.*, RA, ERA.3103.1.3, fol. 473.

27 The Cambridge and Paul Instrument Company Limited to the Institute of Physics, 1 December 1923, *ibid.*, RA, ERA.3103.1.2, fols. 70–73.

28 James B. Macelwane to Masing, 12 January 1927, *ibid.*, fol. 279.

29 Nörlund to Masing, 13 April 1927, *ibid.*, fols. 393–394.

Golitsyn-Vilip seismograph (instruments by Bosch,[30] Wiechert[31] and Milne-Shaw),[32] and the relevant literature provides a few additional examples.[33]

5 Hugo Masing's Marketing Strategies

5.1 *State Trade Deals*

Firstly we should investigate whether Masing benefited from the marketing and trade strategies promoted by the state of Estonia. The industrialisation of the Estonian region began in the second half of the nineteenth century when the area was part of the Russian empire. The industry relied heavily on the large imperial market and its raw materials. When Estonia declared itself an independent country it found itself without raw materials, fuel, market or capital, not only because of its newly-declared independence but also because of the First World War.[34] It took time for companies to reorganize themselves so that their products could be sold in the Western market.[35] The first trade deals were in place by 1924 with several countries including Great Britain, France, Denmark, and the Netherlands.[36] The economic crisis of 1929 naturally had an effect on the economy and economic policy in Estonia.[37] By the second half of the 1920s there were several companies that had moderate success with exporting their goods, such as the chocolate company "Kawe". Its products were sold in Great Britain, the United States, France, Morocco, China and India.[38] From Tartu, more prominent examples are the Tartu Telephone Company (which exported to China, Turkey, Persia, Lithuania and Finland)[39]

30 T. Banachiewicz to (probably) Prof. J. Vilip, 2 October 1926, *ibid.*, fol. 39.

31 "Coching Chu, Institute of Meteorology, National Research Institute of China, Pie-Chi-Ko, Nanking", 13 September 1930, *ibid.*, RA, ERA.3103.1.3, fol. 139.

32 James J. McAsey to Masing, 29 October 1935, *ibid.*, fol. 331.

33 Such as Arnold Romberg, "A Practical Long-Period Seismograph", *Science* 50 (1919), pp. 141–142; Hugo Benioff, "A New Vertical Seismograph", *Bulletin of the Seismological Society of America* 22 (1932), pp. 155–169.

34 Maie Pihlamägi, *Eesti industrialiseerimine 1870–1940* [The industrialization of Estonia, 1870–1940], Ajaloo Instituut, Tallinn, 1999, p. 75.

35 *Ibid.*, p. 77.

36 *Ibid.*, p. 82.

37 *Ibid.*, pp. 115–117.

38 *Ibid.*, pp. 178–180.

39 Aime Kärner, "Tartu tööstuse kajastusi ajalehes Postimees aastatel 1920–1940" [Reports on Tartu Industry in the Newspaper Postimees in the years 1920–1940] in Marge Rennit (ed.), *Uurimusi Tartu tööstusest enne Teist maailmasõda* (*Tartu Linnamuuseumi aastaraamat*, 19), Tartu Linnamuuseum, Tartu, 2013, pp. 25–37: 31–32.

and Johan Lill's bakery (his honey cakes were sold in Europe, the United States and Australia).[40]

Compared to other companies the products of Hugo Masing were very specific, expensive and sold in very small quantities, thus it is difficult to say if at all or how the general trade policies of the young Estonian republic benefited the sale of seismographs. As there was a very small market for seismographs inside Estonia the company did have to export most of those made. Masing did not sell seismographs in large quantities in any country so he may have benefited from some trade deals but he probably did not base his choice of clients on which countries had a trade deal with Estonia. Masing may have slightly benefited from the policy introduced in 1930s that reduced the duty on imported goods if these were used to build goods that were then exported,[41] as Masing used imported galvanometers. Unfortunately, we do not have his detailed financial records to corroborate this theory.

5.2 *Trade Catalogues and Advertisements in Specialist Magazines*

There were specific catalogues for the Golitsyn-Vilip seismograph published in English and German.[42] These were nineteen pages long complete with black-and-white photographs and lengthy explanations in both languages. A world map was included in later editions of the catalogue to advertise the stations already equipped with the new seismographs. At least one person wrote that he was holding the catalogue in his hand whilst inquiring about just one specific part.[43] There were also some advertisements in general catalogues or magazines. Masing advertised in the magazine *Instruments* in 1929 and there were at least six requests for further information coming from these

40 Andres Sepp, "Eesti esimene mesikoogitööstus ja märkmeid pagaritööstusest Tartus" [Johan Lill's first Estonian honey cake manufacture and notes on bakeries in Tartu], in *ibid.*, pp. 49–77: 49. Lill was noted for his new marketing strategies when local market was concerned, see p. 60.

41 Maie Pihlamägi, "Tartu tööstuse areng Eesti Vabariigi tööstuspoliitika taustal kahe maailmasõja vahel" [Evolution of industry in Tartu against the backdrop of the industrial policy of the Republic of Estonia between the two world wars], in *ibid.*, pp. 11–21: 16.

42 "Seismographen für Galvanometrische Registrierung nach Fürst B. Galitzin und Prof. J. Wilip" [Seismographs with Galvanometric Registration after Prince B. Golitsyn and Prof. J. Vilip], in *Hugo Masing Werkstatt für Wissenschaftliche Instrumente. Eesti, Tartu, Lai 17*, Not dated, University of Tartu Museum, Archival Collection, ÜAM 1554:9 Ar. The earlier 1929 edition was without a map and in a smaller format.

43 John P. Delaney to Masing, 9 June 1930, National Archives, *cit.* (n. 1), RA, ERA.3103.1.3, fol. 147.

advertisements.[44] John Tynan from Fordham University in New York sent a letter with a copy of the magazine *Science and Invention* that featured images of the Masing seismograph.[45]

5.3 *Trade Companies*

The majority of the Masing's seismographs seem to have been sold directly from Masing to the buyers. The correspondence, however, contains a few letters in which some companies showed interest in selling the product. Carlowitz & Co. in Hamburg wanted to sell seismographs to China.[46] Masing did sell to China, but there was also a direct personal contact with the observatories in question. Another company showing interest was Spindler & Hoyer in Göttingen.[47] Thus, we cannot say conclusively from the correspondence whether Masing actually sold any instruments through these companies.

5.4 *Personal Offers*

The correspondence suggests that at least in the middle of the 1920s Masing made the first contact with potential buyers by writing personal letters advertising the new seismographs. One of these early examples is a letter addressed to the seismologist, astronomer and director of the Osservatorio Ximeniano in Florence, Guido Alfani.[48] It could be an answer to a query that has not survived.[49] The letter states that the seismograph is improved and has an advantage over the Zöllner pendulum, and also mentions the connection to Prince Boris Golitsyn. In several later letters Masing lists the people, the institutions or the places that have already bought a seismograph and are happy with it or have placed orders recently.[50]

---

44 "Letters about Instruments. Industrial-scientific instrument publishing company", *ibid.*, RA, ERA.3103.1.2, fols. 148–151.

45 John St Tynan to Masing, 8 December 1926, *ibid.*, fols. 534–535.

46 Carlowitz & Co. in Hamburg to Masing, 19 January 1929, *ibid.*, fol. 69. Some of Masing's seismographs ended up in China but it is unclear whether through Carlowitz & Co. or not. Further correspondence hints that there was a misunderstanding or even a conflict.

47 Spindler & Hoyer in Göttingen to Masing, 18 March 1930, *ibid.*, fol. 506. Also N.V. Wed. J. Ahrend & Zoon to Masing, 23 August 1937, *ibid.*, RA, ERA.3103.1.3, fol. 9.

48 I am grateful to Giorgio Strano for his biographical notes on Guido Alfani.

49 Masing to Monsieur le Directeur P. Guido Alfani in Florence, 2 January 1925, *ibid.*, RA, ERA.3103.1.2, fol. 2.

50 Masing to Dr. C.L. Abell, 7 September 1927, *ibid.*, fol. 6.

5.5 *Asking for Recommendations*

Masing also advertised through former clients and asked satisfied clients to recommend his seismographs.[51] He had many helpful clients. James B. Macelwane, Director of the Department of Geophysics at Saint Louis University heard about the prices from the Reverend Manuel Navarro Neumann, the director of the Cartuja Seismological Station in Granada,[52] and had recommended the seismographs to a certain Mr. L.W.H. Swinger in Nauru Island (Central Pacific).[53] Macelwane also wanted to use photographs of the seismographs in his Lowell lectures publications.[54] Masing asked the Rector of St. Xavier College in Cincinnati, Hubert J. Brockmann, if he could possibly recommend them to his colleagues.[55] Dr Lunkenheimer of the Observatorio Astronómico de la Universidad Nacional in La Plata, Argentina, provided the contacts for Pierre Marie Descotes, Director of the Observatory of San Calixto in La Paz, Bolivia.[56] John Tynan from the Seismic Station of Fordham University wrote to Masing saying that there was to be a meeting of the United States Seismological Society East Wing in Fordham in 1929 and the participants would be able to see Masing's seismographs, hence if he wished to send any brochures he would be very welcome to do so.[57] Masing showed his gratitude by sending thank-you letters to the people who had recommended the seismographs.[58]

5.6 *The Inventor, Johan Vilip*

Johan Vilip published some of his articles in the Tartu University Press in English.[59] It appears that Masing used Vilip's articles to advertise his

51 For example, Masing to James B. Macelwane, 21 December 1931, *ibid.*, fol. 224, and 6 April 1934, *ibid.*, fol. 210.

52 James B. Macelwane to Masing, 27 September 1927, *ibid.*, fol. 286.

53 Masing to James B. Macelwane, 24 November 1937, *ibid.*, fol. 204.

54 Masing to James B. Macelwane, 1936, *ibid.*, fols. 208–209.

55 Masing to Mr. Hubert J. Brockmann, *ibid.*, fols. 25–27.

56 Masing to P.M. Descotes, 29 March 1928, *ibid.*, fol. 122. Another example is N.V. Wed J. Ahrend & Zoon who got the address through the Koninklijk Nederlands Meteorologisch Instituut, see N.V. Wed. J. Ahrend & Zoon to Masing, 23 August 1937, *ibid.*, RA, ERA.3103.1.3 fol. 9.

57 John St. Tynan to Masing, 8 December 1926, *ibid.*, RA, ERA.3103.1.2, fols. 534–535.

58 For example, Masing to Monsieur le Prof Manuel Ma[ria] S. Navarro-Neuman Grenae, 28 December 1926, *ibid.*, fol. 290.

59 One institution that seems to have developed an interest because of the Vilip article is the Koninklijk Nederlands Meteorologisch Instituut, see "Koninklijk Nederlands Meteorologisch Instituut to Masing", 2 February 1927, *ibid.*, fol. 615.

seismographs. At times it was the potential buyer or a happy customer who asked for Vilip's scientific articles to be sent.[60] At other times the initiative to send publications came from Masing.[61]

In other letters only Vilip's name is mentioned and we cannot ascertain how the initial contact was made.[62] The "ERDA" Aktien-Gesellschaft Institut für angewandte Geophysik of Göttingen contacted Masing through Vilip and inquired about a new component for an older Golitsyn seismograph.[63] Vilip forwarded letters and inquiries addressed originally to him to Masing.[64] He also forwarded business cards to Masing, who sent out further information. One such example is a card from Trajan D. Gheorghiu, vice president of the Société de Physique de Roumanie and a member of the Academy of Sciences of Romania, associated with the University of Bucharest.[65]

Vilip was not the only scientist to recommend the Golitsyn-Vilip seismographs. A colleague of Vilip in Tartu, Karl Frisch (1892–1953, later known as Kaarel Kirde) had also given the contact details of a potential customer to Masing.[66]

It has to be remembered that both Vilip and Masing had been collaborating with Golitsyn on his seismographs. It should not be surprising that those with favourable experience of the previous version would turn to the new one,[67] but this does not occur very often. Probably there were few customers who had need of a new seismograph after only ten or twenty years.

60 For example, Dr. C.L. Abell to Masing, 7 September 1927, *ibid.*, fol. 7.

61 For example, Masing to Carlowitz & Co. in Hamburg, 25 January 1929, *ibid.*, fol. 69.

62 For example, Masing to Mr. Donald C. Barton in Houston, Texas, 8 September 1927, *ibid.*, fol. 10.

63 Aktien-Gesellschaft Institut für angewandte Geophysik "Erda" to Hugo Masing, 17 April 1925, *ibid.*, fol. 126.

64 T. Banachiewicz to [probably] Prof J. Vilip, 2 October 1926, *ibid.*, fol. 39. Tadeusz Banachiewicz had worked at Tartu Observatory between 1915 and 1918 and defended his master's thesis in Tartu, see Želnin, op. cit. (n. 9), p. 50.

65 Business card of Trajan D. Gheorghiu, National Archives, *cit.* (n. 1), RA, ERA.3103.1.3, fol. 10; Masing to Dr. Trajan D. Gheorghiu, 28 October 1937, *ibid.*, RA, ERA.3103.1.4, fol. 11.

66 Masing to S. Dlusri, Gdynia, Observatoire Maritime, 14 July 1933, *ibid.*, RA, ERA.3103.1.3, fol. 144.

67 Geodätisches Institut to Masing, 3 July 1925, *ibid.*, RA, ERA.3103.1.2, fol. 147. Another old client was Father Francis Tondorf, see F. Tondorf to Masing, 31 October 1926, *ibid.*, fol. 523. A third mention of the old Golitsyn seismographs in Fr. E. Gherzi to Masing, 11 August 1930, *ibid.*, RA, ERA.3103.1.3, fol. 228. A fourth is in a letter from Okura & Company to Masing, 12 February 1933, *ibid.*, fol. 474.

6 Conclusions

Although Prince Boris Golitsyn did not teach seismology while he was in Tartu (unlike Grigori Levitski and Aleksandr Orlov) he was instrumental in bringing together Johan Vilip and Hugo Masing, who later formed the team behind the Golitsyn-Vilip seismograph constructed in Tartu.

The reinstating of the university in Tartu in 1802 had created a demand for fine mechanical workshops in the city. There were several workshops in Tartu at the time, but Hugo Masing's workshop was perhaps the one that concentrated most on complex instruments. Masing co-operated successfully with two scientists in particular, Alfred Fleisch and Johan Vilip, and from their collaboration came international recognition that benefited both parties. It could be said that perhaps Masing showed more entrepreneurship in developing products that could be sold elsewhere while other precision mechanics did 'odd mechanical jobs' in specific institutes and produced simpler instruments for sale.

Looking at Masing's marketing strategies, it seems that rather than relying on trade catalogues he preferred a personal approach. It is likely that the reason why Masing's seismographs ended up on five continents was not because he and Vilip developed the best seismograph in the world (as the Estonian historiography seems to imply) but because they created a very good seismograph and were both very pro-active with their marketing strategies. In marketing the seismograph Vilip and Masing formed a team. It was in Vilip's interest that his improved seismographs sold widely.

There are still questions about Masing's seismographs. Were there enough orders? That is, did the demand satisfy Masing? Was his marketing strategy successful? As the making and calibrating of each seismograph took several months we may question whether Masing wanted more orders. He could have perhaps hired more labour, but could he find mechanics skilled enough to work on the finer parts of his seismographs? These questions are left unanswered for lack of appropriate sources.

CHAPTER 10

Instruments and Laboratories in the Schools of the Greek Community of Istanbul, 1850–1960

Panagiotis Lazos, George N. Vlahakis and Constantine Skordoulis

1 Introduction

After the fall of the Byzantine Empire in 1453 the Greek community of Istanbul flourished for over four centuries under the political umbrella of the Ottoman Sultans and the cultural and religious leadership of the Ecumenical Patriarchate. From the late seventeenth century and throughout the eighteenth century there was a powerful movement towards the scientific education of the Greek population of the Ottoman Empire, though there was often strong and persistent opposition from some theological circles which considered that science was opposed to Greek Orthodox religious dogma. Educated Greeks were almost always close to the *Sublime Porte*[1] as diplomats or high-ranking officials, and Greek merchants were highly influential in the financial life of Istanbul. It was common knowledge that Greek bankers usually supported the military and other enterprises of the Sultan with loans. The Greek community of Istanbul had always played a considerable role within the multicultural milieu of the Ottoman Empire. After the formation of the independent Greek state in 1830 the picture became more complicated. Though still outside the main Greek state, the Greek community experienced strong growth in commercial and cultural influence in the second half of the nineteenth century, which lasted until the 1920s. Then, due to the defeat of the Greeks in the Greco-Turkish war of 1919–1922,[2] and the destruction of Smyrna, a large part of the Greek population of Istanbul was forced to leave their homes and go to Greece as immigrants.

The period of reforms known as *Tanzimat* that began in 1839 and ended in 1876 to modernize the Ottoman Empire played an important role in this, as well as in improving the broader political and economic conditions of the time.

1 The *Sublime Porte* was a term used to describe the buildings and personnel of the government of the Ottoman Empire.

2 The Greco-Turkish war is also known as the Western Front of the Turkish War of Independence, mainly in Turkish historiography.

 | DOI:10.1163/9789004412842_011

Though reforms continued after 1876, it is acknowledged that *Tanzimat* was the response of the political leaders of the empire to its decline, which started early in the nineteenth century after the defeat of the Ottoman Navy in the battle of Navarino (1827). After the battle some rights were given to Christians, and opportunities for economic and cultural development were created. The Greek community of Istanbul, which was close to the centre of power, was active in many fields, especially education. Through the Patriarchate, but also through a series of societies (mainly the Greek Philological Society of Constantinople) elementary schools and high schools, called Γυμνάσια (Gymnasiums) in Greek, were established wherever Greek populations existed in the empire. Many of these schools were supported financially by wealthy Greek merchants.

What is of interest is that these schools started a trend towards secular education. Also, a number of schools for girls were established, at a time when in the independent Greek state girls were practically excluded from the educational system.

The secular character of these schools was critical for the curriculum. In the schools controlled by the clergy Aristotelian natural philosophy was still taught, but in the schools administrated by civilians modern sciences were taught. It is significant that not only the content of the lessons but also the books and the educational methods followed in these schools were up-to-date and progressive. As a result, many of these schools used books already popular in Europe, and acquired instruments to establish laboratories for the experimental teaching of natural sciences. The books were translated from (or based on) mainly French and German texts such as Ganot's *Traité élémentaire de Physique*[3] and Jochmann's *Grundriss der Experimentalphysik*.[4] The instruments were imported from Europe (mainly France in the nineteenth century) and the cost of their acquisition was usually covered by rich Greeks of Istanbul or the Greek diaspora. Almost every Greek high school in Istanbul had at least a modest collection of minerals to establish a geological cabinet, a collection of plants to study botany, and physics and chemistry instruments to equip educational laboratories. The experiments performed by the Greek teachers were usually those described in the books they used as sources for their lessons but

3 Adolphe Ganot, *Traité élémentaire de physique expérimentale et appliquée et de météorologie*, Chez l'auteur, Paris, 1851, and many other editions.

4 Emil Jochmann, *Grundriss der Experimentalphysik*, Winckelmann & Söhne, Berlin, 1874 and other editions. Professor Andreas Spatharis based his book *Στοιχεία Πειραματικής Φυσικής* [Elements of experimental physics], εκδόσεις Ανέστη Κωνσταντινίδου, Athens, 1886, on Jochmann's book, according to the editor's note in the first and only edition of the book.

sometimes the teachers of physics modified them to make them simpler or more spectacular.

Today, three high schools for the Christian orthodox citizens of Turkey continue to operate in Istanbul: Phanar Greek Orthodox College (known in Greek as Great School of the Nation), Zappeion and Zografeion.[5] They all hold valuable and rich collections of scientific instruments.[6] Another collection is held in the Theological School of Halki, which has been closed for political reasons since 1971.

According to the Principals of the three schools, no documents relating to the acquisition of the scientific instruments, nor any descriptions of them, have been found in the school archives. The archive of the Theological School of Halki is currently not accessible to researchers. Access to such archives might offer definitive answers to questions about how, when and where the instruments were purchased, how they were used in teaching practice, and how they were maintained and repaired. Fortunately, some of the gaps can be covered by the fairly regular annual operating reports of some of the schools, and from other sources.

This chapter presents a brief history of these institutions and their scientific instrument collections, along with some information about the evolution in the teaching of natural science in them.

2 A Brief History of the Schools

The oldest of the four educational institutions is the Phanar Greek Orthodox College, which was founded in 1454, just one year after the Ottomans captured Constantinople. Its founder was Patriarch Gennadios Scholarios. The school has continued in operation up to the present day, apart from a short interruption in 1821 caused by the Greek revolution against the Ottoman Empire. Until the early nineteenth century the overwhelming majority of the graduates of the school were destined to become clergy. This is reflected in the curriculum of the time, which was oriented towards classical studies, with sporadic physics teaching.[7] There is, however, a reference to the existence of some scientific

5 Phanar Greek Orthodox College and Zografeion were initially schools for boys. Zappeion was a school for girls. They all became mixed schools in the 1990s.

6 There used to be other public girls high schools and private boys high schools, but not much is known today about the possible existence of collections of scientific instruments.

7 Κωνσταντίνος Κούγκας [Konstantinos Kougas], "Οι φυσικές επιστήμες στο πρόγραμμα διδασκαλίας της Μεγάλης του Γένους Σχολής [The natural sciences in the curriculum at the Great School of the Nation]", in *Πρακτικά Επιστημονικής Ημερίδας "Η Πατριαρχική Μεγάλη του Γένους Σχολή, Ιστορία και Προσφορά"* [Proceedings of the scientific conference on "The

instruments at the beginning of the nineteenth century, without specifying their type and number. This collection had almost completely disappeared by 1860.[8]

The orientation of the school changed gradually after 1850 and subjects such as physics, cosmography[9] and chemistry acquired a permanent place in the curriculum. The teaching of physics and chemistry were introduced in the school year 1851–1852.[10] This does not mean that classical studies lost their primacy, but that natural sciences were considered to be necessary weapons in the students' armoury. Through a series of donations (in money or in instruments), in less than two decades (1865–1880) the school acquired a significant collection of instruments and devices, numbering more than 250 items. As a result there was an increase in the time devoted to the teaching of physics and chemistry. It is not a coincidence that a group of students under the supervision of Professor Andreas Spatharis took first prize in the Zappios Science Contest held in Athens in 1882.[11]

The contribution of Andreas Spatharis (1837–1901) in the progress of the teaching of science in the College is remarkable. He taught experimental physics, chemistry, cosmography and mathematics for almost forty years. Spatharis initially studied architecture at the Academy of Architecture in Berlin and then attended general courses at the University of Berlin. He was a charming speaker with a wide education, an excellent teacher and an expert in experimental physics and chemistry. He was first appointed to the Commercial School of Halki in 1859 and, in addition to the Phanar Greek Orthodox College, he taught at Zappeion School for girls, at the Greek Lyceum of Chatzihristou and at the Pallas School for girls.

Indicative of Spatharis's commitment to science and teaching was his participation for many years in the public lectures of the Greek Philological

Patriarchal Great Nation School. History and Offering"], Εταιρεία Μελέτης της καθ'ημάς Ανατολής (ΕΤΜΕΛΑΝ), Athens, 2004, pp. 167–183.

8 Ευστάθιος Κλεόβουλος [Efstathios Kleovoulos], *Λογοδοσία περί του σχολικού έτους 1865–1866 της εν Κωνσταντινουπόλει Μεγάλης του Γένους Σχολής* [Report for the school year 1865–1866 about the Great School of the Nation in Constantinople], Ανατολικός Αστέρας, Constantinople (Istanbul), 1866, p. 30.

9 The Greek schools followed French practice and used the word 'cosmography'. In English in this context the usual word would have been astronomy, which does not have exactly the same meaning.

10 Τάσος Γριτσόπουλος [Tasos Gritsopoulos], *Πατριαρχική Μεγάλη του Γένους Σχολή* [Patriarchal Great School of the Nation], vol. 2, re-issued by the Association of Greek Orthodox College Alumni, Athens, 2004, p. 176.

11 "Δελτίον Μεγαλοσχολιτών", Σύνδεσμος των εν Αθήναις Μεγαλοσχολιτών, Αθήνα [Bulletin of the Association of Alumni of the Great School of the Nation] 2 (Dec. 1949–Jan. 1950), p. 67.

Society of Constantinople. Among his speeches were those on physics (electricity, the barometer, acoustics), which were accompanied by demonstration experiments using the scientific instruments in the Society's collection. A report of a public lecture he gave in 1865, at the age of 28, with the title "On the Air" notes that Spatharis was able to present his subject clearly through an excellent selection of experiments, and that he received the admiration of the audience. It is not surprising that Spatharis was responsible for the physics cabinet of the Society for five years (1875–1880). It is clear that he was a professor with great experience and special interest in demonstration experiments and scientific instruments.

One incident described by one of Spatharis's students, S. Hudadoroglou, is illuminating:

> When we climbed the tower of the school for a chemistry or physics lesson there was much noise and disorder. In a lesson on fluoride, Spatharis, demonstrated how the fluoride attacked glass where it was not protected by a layer of wax by etching the words 'Do not make a noise while coming into the laboratory' into the glass. This was enough to make us to behave in the right way.[12]

Spatharis's immediate and intelligent treatment of the misbehaviour of his students reveals not only an experienced and tolerant teacher, but also the confidence with which he handled demonstration experiments in the laboratory. Another of Spatharis's students, Kleonimos Stylianidis, writes about his teacher: "It was very impressive that his teaching and his physics and chemistry experiments were always conscientiously prepared".[13] Spatharis once remarked to his son: "your father has taught the same courses for many years, but never comes to a lesson without having prepared carefully for the teaching and for the experiments".[14] Finally, as the Head of the school, Grigorios Palamas, mentions: "[Spatharis] during this year spent many hours and days on the preparation of experimental physics and chemistry courses and on the cleaning and arranging of each of the school's instruments".[15]

12 *Ibid.*, p. 48.

13 *Ibid.*, 3 (Feb. 1952), p. 74.

14 *Ibid.*

15 Γρηγόριος Παλαμάς [Grigorios Palamas], *Έκθεσις της καταστάσεως της Μεγάλης του Γένους Σχολής κατά το σχολικόν έτος 1878–1879* [Report of the condition of the Patriarchal Great School of the Nation during the school year 1878–1879], Ανατολικός Αστέρας, Constantinople (Istanbul), 1879, pp. 67–68.

During the last third of the nineteenth century the great growth of the Greek community in Istanbul was accompanied by an increased need for education in economic and scientific matters. Greek merchants, doctors, pharmacists and lawyers wanted a more Western-European oriented education for their children. Despite the efforts of Spatharis, the College could not meet these needs as it continued to provide a programme focused on classical education. In addition, the commercial activity of the community took place mainly in the district of Pera, but the grandiose building of the College was far away in the district of Fanari or Phanar (modern Fener), near the Patriarchate.

The gap was covered by the founding of Zografeion high school in 1883 in Pera. The School of the Virgin Mary had been in operation in that area since 1833, initially as an elementary school, but later a high school was added.[16] In 1893 the school acquired a spacious four-storey building with both physics and chemistry laboratories (*Fig. 10.1*) and was renamed Zografeion Gymnasium after Christakis Zografos (1820–1898), who had made a generous donation. The building was designed by Pericles Fotiadis (1859–1959). The architect was chosen by competition.

The curriculum of Zografeion high school is clearly focused on the teaching of natural sciences as proven by the time devoted to the teaching of science. According to the 1903 regulations experimental physics was to be taught in the last two grades of the elementary school and the last three grades of the high school. Similarly, chemistry was to be taught in the last grade of the elementary school and in the last three grades of the high school. Each subject was taught for two or three hours a week.

In 1875, almost twenty years before the founding of Zografeion, a famous girls' school, Zappeion, had been founded in Pera. It is remarkable that the Greek community in Istanbul pioneered the education of girls. In 1874 the Association for Women's Education was founded with the aim of establishing a girls' school and a school for training teachers. The search for the necessary donors did not initially produce the required results, as offers from Georgios Zarifis and Christakis Zografos were not enough to cover the required amount. Finally, very generous sponsorship from a rich entrepreneur from Epirus, Constantine Zappas, made it possible to establish the school, which was named Zappeion in honour of the benefactor.

During the first years of its operation (1875–1885) the girls' school was housed in a rented building. The Greek Kingdom recognized the high school

16 Πολύβιος Στράντζαλης [Polyvios Strantzalis], *Η Σχολή της Παναγίας (1833) και το Ζωγράφειο Γυμνάσιο (1893)* [The Virgin Mary School (1833) and Zografeion High School (1893)], Private publication, Athens, 2003, p. 92.

FIGURE 10.1 The physics laboratory of Zografeion School in about 1910 (from *Lycée Zographeion*, Pera, Constantinople [Istanbul], 1907–1908). There was a separate chemistry laboratory.

as being equal to the famous Arsakeion Girls' School in Athens in 1879.[17] This move had a double benefit as it strengthened the prestige of Zappeion School and at the same time gave its graduates the right to work as teachers in elementary schools in Greece. This led to the establishment of a facility for training teachers for the graduates who chose to become teachers. The training course took two years and created a need for a more spacious building.

Between 1880 and 1885, an impressive new building designed by Ioannis Ioannidis was erected on a plot of land next to the church of Agia Triada (the Holy Trinity) in Istanbul. Zappas bore the cost again. The inauguration took place in 1885. The audience was stunned into silence at the sight of the

17 *Κανονισμός του Ζαππείου Ανωτάτου Ελληνικού Εκπαιδευτηρίου του θηλέος φύλλου εν Κωνσταντινούπολει* [Regulations of Zappeion National School for Girls in Constantinople], τύποις Σ. Ι. Βουτυρά, Constantinople (Istanbul), 1879, pp. δ'-ε'.

magnificent building. The four floors, with a total height of 32 metres, the halls, workshops and equipment were etched into the memories of visitors.[18]

The provision for science teaching was described as follows:

> On either side of the ceremonial hall there are two halls containing the instrument cabinet, with all the necessary equipment needed for the teaching of physics, chemistry, zoology, astronomy, mineralogy, and all other departments of practical human knowledge. These halls [...] constitute valuable museums for whoever looks into them and wants to learn even just by a simple conversation, as they give them useful skills and expertise.[19]

The provision in the school of so much space for science teaching and the necessary apparatus would have been impressive in any case, but especially so in a girls' school in the late nineteenth century in Istanbul. This was pioneering, at least within the Ottoman Empire in that era.

According to the school regulations of 1879,[20] physics was taught in the high school in the third and fourth grades and chemistry was taught in the fifth and last grade,[21] both for an hour per week. In 1906,[22] the numbering of elementary and high school classes was unified and physics was taught in the new eighth and ninth grades,[23] as was cosmography. Chemistry was no longer taught in the high school, but had been transferred to the first year of the teacher training school, as were some topics in physics (optics, meteorology and climatology).

The Theological School of Halki was founded by Patriarch Germanos IV in 1844. The increased need of the Patriarchate for clergy "with a good general and theological education" led to this decision. The school was installed in the previously ruined monastery of Agia Triada (the Holy Trinity), located at the top of the hill of Elpida (Hope), one of the two hills of the island, offering a splendid view and the necessary isolation for the monks, teachers and students of the Theological School (*Fig. 10.2*).

18 "The Opening of the Zappeion", *Neologos* [an Istanbul newspaper], 10 July 1885.

19 *Ibid.*

20 *Κανονισμός του Ζαππείου* [Regulations of Zappeion], cit. (n. 17), pp. 52–54.

21 The elementary school had four grades and the high school five grades.

22 *Κανονισμός του Ζαππείου Εθνικού Παρθεναγωγείου εν Κωνσταντινούπολει,* [Regulations of Zappeion School for Girls in Constantinople], Τύποις Αδελφών Γεράρδων, Constantinople (Istanbul), 1906, p. 17.

23 The elementary and the high school had a total of nine grades, so the new regulation moved the teaching of physics one grade later.

FIGURE 10.2 The building of Theological School of Halki. Drawing by the architect who designed it, Perklis Fotiadis, around 1894 (P. Fotiadis personal archive, Centre for Asia Minor Studies, Athens).

The Faculty of Theology initially offered a three-year course, but it was soon extended by another year. It is noteworthy that the earliest regulations for the course included "elements of physics" in the second year of study. It seems, however, that this was not taught, at least during the early years of the school. The first clear reference to the teaching of physics is for the school year 1860–1861, but it is not certain that it continued in the following years.

There was no laboratory. According to the reports and correspondence from the Head of the school, Metropolitan Bishop of Stavropolis Konstantinos Tipaldos (1795–1867), the only scientific instrument that the school had until 1864 was an "accurate" barometer-thermometer. Nevertheless, in his report for the school year 1860–1861 he states that Professor Kyriazidis taught experimental physics, but without experiments because of the lack of instruments. It is not surprising that in the school regulations of 1867 experimental physics was no longer included in the lessons to be taught and was not included in the next regulations in 1874.

Experimental physics reappears in the regulations which took effect in the school year 1897–1898 and was taught in both the third and the fourth years.[24] It is unlikely that the teaching of physics began without the provision of a laboratory and apparatus. Therefore it is safe to assume that by 1897 at the latest, and probably sooner, the school had acquired suitable equipment. In the school year 1903–1904 new regulations included courses in cosmography and

24 Απόστολος Μέξης [Apostolos Mexis], *Η εν Χάλκη Ιερά Θεολογική Σχολή: Ιστορικά σημειώματα (1844–1935)* [The Holy Theological School on Halki: Historical notes (1844–1935)], Τύποις « Φαζιλέτ », Istanbul, 1935, p. 152.

chemistry,[25] but chemistry and cosmography are not mentioned anywhere in the school timetable. Physics, on the other hand, seems to have been taught in both the third and the fourth years for two hours a week. It appears that the chemistry and the cosmography were taught within the time allocated for physics. It is clear that, at least in science lessons, there was at times a degree of flexibility in the application of the regulations. Physics was taught as early as 1872 by Georgios Lianopoulos, and this was probably made possible by the acquisition, by purchase or donation, of appropriate scientific instruments. Unfortunately, lack of access to the school's archives prevents further investigation.

The years 1910–1919 were particularly difficult, with the buildings requisitioned for troops or refugees and the temporary transfer of the school to the Monastery of Saint George of Kremnos (the Cliffs) in Halki. It is easy to imagine what damage and loss this probably inflicted on the collection of scientific instruments.

In the period 1919–1923, the school became a Theological Academy, and as a result physics and chemistry were removed from the curriculum.[26] In the 1923–1924 school year a significant decrease in the number of students led to the return to the 1903 regulations. From 1925 the teaching of physics and chemistry, as well as of mathematics, was undertaken by Alexandros Iatropoulos (1892–1970), a chemical engineer. The following year Stylianos Stassinopoulos (1865–?), a mechanical engineer, took over the teaching of physics and cosmography.[27]

In 1953 the school's latest regulations were published, according to which students could attend either a three-year high school course or a four-year course in the theological department. In the curriculum of the high school both physics (two hours per week in each class) and chemistry (two hours per week in the first grade and one hour in the next two classes) are listed.[28]

In the decade 1950–1960 a series of projects was completed aimed at improving the functioning of the school. Among other things there was a rearrangement of the physics and chemistry laboratories (*Fig. 10.3*) and an "enrichment"

---

25 *Κανονισμός της εν Χάλκη Ιεράς Πατριαρχικής Θεολογικής Σχολής της του Χριστού Μεγάλης Εκκλησίας* [Regulations of the Holy Patriarchal Theological School of the Great Church of Christ at Halki], Πατριαρχικό Τυπογραφείο, Constantinople (Istanbul), 1903, p. 25.

26 Μέξης [Mexis], op. cit. (n. 24), pp. 177–178.

27 *Ibid.*, p. 197.

28 *Κανονισμός Ιεράς Θεολογικής Σχολής Χάλκης* [Regulations of the Holy Theological School of Halki], Πατριαρχικό Τυπογραφείο, Istanbul, 1953, p. 30. This book has also been published in Turkish.

FIGURE 10.3 The chemistry laboratory of the Theological School of Halki in 2016 (Photo: Panagiotis Lazos)

(*εμπλουτισμός*) of the collection of instruments.[29] The school had, after decades, acquired new equipment for teaching physics. The school was closed in 1971 after the Turkish government passed a law banning the operation of private higher education.

3 The Instruments and the Makers

The schools obtained most of their instruments during the period 1860–1910, which was followed by a long period of inactivity. There was almost no purchase of instruments between 1910 and 1950. These variations are just a reflection of the general social and economic situation of the Greek community of Istanbul. The instruments obtained up to 1910 are described below.

3.1 *Phanar Greek Orthodox College*

The richest of all the collections is the one in the Phanar Greek Orthodox College. This was not the case until 1865 but after that the school obtained

29 Βασίλειος Σταυρίδης [Vasileios Stavridis], *Η Ιερά Θεολογική Σχολή της Χάλκης. Τόμος 2ος 1923-σήμερον* [The Holy Theological School of Halki. 2nd tome: 1923 to the present], Τσιβεριώτης Ο.Ε., Athens, 1968, p. 14.

many instruments in five phases. As a result, the collection contained more than 350 instruments by 1910.

The first twenty-two instruments were donated to the College in 1865 by Chatzi-Stefanis Gaitanakis (1821–1880), a Greek architect.[30] They were used initially in the school he had founded in 1857 in his home-city, Madytos. It is not clear why Gaitanakis decided to donate the instruments. Perhaps, the reason was that by doing this he was permitted to send a student from Madytos to the College for free every year. The instruments are not signed but they were listed both in Greek and in French in the annual report of the Principal of the College, so it is most probable they were made in France. Most of them were for static electricity and topography. The most impressive instruments were a planetarium, two parabolic reflectors for experiments on radiant heat and a battery of six Leyden jars (*Fig. 10.4*).

FIGURE 10.4 A battery of six Leyden jars in a wooden case, in the collection of the Phanar Greek Orthodox College (Photo: Panagiotis Lazos, 2016)

30 Κλεόβουλος [Kleovoulos], op. cit. (n. 8), p. 30.

The next year the College obtained almost ninety new instruments,[31] including a double-barrelled air pump, a Hero's fountain, a fountain in vacuum, a centrifugal machine, Magdeburg hemispheres, a Morse telegraph and an interesting device for simulating the propagation of longitudinal waves. There is no information about their makers and the instruments that have survived are not signed. The money to pay for the instruments was collected through various donations. With this purchase the College had a rich collection for teaching the mechanics of solids, gases and liquids, and optics, acoustics and electricity.

In 1878 the College purchased a group of about sixty second-hand instruments.[32] Professor Spatharis was informed that the private *Ελληνικό Εκπαιδευτήριο* (Greek School) was going to close and he suggested to the principal that the College should purchase the instruments because they would be obtained at a bargain price. A Carré electrical machine made by Adolphe Gaiffe, a Gay-Lussac barometer, a gasometer for measuring the quantity of a gas, a sonometer and a cathetometer were the most significant instruments. Nevertheless, the collection was endangered because of the poor condition of the old building of the school and the lack of space in the cabinet.[33]

The biggest enlargement of the collection occurred in 1880,[34] when more than 140 instruments of high quality were acquired in order to enrich the physics cabinet of the new College building (*Fig. 10.5*). The significant amount of money needed was collected from donors. The instruments were made by various French makers and were bought through André Mandeleney, a French pharmacist in Istanbul. They were chosen by Professor Spatharis from a Hachette catalogue given to him by Mandeleney. They included a beautiful Melloni's apparatus and an elaborate Atwood machine with seconds pendulum both made by Breton Frères. Many instruments, mainly but not only for acoustics, appear to have been made by Rudolph Koenig, but they are signed L.B.F.G., which is a puzzling set of initials. Other instruments worth mentioning include a Helmholtz double siren, a set of ten Helmholtz resonators, a

31 Ε. Κλεόβουλος [E. Kleovoulos], *Λογοδοσία περί του σχολικού έτους 1866–1867 της εν Κωνσταντινουπόλει Μεγάλης του Γένους Σχολής* [Report for the school year 1866–1867 about the Great School of the Nation in Constantinople], Ανατολικός Αστέρας, Constantinople (Istanbul), 1867, pp. 40–42.

32 Παλαμάς [Palamas], op. cit. (n. 15), pp. 145–150.

33 *Ibid.*, p. 114.

34 Γ. Παλαμάς [G. Palamas], *Έκθεσις της πνευματικής και υλικής καταστάσεως της Πατριαρχικής Μεγάλης του Γένους Σχολής κατά το σχολικόν έτος 1880–1881* [Report of the spiritual and physical condition of the Patriarchal Great School of the Nation during the school year 1880–1881], Πατριαρχικό τυπογραφείο, Constantinople (Istanbul), 1881, pp. 168–174.

FIGURE 10.5 The building of Phanar Greek Orthodox College erected in 1882, also known as the "Red Castle" (Photo: Panagiotis Lazos, 2016). On the top of the tower there is a small refracting telescope.

Coulomb torsion balance, a series of organ pipes, a spectrometer, a Papin digester, a Haldat apparatus for teaching hydrostatics, and an Archimedes screw. After 1880 the College's instrument collection compared well with collections in most high schools in Europe.

Finally, in the period 1900–1910 an unknown number of instruments made mainly by the German maker Max Kohl was purchased. A few surviving instruments made by the Austrian maker W.J. Rohrbeck's Nachfolger were probably bought during this period. An electrical machine on a sturdy metal base, the most bulky instrument in the collection, was also purchased (*Fig. 10.6*). It was made by A. Gaiffe and has four discs. Gaiffe exhibited a very similar machine at the *Exposition Universelle* in Paris in 1900,[35] but it was rotated by an electric motor whereas the machine at the school is rotated by a hand crank.

3.2 *Zografeion*

In the collection of Zografeion School there are about 300 instruments including an extensive series of barometers and another of cathode ray tubes.

35 Édouard Hospitalier, Jules Armand Montpellier, *L'électricité à l'exposition de 1900*, Ch. Dunod, Paris, 1902, pp. 3–7.

FIGURE 10.6 The Bonetti electrical machine of the Phanar Greek Orthodox College, made by A. Gaiffe (Photo: Panagiotis Lazos, 2016)

Because the school was founded later than the other Greek schools there are only a few French instruments, all made by Eugène Adrien Ducretet of Paris. The most significant is an impressive tellurium, called a "*cosmographe de Girod*" in the 1905 trade catalogue of the maker (*Fig. 10.7*).[36] Most of the instruments were made by the German firm Kohler & Volckmar (KV), the Austrian firms Noir & Forster and W.J. Rohrbeck's Nachfolger, both of Vienna, the Belgian firm Robert Drosten of Brussels, and the English firm Griffin of London. Zografeion used the widest variety of makers.

3.3 *Zappeion*

The collection of Zappeion school includes more than 150 instruments. The older ones are part of the original equipment of the school and were acquired in about 1890, at the same time as the building was erected. Like other schools

36 Eugène Ducretet, *Catalogue raisonné des instruments de précision de E. Ducretet*, Ducretet, Paris, 1905, p. 307.

FIGURE 10.7 The "Cosmographe de Girod" or tellurion of Zografeion School, made by E. Ducretet (Photo: Panagiotis Lazos, 2016)

at the time, Zappeion purchased French-made instruments. It is notable that at least seventeen instruments were purchased second-hand. They have labels on them reading "Ch. Duterme, Instruments de Physique, d'occasion, 11 Rue Blainville, Paris." Duterme was a retailer of scientific instruments, both new and used, in the last third of the nineteenth century. Some of them are signed by their maker, such as A. Gaiffe on a Ruhmkorff coil and Dumotiez on a broken differential thermometer. Dumotiez Frères was not an active business after about 1820, so it would be very interesting to know the route by which this instrument reached Zappeion almost seventy years later.

The instruments in the collection are, in general, in very good condition. The most interesting ones are a Ramsden machine, a Carré machine, an armillary sphere, two telluriums, a phonograph, a double-barrelled air pump signed Loiseau, Paris, a sonometer, a small refracting telescope and a Hero's fountain.

Indicative of the high quality of teaching methods is the existence of almost 250 slides about science, which were projected with a magic lantern, made by Elie Mazo of Paris.

3.4 *Theological School of Halki*

The collection of instruments of the Theological School of Halki is kept in the so-called chemistry laboratory of the building. It is a room with four tiers of benches, a table for demonstration experiments, a fume cupboard and instrument shelves. About forty-five physics instruments and several pieces of chemical apparatus are preserved, and there are indications from the numbering on the labels on some instruments that the collection once contained at least one hundred items.

The earliest instruments date back to the last quarter of the nineteenth century. Some of them are signed by the well-known French manufacturer Antoine Molteni of Paris. The rest are unsigned. Among them are a solar microscope, a hydrostatic balance, an electric egg, a Volta's electric lighter,[37] a Babo device,[38] two vacuum fountains, two Leyden jars and a beautiful electric whirl with seven luminous tubes (*Fig. 10.8*).

The presence of static electricity devices such as an electric egg and Leyden jars shows there used to be an electrical machine. Similarly, the existence of the vacuum fountains shows that there used to be an air pump. No such instruments have been found. This is not surprising if we consider the difficulties the Theological School has faced, which leads us to believe that only a small part of the nineteenth-century collection has survived.

4 The Collections after 1945

All the collections include instruments made by the American company CENCO. These were purchased between 1948 and 1960. Trade catalogues of the company have been found in Zografeion and Zappeion schools. One of them bears the stamp of Mehmet Vasfi (1900–1975), the representative of CENCO

37 Paolo Brenni, "Volta's electric lighter and its improvements: The birth, life and death of a peculiar scientific apparatus which became the first electric household appliance", in Marco Beretta, Paolo Galluzzi, Carlo Triarico (eds.), *Musa Musaei: Studies on Scientific Instruments and Collections in Honour of Mara Miniati*, Olschki, Florence, 2003, pp. 371–394.

38 William Jensen, "Of Beehives and Babo Generators: The Adventures of a Museum Curator", *Bulletin for the History of Chemistry* 8 (1990), pp. 34–35.

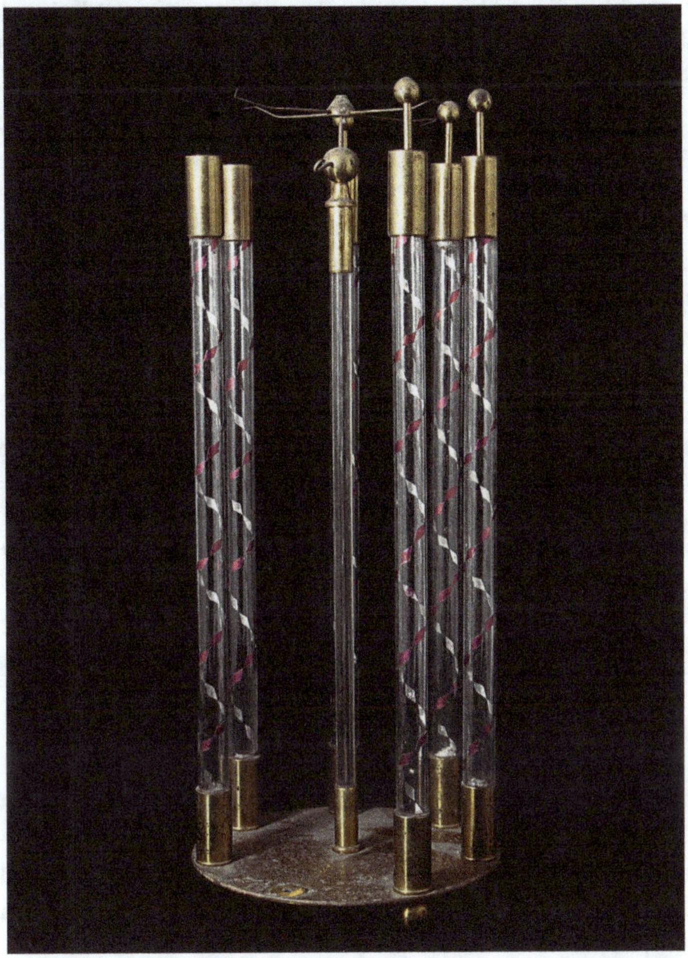

FIGURE 10.8 The electric whirl with seven luminous tubes of the Theological School of Halki (Photo: Panagiotis Lazos, 2016)

in Turkey.[39] It is not clear why Greek schools purchased instruments only made by CENCO after World War II. In the case of the Phanar Greek Orthodox College the scientific instruments were donated by a Greek ship-owner living in the USA and perhaps this played a role for the other schools as well.

Between 1948 and 1950 the Phanar Greek Orthodox College purchased about thirty instruments plus a periodic table showing the chemical elements

39 We thank Tuğba Canturk, MA student in History of Science in Istanbul University, for the information about Mehmet Vasfi.

known in 1947. Zografeion School has almost twenty instruments, including a thermoelectric magnet, a Zeleny electroscope, a D'Arsonval galvanometer and an electrically-maintained tuning fork. In Zappeion School there are ten instruments, among them a Kundt's tube for acoustic experiments, a Hartl disc for optical experiments and an oscilloscope. There is also a group of ten instruments in the collection of the Theological School of Halki acquired in the 1950s and 1960s. Among them is a high-frequency Tesla apparatus, a thermoelectric magnet and an electromagnetic device of Elihu Thomson. There is also a spectroscope made by B & L (Bausch & Lomb) but sold through the CENCO network.

Finally, there are some instruments in all the schools dating from the 1960s and 1970s made by the Instruments Factory of the Greek Ministry of Education. Those instruments were given to the schools by the Greek State.

An interesting fact about all the collections is that there are almost no instruments made in Istanbul. The one and only exception is a thermometer in Zografeion made by the maker and optician J. Verdoux. That suggests that the quality and quantity of production of scientific instruments in Istanbul was poor.

5 Conclusion

Our research findings show that the Greek schools of Istanbul made big steps towards the teaching of physics and chemistry by experimental methods during the period under examination. A large amount of money was invested in purchasing rich collections of scientific instruments made by well-known makers. The laboratories were spacious and used only for science lessons. Experimental physics was taught even in the Theological School of Halki and in Zappeion, a school for girls. Both were forward-looking decisions for the era and the place.

The process may be regarded as one aspect of the urbanization and expansion of the Greek community in Istanbul, but this would be the subject for further research.

CHAPTER 11

From the Ottoman Empire to Canada

George Petrovic's Metrological Instruments in the Canada Science and Technology Museum

Hasan Umut and David Pantalony

1 Introduction

In 1980 the National Museum of Science and Technology in Ottawa Canada (now Ingenium: Canada's Museums of Science and Innovation) acquired a collection of mathematical and metrological instruments that stood out from the traditional holdings of the museum. A man named George Petrovic (in his native Serbian, Djordje Petrović, 1927–2007), who emigrated from Yugoslavia to Montreal in the mid-1970s when he was in his late forties, sold the collection to the museum (*Fig. 11.1*). At first glance Petrovic's immigration story seems unexceptional in the Canadian context, given the fact that every year Canada welcomes thousands of people from all around the world. Petrovic's immigration, however, meant more than just setting up a new life in a new cultural and social environment. It was also a migration of the instruments which he had collected over many years from one of the most multicultural regions of the Old World, namely from the Eastern Mediterranean region, and which had mostly been produced or used in the Ottoman territories, to one of the oldest and most multicultural cities of the New World, that is Montreal. In other words, Petrovic's collection marks an intriguing journey of instruments between East and West.

In this chapter we look at the history of this collection as a whole, its formation and changing context, and the background and motivations of its collector. We shall also explore some examples from the collection and their original places, dates and uses. Through this broader collection history—from the Eastern Mediterranean to Yugoslavia to Montreal—we gain a critical understanding of the changing uses of these instruments over time from technical to cultural objects, which has an impact on the kinds of stories we can tell about these objects.

 | DOI:10.1163/9789004412842_012

FIGURE 11.1 George Petrovic at the opening for his exhibition "A Vision of the Future", 1986, at the National Museum of Science and Technology in Ottawa
LIBRARY AND ARCHIVES, INGENIUM: CANADA'S MUSEUMS OF SCIENCE AND INNOVATION

2 The Petrovic Collection

The Petrovic collection connects various regions and periods of time under the theme of metrology, having 130 artefacts including rulers, compasses, levels, plumb bobs and other kinds of measuring instruments dating from the twelfth to the nineteenth centuries (*Tab. 11.1*), and covering a vast territory in which more than twenty nation-states are located today.[1] They mostly derive from building and other trades, representing fairly routine mathematical practice and culture within those communities. Some of the examples, however, represent more decorative, ceremonial instruments. Approximately eighty of

1 George Petrovic, "Collection of East Mediterranean Measuring Rods and Instruments from the 15th to 19th Centuries", Ottawa, Ingenium: Canada's Museums of Science and Innovation, George Petrovic Collection File, Petrovic Collection, 1980.0610-1980.0739.

TABLE 11.1 Number of instruments in the Petrovic collection by date of production

Date	Number of instruments
12th Century	2
15th Century	5
16th Century	10
17th Century	17
18th Century	39
19th Century	57
Total	**130**

the instruments were collected by Petrovic when he was in Yugoslavia from the regions that were once ruled by the Ottomans. Most of them were made during the Ottoman period, except two plumb bobs from the twelfth century, of which one is attributed to Gamzigrad and the other came from Split, the former in Serbia and the latter in Croatia. Petrovic referred to the broader collection as 'East Mediterranean'. Having said this, in order to express the historical continuity in the use (and survival) of 'non-Ottoman' instruments during the Ottoman period from the fourteenth century onwards, this chapter also assumes them under the 'Ottoman collection'. The rest of the instruments, mostly dating from the nineteenth century, originated from countries across the world, indicating that Petrovic continued collecting instruments after he emigrated to Canada.

One of the more striking aspects of the collection is that Petrovic attributed the instruments to a wide range of regions and periods, which appears to have been based on his determination of where the objects were found and used.[2] There are instruments from the former Yugoslavia (the largest number is twenty from Serbia), Greece, Egypt, Israel, Lebanon, Cyprus, Bulgaria, Libya, Turkey, and Tunisia. Petrovic also collected over thirty instruments from North America following his immigration. As we will see below, some attributions related to instruments from the Eastern Mediterranean need further verification: Petrovic seemed to have based them on his own research and collecting, and attributions from his network of collectors.[3] Similar to the attributions of

2 *Ibid.*

3 These attributions are the subject of further research that needs to be done on the nature of Petrovic's studies in Belgrade. We are presently working on a full translation of his PhD

place, further research into Petrovic's studies in Belgrade is needed to verify some of his claims about dates. Of the instruments in the 'Ottoman Collection', there are two from the twelfth century, five from the fifteenth century, ten from the sixteenth century, seventeen from the seventeenth century, twenty-seven from the eighteenth century, and twenty from the nineteenth century. Below, we shall highlight some physical tests that have both confirmed and challenged some of these dates. While the questions about dates and places continue to pose historical problems about the earliest use of these instruments, at the same time these questions point to historical opportunities related to the complex context of the collector and his collecting activities and research, and the significance of the collection and its recent provenance as a window into post-Ottoman history and instrument migrations.

At the time of purchase in 1980 there was an internal debate about the authenticity, originality and importance of the collection. Archival documents in the Canada Science and Technology Museum show that, before purchasing the collection, curators corresponded with some of those who were familiar with the Petrovic collection. For instance, a letter from Peter Swann, the President of the Association of Cultural Executives in Montreal, describes the instruments he had once seen as "being of good quality" and that "obviously a great deal of time and study had been expended on them by the collector". Swann mentioned that Petrovic was asking $250,000 for the collection, which clearly indicated that Petrovic was exceedingly confident about its uniqueness. However, Swann stated that the real value of the collection should be $25,000.[4] Another acknowledgement came from Silvio A. Bedini, the well-known historian of technology and scientific instruments from the National Museum of American History, Smithsonian Institution.[5] In his letter Bedini mentioned his familiarity with the Petrovic collection, stressing that "the acquisition of the collection for a museum of science and technology would be most desirable".[6] In his catalogue, Petrovic pointed out that his collection had been exhibited four times,

---

dissertation, as well as tracking down his papers in Belgrade.

4 Peter C. Swann, "Letter to Dr David Baird of the National Museum of Science and Technology", 19 December 1980, Ottawa, Ingenium: Canada's Museums of Science and Innovation, George Petrovic Collection File, Petrovic Collection, 1980.0610-1980.0739.

5 For a general account of Bedini's importance in the fields of history of technology and museology, see Robert C. Post, "Silvio A. Bedini: 17 January 1917–14 November 2007", *Proceedings of the American Philosophical Society* 156 (2012), pp. 425–433.

6 Silvio A. Bedini, "Letter to Dr David M. Baird of the National Museum of Science and Technology", 6 March 1981, Ingenium: Canada's Museums of Science and Innovation, George Petrovic Collection File, Petrovic Collection, 1980.0610-1980.0739.

one of which was at the First International Congress for Historical Metrology held in Zagreb in 1975.[7] Even with these supporting materials and testimonials, questions remained about the provenance and value of the collection.[8]

In 1980 the National Museum of Science and Technology in Ottawa did purchase the collection for $35,000. Even though the authenticity and value of the collection is still an open research question, there is much to learn from this unusual collection of instruments, and the story of the collector George Petrovic, his scholarship and the transmission of instruments through several countries, regions, time periods and owners.

3 The Collector

George Petrovic's career as a collector is an integral part of the whole story of the formation and transmission of his collection across continents. He was born in Zaječar on 24 September 1927, a city which is in Serbia today. After receiving his undergraduate degree in the Faculty of Architecture at the University of Belgrade in 1952 he worked as a teaching assistant in the same faculty between 1952 and 1959, followed by his professorship in architecture between 1959 and 1975. He received his PhD in the same faculty in 1971. He moved to Montreal in the mid-1970s where he started working as a consultant for an architectural office.

Petrovic's dissertation submitted in Serbian is central to the history of the collection.[9] It reflects an interdisciplinary approach to architecture and its connections to metrology. By analyzing the proportions, dimensions and metrics of the wooden skeleton system used for traditional Serbian houses, Petrovic's dissertation deals with the applications of the *arshin*,[10] a traditional

7 Petrovic, "Collection of East Mediterranean …", cit. (n. 1).

8 Personal communication with former curator at Canada Science and Technology Museum, July 28, 2016.

9 Djordje Petrović [George Petrovic], *Aršin, Dimenzionalna, Proporcijska i Metrološka Analiza Aršinom Alemenata Konstruktivnog Sklopa Drvenog Skeletnog Sistema (Bondruka) u Folklornoj Arhitekturi Srbije XIX veka*, PhD dissertation, University of Belgrade, Faculty of Architecture, 1968. We are grateful to the University of Belgrade Library for kindly sharing with us a digital copy of this dissertation.

10 For the evolution of *arşın* (*arshin*), especially in the Ottoman context, and its technical explanation see Halil İnalcık, "Introduction to Ottoman Metrology", *Turcica* 15 (1983), pp. 311–348; Alpay Özdural, "Sinan's Arşın: A Survey of Ottoman Architectural Metrology", *Muqarnas* 15 (1998), pp. 101–115; Neslihan Sönmez, "Osmanlı Mimarlığında Kullanılan Uzunluk Ölçü ve Birimleri", in Güler Eren (ed.), *Osmanlı: Kültür ve Sanat*, Yeni Türkiye

measurement unit, as well as the name of some types of rules that were once used extensively in the Ottoman territories including the Serbian region. In the dissertation he also describes the ways in which measuring instruments calibrated in this unit were used to build a Serbian house in the nineteenth century. In other words, his interest seems to have been focused on folk architecture and theories of proportion. The dissertation also stresses the anthropometric character of these units. Hence his dissertation can be characterized as a landscape of intersecting architectural theory, scientific practice, material culture and relevance to a particular social and historical context.

Most importantly, Petrovic used a collection of historic metrological instruments to study these connections. For his dissertation research he examined a number of rules some of which are now in the museum's collection, using them as one of his main historical and architectural sources from this subject and period of research.[11] The nature and origins of the collection, therefore, is important. In his catalogue Petrovic stated "the authenticity of each piece of this collection from thirteen Eastern Mediterranean countries has been verified scientifically (the collection served as material for a doctoral thesis on the history of metrology)".[12] The collection appears to comprise a number of smaller collections. But there are still questions about how he obtained these instruments (and smaller collections) and his methods of attributing date and place, which seemed to derive from his subject expertise and a network of collectors with similar interests (one of whom is listed as the architect Le Corbusier).[13]

As a scholar and professional architect George Petrovic worked on and contributed to a wide range of topics, being involved in many architectural projects. He won several architectural competitions in the former Yugoslavia. In addition to historical metrology and its application to traditional Serbian architecture, his scholarship also covered such fields as modern architectural theories, their applications, and urban planning. Additionally, he published more than forty articles or books, some of which he entitled *National*

Yayınları, Ankara, 1999, pp. 125–133. Petrovic also gives a short summary of the *arshin* in the Serbian context; see, Djordje Petrović [George Petrovic], "Aršinski sistem mera u Srbiji", *Zbornik Odsjeka za povijesne znanosti Zavoda za povijesne i društvene znanosti Hrvatske akademije znanosti i umjetnosti* 7 (1974), pp. 273–275.

11 Comparison of the list of the instruments Petrovic examined for his PhD research with the instruments in Ingenium: Canada's Museums of Science and Innovation shows that many of them are now in the museum. See, Petrović, *Aršin* ..., cit. (n. 9), pp. 127–130; Petrovic, "Collection of East Mediterranean ...", cit. (n. 1).

12 *Ibid.*

13 Though this paper does not provide a detailed technical analysis of Petrovic's dissertation, we have plans to do further research and critical analysis after its translation.

Architecture of Serbia, *Compositions of Architectonic Forms*, and *Visual Research on the Environment*, all written in Serbian.[14]

In addition to his long-term research on the theory and practice of measuring instruments and their architectural and historical implications, Petrovic was also involved in the institutionalization of the history of metrology in the international arena. An official letter sent to Petrovic on 27 October 1975 by Dr. Zlatko Herkov, the Secretary General of the *Comité International pour la Métrologie Historique*, who would later become the honorary president of the committee,[15] confirmed Petrovic's membership of the committee as a result of his work in historical metrology.[16] Among the members of the committee at the time were Jean-Claude Hocquet, Frank Greenaway, Helmut Grotzsch, Otto Krätz and René Taton.[17] Petrovic presented papers at two meetings of the committee.[18] Further research will be needed to clarify the depth of Petrovic's involvement in this committee, and the impact this involvement had on his research and collecting.

Petrovic was also an artist, and his art reveals much about the context of his life and studies.[19] In particular, he expressed his interest in science and

14 English translations of the titles of the works in Serbian are given in George Petrovic, "Curriculum Vitae", Ottawa, Ingenium: Canada's Museums of Science and Innovation, George Petrovic Collection File, Petrovic Collection, 1980.0610-1980.0739. For some of his works published in Serbian, see Petrović, *Narodna Arhitektura: Doksati i Čardaci*, Građevinska knjiga, Belgrade, 1955; Petrović, *Teoretičari Proporcija*, Vuk Karadžić, Belgrade, 1967; Petrović, *Vizuelna Istraživanja Čovekove Sredine i Urbani Dizajn*, Univerzitet; Beogradski izdavačko-grafički zavod, Belgrade, 1972; Petrović, *Kompozicija arhitektonskih oblika*, Naučna knjiga, Belgrade, 1972.

15 For the importance and a short history of the committee, as well as Zlatko Herkov's role in it, see Jean Claude Hocquet, "Bilan et Développement de Vingt ans d'Activités du CIMH", *Cahiers de Métrologie* 11–12 (1993–1994), pp. 11–26.

16 Zlatko Herkov, "Letter to George Petrovic, 27 October 1975" (in Serbian), Ottawa, Ingenium: Canada's Museums of Science and Innovation, George Petrovic Collection File, Petrovic Collection, 1980.0610-1980.0739.

17 "Comité International des Sciences Historiques: Les Adresses des Membres Du Comité", *ibid.*

18 Petrović, "Aršinski Sistem Mera u Srbiji (Das Aršin-Massystem in Serbien)", in Zlatko Herkov, Miroslav Kurelac (eds.), *Travaux de la 2e Conférence Internationale sur la Métrologie Historique (Rijeka 1973)*, Institut d'histoire de l'Académie yougoslave des sciences et des arts, Zagreb, 1974, pp. 273–275; Petrović, "Dimenzionalne Veze Između Neimarskog Aršina i Starih Graditeljskih Instrumenata (Dimensional Connections between the Builder's Ell and the Ancient Building Instruments)," in Zlatko Herkov (ed.), *Radovi 1. Međunarodnog Kongresa za Povijesnu Metrologiju, Zagreb, 28–30 Listopada 1975 (Travaux du 1er congrès international de la métrologie historique, Zagreb, 28–30 Octobre 1975)*, Jugoslavenska Akademija Znanosti i Umjetnosti, Historijski Zavod, Zagreb, 1975, pp. 134–139.

19 In fact, when we contacted the Serbian Embassy in Ottawa about Petrovic they knew of him through his art.

FIGURE 11.2 Exhibition poster for "A Vision of the Future", 1986
LIBRARY AND ARCHIVES, INGENIUM: CANADA'S MUSEUMS OF SCIENCE AND INNOVATION

architecture through a series of futuristic and utopian paintings. Six years after the acquisition of his instruments, the National Museum of Science and Technology in Ottawa exhibited his work "A Vision of the Future" between 18 September and 15 November 1986 (*Fig. 11.2*). The exhibit booklet was effusive about his work and methods, stating:

> Petrovic's meticulously prepared paintings are testimony to the imagination and passion with which he pursues his métier as an artist. Through

> years of diligent practice and experimentation, he has evolved a rare original technique using cotton paper and China ink prepared to his own specifications.[20]

His art work deserves critical study on its own terms, but just as intriguing are the connections between the cosmopolitan ideals of his utopian paintings, the nationalistic undertones of his research into traditional Serbian architecture, and the context of his life and work in communist Yugoslavia in the 1960s and 1970s. This broader context of the history of his studies and collecting activities points to his collection's evolving function in a post-Ottoman context, and therefore deserves further study in its own right.

As pointed out above, Petrovic collected the majority of his instruments when he was affiliated to the University of Belgrade. He added to his collection a number of Canadian-oriented instruments from various cities around the country, including Montreal, Toronto, Winnipeg, Vancouver, and Halifax. This is important because it shows that Petrovic continued collecting metrological instruments even after his arrival in Canada, from the mid-1970s until 1980 when the museum purchased the collection. The artefacts collected in North America are largely products of the nineteenth century.

The Ottoman instruments, however, have the most intriguing histories. The remaining part of this chapter introduces some of the instruments in this collection which are thought to have been used or produced in the Ottoman territories. The aim is to provide a few examples that demonstrate use and manufacture in a wide range of historical periods and geographies, and also to highlight some significant challenges related to tracing the early history of these instruments, and how these problems raise instructive museological and historiographical issues for the collection as a whole, and issues in the history of science in general and Ottoman studies in particular.

4 The Instruments

The Ottoman instruments in the Petrovic collection include rules, plumb bobs, levels, compasses, and a scale. The rules are geographically, temporally, and practically complex, diverse, and hybrid. They are generally fairly common, but have clearly been designed for the different social, economic and cultural needs of the regions where they were produced and used (*Fig. 11.3*).

20 "A Vision of the Future: An exhibition of paintings by Georges Petrovic", National Gallery of Canada, Petrovic File, Record Number: 28936, p. 2.

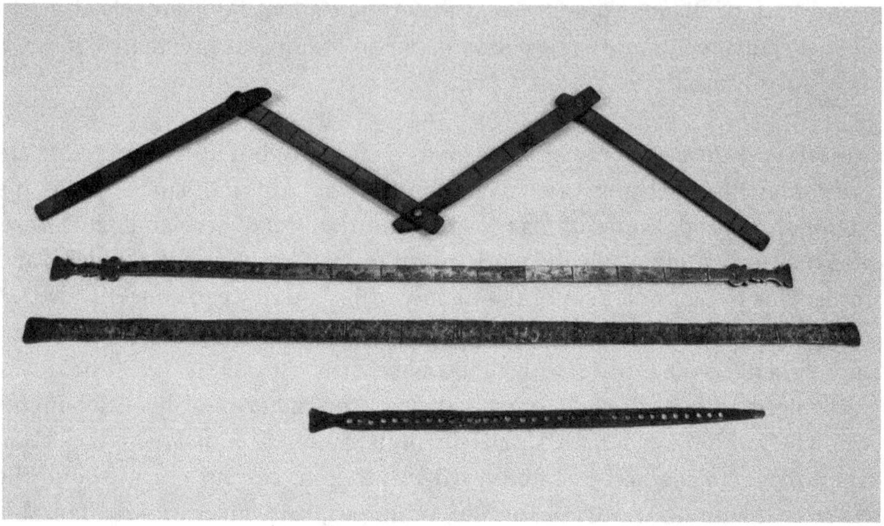

FIGURE 11.3 Four rules (top to bottom): eighteenth-century wooden builder's rule, attributed to Thessaloniki, Greece, art. no. 1980.0615 Ingenium; sixteenth-century mercantile rule attributed to Prizren, in Kosova (in former Yugoslavia), art. no. 1980.0641 Ingenium; seventeenth-century mercantile rule, attributed to Cairo, Egypt, art. no. 1980.0634 Ingenium; seventeenth-century weaver's rule attributed to Bicaj, Albania, art. no. 1980.0701 Ingenium
PHOTO: ANDRÉ LÉVESQUE, INGENIUM: CANADA'S MUSEUMS OF SCIENCE AND INNOVATION

The different types of rule can be classified according to their uses and lengths. These types were used not only as the names of the corresponding rules, but they were also considered as measurement units themselves. Those of the first type are called *endaze* and were basically used for measuring expensive cloths. The collection has eight rules of this type. Most of them are approximately 61 cm in length, 0.5 cm in height, and 1.3 cm in width. Most of them have 16 sub-divisions (*kirah*) indicated by alternating double and single lines. Most of them are double-sided. They are made of a ferrous alloy except for one which is made of wood.[21]

The second type of rule is the *çarşı arşını* (*arshin* of the market) used to measure ordinary or cheaper cloths. The collection includes seventeen *arshin* rules, some of which are folding ones. The fixed rules are approximately 68

21 Sandra Judge-Gervais, "Metrological Instruments of the Petrovic Weights and Measures Collection: Twelfth to Nineteenth Century", 1994, Ottawa, Ingenium: Canada's Museums of Science and Innovation, George Petrovic Collection File, Petrovic Collection, 1980.0610-1980.0739, p. 8.

cm in length, 0.5 cm in height and 1.5 cm in width, and the folding ones are approximately 66 cm in length, 0.5 cm in height and 1.4 cm in width. They are subdivided into 16 equal parts (*kirah*)[22] indicated by alternating double and single lines. Like the *endaze* rules, these are also double sided. They are all made of ferrous alloy.[23] *Endaze* rules and *çarşı arşını* are often confused. Their length varied depending on the period and region where they were used.[24]

The third type is an *arshin* for builders. It is known as *mimar arşını* (*arshin* of the architect) in Turkish or *zira-i mimari* (*zira* of the architects) in Arabic. Both terms were used in the Ottoman Empire. There are four instruments in this category. They measure approximately 76 cm in length, with 24 subdivisions.[25] Two of them are made of ferrous alloy, the other two are of wood.[26] It is interesting that although Petrovic's main research interest for his dissertation was on architectural rules, there are very few of this type of rule in the collection.

The fourth type of rule in the collection is a weaver's rule. There are fourteen in this category, all made of ferrous metal. A weaver's rule has two adjustable parts which are kept together by a hook and a ring. The hook in one part of the rule can be placed into holes in the other part. One end of each part is serrated, with varying numbers of teeth. Some are missing one of the two parts.[27]

The fifth type of rule is what Petrovic calls 'yards'.[28] There are two examples in this category.[29] To the best of our knowledge, the yard was not used as a measurement unit in the pre-modern Ottoman period.[30] Given the fact that these rules, whose places of origin are given as Alexandria, Egypt, and Haifa, Israel, have been attributed to the eighteenth and nineteenth centuries respectively, with the latter including both Arabic and Latin numerals, the yard seems

22 Both the *endaze* and the '*arshin* of the market' are divided into 16 *kirah*. The *kirah* of the *endaze* measures roughly 4 cm while the *kirah* of the '*arshin* of the market' measures 4.25 cm. See Feza Günergun, "Introduction of the Metric System to the Ottoman State", in Ekmeleddin İhsanoğlu (ed.), *Transfer of Modern Science and Technology to the Muslim World*, IRCICA, Istanbul, 1992, pp. 297–316: 310.

23 Judge-Gervais, op. cit. (n. 21), pp. 10 and 13.

24 For measurement units that were used in the Ottoman period, and their modern equivalents, see *ibid.*, pp. 310–311.

25 The subdivision of the '*arshin* of the architect' is named *parmak* in Turkish or *usbu'* in Arabic, and it is about 3.2 cm. See *ibid.*, p. 310.

26 Judge-Gervais, op. cit. (n. 21), p. 16.

27 *Ibid.*, p. 18.

28 A yard is an English unit of measurement still in customary use in number of countries. It is now defined as 0.9144 meter.

29 *Ibid.*, p. 16.

30 Günergun, "Introduction of the Metric System …", cit. (n. 22), pp. 309–311.

to have been used as a unit in the Ottoman lands from the eighteenth century onwards, as the encounter with European scholarship intensified. They measure approximately 91.0 cm with 36 subdivisions.

Finally, there is a single cubit rule 50.9 cm in length, 1.1 cm in width, and 0.3 cm in height. Having attributed this one to Dubrovnik in today's Croatia, Petrovic dated it to the fifteenth century.[31]

In the collection there is a rare example of a rule which shows both the transmission of a scientific object within the Ottoman Empire and the transmission of scientific knowledge between continents. It is an *arshin* of the market, 66.2 cm in length. Petrovic attributed it to Benghazi, Libya, and dated it to the nineteenth century. It is double sided and both sides have 16 subdivisions indicated by alternating single and double lines. On one side Arabic numerals are punched into the metal at 10 cm intervals.[32] On the other side European numerals are punched into the metal. At the left-hand end of the side marked in Arabic script, the following is written from right to left: *sene-yi hicriyye* (*hijri* year), *sene-yi kameriyye* (lunar year) 1299, and *sene-yi şemsiyye* (solar year) 1260. The words *sene-yi hicriyye* indicate that the dates that follow are counted from the year of the Prophet Muhammad's migration from his home city of Mecca (622 CE) in the lunar calendar. 1299 is the number of lunar years and 1260 is the number of solar years since that date. Both dates correspond to 1882 CE.[33]

On the side bearing the dates, 'Tophane' is written in Arabic script, which is the presumed place of production of the rule. Tophane was and is still a district in Istanbul where the imperial foundry *Tophane-i Amire* was located, so this piece was most-likely produced in Istanbul (*Fig. 11.4*).[34] If Petrovic's attribution is correct, how it came to be in Benghazi is unknown.

---

31 Though the extent to which cubit rules were used in the Ottoman territories need further research, there is evidence that rules with an average length of 54.1 cm were used in Ottoman Egypt. See, Özdural, op. cit. (n. 10), p. 104.

32 Judge-Gervais, op. cit. (n. 21), p. 12.

33 For an account of the history of calendars in Islamic societies, see F.C. de Blois *et al.*, "Ta'rīkh", in P. Bearman, Th. Bianquis, C.E. Bosworth, E. van Donzel, W.P. Heinrichs (eds.), *Encyclopaedia of Islam, Second Edition*, online at http://dx.doi.org/10.1163/1573-3912_islam_COM_1184 (accessed 23 April 2018).

34 According to the Ottoman state decree that was issued in 1229 H (1882 CE), instruments produced in Tophane in Istanbul were supposed to be stamped with this year and the word "Tophane." For an account of the Ottoman state's involvement in the changing of the measurement system, see Feza Günergun, "Du 'Zira' au 'Mètre': Une Transformation Métrologique dans l'Empire Ottoman", in Patrick Petitjean, Catherine Jami, Anne Marie Moulin (eds.), *Science and Empires: Historical Studies about Scientific Development and European Expansion*, « Boston Studies in the Philosophy of Science 136 », Springer, Dordrecht & Boston, 1992, pp. 103–110. More importantly, a weight in the Pera Museum in Istanbul has exactly the same date and place name as those in the rule in question,

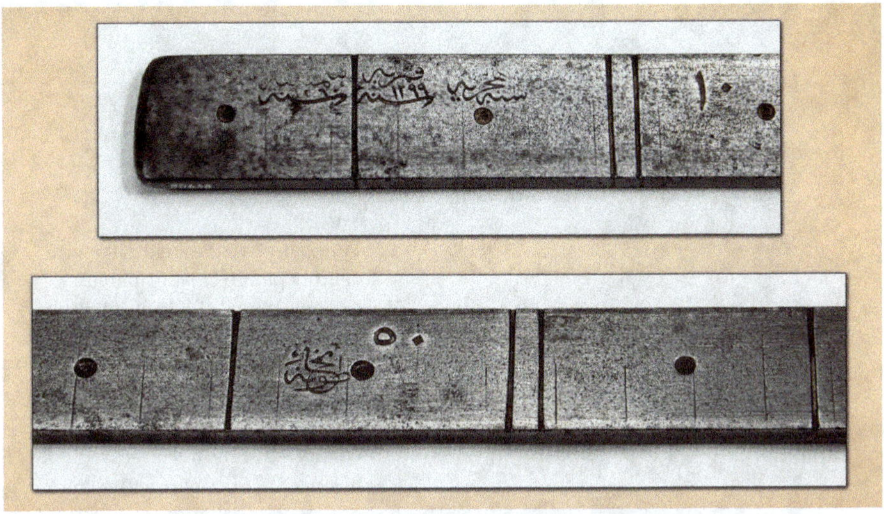

FIGURE 11.4 Details from a nineteenth-century mercantile rule attributed to Benghazi, Libya. On the Arabic side there is the date 1299 H, corresponding to 1882 CE. There is also the inscription in Arabic for "Tophane", a district of Istanbul where it was made, art. no. 1980.0638 Ingenium
PHOTOS: ANDRÉ LÉVESQUE, INGENIUM: CANADA'S MUSEUMS OF SCIENCE AND INNOVATION

There are eleven plumb bobs in the Petrovic collection that could be considered Ottoman (as we defined earlier, through origins, survival, and continued use). According to Petrovic's records their dates range from the twelfth to the eighteenth centuries. The Ottomans called them *şakül.* They are diverse in materials and style, and were produced in various regions of the empire and at different periods of time. They were used as vertical datum lines for measuring heights and angles. They are mostly made of bronze with varying amounts of decorative features. There are two dated back to the twelfth century by Petrovic, so they are credited with being the oldest artefacts in Ingenium: Canada's Museums of Science and Innovation. One of them measures 8.1 cm in length and 2.1 cm in diameter. It seems to have been cast in one piece. It has a decorative shape and decorative details. Petrovic attributed this piece to Gamzigrad, which is today in Serbia. He attributed the other twelfth-century plumb bob to Split, which is now in Croatia. Its length is 3.2 cm and it is 2.0 cm in diameter. Unfortunately, it is in poor condition so it is not possible to be sure of its original shape.

---

revealing that the rule was produced in Istanbul in the late nineteenth century. See Garo Kürkman, *Anadolu Ağırlık ve Ölçüleri*, AKMED, Istanbul, 2003, p. 90.

FIGURE 11.5
Plumb bobs (left to right): twelfth-century plumb bob attributed to Gamzigrad, in Serbia (in former Yugoslavia), art. no. 1980.0657 Ingenium; fifteenth-century plumb bob attributed to Istanbul, Turkey, art. no. 1980.0656 Ingenium, which may be connected to the *havai terazi* (aerial balance) in *Fig. 11.6*; twelfth-century plumb bob attributed to Split, in Croatia (in former Yugoslavia), art. no. 1980.0679 Ingenium.
PHOTO: ANDRÉ LÉVESQUE, INGENIUM: CANADA'S MUSEUMS OF SCIENCE AND INNOVATION

Perhaps one of the more intriguing examples is a plumb bob that Petrovic attributed to Istanbul and dated back to the fifteenth century. This one has a hinged, decorative medallion attached to the head (*Fig. 11.5*, centre), which is unusual for plumb bobs due to the fragile nature of the hinge. Therefore, this plumb bob may have served a ceremonial purpose rather than a functional one.[35] Alternatively it may have been connected to the *havai terazi* (aerial balance, also known as a plumb bob level) in the collection (see below and *Fig. 11.6*) for showing whether two points are at the same height.[36] Similar types of plumb bob can be seen in the Pera Museum in Istanbul. The catalogue of that collection dates them between the sixteenth and nineteenth centuries, which might lead one to reconsider Petrovic's dating of his plumb bob to the fifteenth century.[37]

35 Assia Kaab, "Metrological Instruments of the Petrovic Collection", 2017, Ottawa, Ingenium: Canada's Museums of Science and Innovation, George Petrovic Collection File, Petrovic Collection, 1980.0610-1980.0739, p. 11.

36 *Ibid.*, p. 16.

37 Pera Museum in Istanbul hosts a collection with more than 500 instruments, mainly of weights that were used from the earliest times to the nineteenth century in Anatolia. As mentioned, some instruments in the museum are of the same types as those in the

FIGURE 11.6 Fifteenth-century level attributed to Istanbul, Turkey, art. no. 1980.0655 Ingenium. It may have been used with the Istanbul plumb bob in *Fig. 11.5*.
PHOTO: ANDRÉ LÉVESQUE, INGENIUM: CANADA'S MUSEUMS OF SCIENCE AND INNOVATION

Among the Ottoman instruments in the Petrovic collection are fourteen levels, dated by Petrovic between the fifteenth and eighteenth centuries. They are generally in good condition. Some of them are plain and the others are elegantly shaped, and decorative in diverse ways. One intriguing example is an instrument which Petrovic attributed to Istanbul and dated to the fifteenth century, which makes it the oldest in the collection (*Fig. 11.6*). It may have been used with the Istanbul plumb bob described above. It was known as a *havai terazi* (aerial balance), especially in Turkish-speaking Ottoman regions.[38]

Petrovic collection, and deserve comparative research that would enrich understanding of the dissemination and provenance of such scientific instruments. See Kürkman, *Anadolu Ağırlık*, cit. (n. 33), p. 103. For more examples of Ottoman weights and measures, see Garo Kürkman, *Osmanlılarda Ölçü ve Tartılar* (*Ottoman Weights and Measures*), Türk ve İslam Eserleri Müzesi, Istanbul, 1991.

38 For some other examples of *havai terazis*, see Kürkman, *Anadolu Ağırlık*, cit. (n. 34), p. 102.

Basically, the *havai terazi* was hung on a nominally horizontal string between two points. The plumb bob was suspended from the centre of the level and if the two points were at the same height then the plumb line would pass over the point of the level. This level measures 20.7 cm in length, 17.2 cm in width and 0.4 cm in depth and it was cast in bronze. It is larger than many levels, with elaborate decorative features some of which indicate that it might have belonged to an architect as a ceremonial piece rather than to a builder.[39]

5 Questions about the Collection

As well as the locations where the objects were made or found, questions remain about some of the production dates attributed to the instruments by Petrovic. In 1994, the Canada Science and Technology Museum sent 26 pieces including twelve levels, ten rules and four plumb bobs to the Canadian Conservation Institute for tests. The report that came back from the Institute was quite helpful. Of the seventeen pieces with metallurgical test results eleven were assigned the same date as Petrovic gave them. However, the metallurgical study did not confirm that the supposedly fifteenth-century level purchased in Istanbul was from that century. Additionally, having confirmed the date of the sixteenth-century level attributed to Plovdiv in Bulgaria, the Institute reported that another level which was attributed to Benghazi and also dated by Petrovic as sixteenth century was actually from the nineteenth century. Obtaining confirmation of the dates of the instruments through physical, stylistic and historical methods is still a pressing research goal.[40]

A new project is being undertaken on the collection by Assia Kaab, a research intern at the Canada Science and Technology Museum. She is analyzing representative instruments in the collection from the point of view of architectural and building practices. She rightly points out that in the history of Islamic architecture measuring and drafting tools have attracted less attention than they deserve. In order to make sense of the tools and determine how they were used she will examine the pieces individually and make replicas, which will help to clarify how the original pieces functioned in measuring and drawing during the Ottoman period.[41]

---

39 Kaab, op. cit. (n. 35), p. 6. See also the chapter by Abdeljaouad and Ageron, *infra*, pp. 23–25 and Fig. 2.3.

40 Judge-Gervais, op. cit. (n. 21), pp. 26–28.

41 For a general description of the project, see Kaab, op.cit. (n. 34).

6 Conclusion

On the most basic level related to the earlier history and use of these instruments, this collection has the potential to tell us about routine instruments and practice related to metrology and activities such as building and architecture. Many pieces in the collection, such as plumb bobs, builders' levels and weaver's rules, are valuable sources for understanding scientific and technical practice as well as the mathematical culture from 'below', from the perspective of 'ordinary' people, not just sultans' courts. Most of the objects in the Petrovic collection are quite common and routine, not necessarily used by architects or people of higher status. In his research and collecting, Petrovic was clearly interested in the tools and processes used for building, surveying and construction, and how the metrological elements of those tools shaped larger architectural traditions, practice and products. Many other instruments in the collection such as the weaver's rules are common in these same ways. On the other hand, some of the Petrovic instruments reveal levels of status between builders and architects, especially those that are more decorative or ceremonial in nature. The place and date attributions of so many similar objects point to shared mathematical practices across several regions and time periods. These material, practical and transmission dimensions of the Petrovic collection have much potential for further research and can nicely complement other metrology collections in the Eastern Mediterranean and former Ottoman regions.

Aside from the actual instruments, one of the more valuable dimensions of the Petrovic collection is the history of the collector and the development and movement of the collection. By exploring the deeper provenance of the Petrovic instruments we follow their path from technical instruments to cultural artefacts.[42] Through their recent history in particular, we see the value of studying the instruments as artefacts of the collector's context as much as from their earlier manufacture and use. This long history of the instruments

42 David Pantalony, "Collectors, Displays and Replicas in Context: What We Can Learn from Provenance Research in Science Museums", in Jed Buchwald, Larry Stewart. (eds.), *The Romance of Science: Essays in Honour of Trevor H. Levere,* « Archimedes, New Studies in the History and Philosophy of Science and Technology 52 », Springer, Cham, 2017, pp. 255–275. For a classic study of object biography as a window into collections and museums, see Samuel Alberti, "Objects and the Museum", *ISIS* 96, 4 (Dec. 2005), pp. 559–571. The study of scientific instruments in museums as assemblages was the focus of a Cain Conference workshop "Shaping Scientific Instrument Collections" at the Science History Institute in Philadelphia, April 18–21, 2018, organized by Samuel Alberti and Carin Berkowitz.

both enriches but also problematizes the history of the Petrovic instruments, a situation which is amplified by the ambiguity surrounding some of the instruments' original history of manufacture and use.

The Petrovic collection therefore raises issues for both curators and historians. In general, this history shows that we must be cautious about the kinds of stories we tell about these instruments and similar collections. At the same time, we must see opportunities within these concerns. For example, a key issue related to the Ottoman dimensions of this history is that Petrovic collected these instruments during a period when nationalism was important in many aspects of social, political, cultural and intellectual life. One of his initial goals related to retracing the roots of traditional Serbian architecture. On the other hand, he clearly saw that this collection had the potential to transcend national, temporal and spatial boundaries, revealing a shared and common culture of practical mathematics in the Ottoman and Mediterranean regions.[43] The resulting collection reflects both of these contexts, and should be explored with both perspectives in mind.

Instruments and their provenance can be a powerful way to navigate these seeming contradictions. A number of studies have been produced on various aspects of the Islamic arts, architecture and science, but artefacts and their practical applications, as well as their ownership history, have attracted less attention than they deserve. As Suraiya Faroqhi, a leading historian of the Ottoman Empire, points out, Ottoman historiography is generally based upon archival research, but "artefacts can tell us things that written sources never mention".[44] Collections such as that of Petrovic has the potential to reveal some of the complexities of the mathematical, architectural and trading cultures in the pre-modern and early modern world, as well as the ways we have collected this material history and constructed narratives around these collections.

---

43 As well as works that deal with the political history of the Ottoman Empire as an important part of the history of the Mediterranean region, studies that investigate Ottoman science in the wider Mediterranean context, especially in the early modern period, have started to emerge. See, among others, Avner Ben-Zaken, *Cross-Cultural Scientific Exchanges in the Eastern Mediterranean, 1560–1660*, Johns Hopkins University Press, Baltimore, 2010; Robert Morrison, "A Scholarly Intermediary between the Ottoman Empire and Renaissance Europe", *ISIS* 105, 1 (Mar. 2014), pp. 32–57; and Nükhet Varlık, *Plague and Empire in the Early Modern Mediterranean World: The Ottoman Experience, 1347–1600*, Cambridge University Press, Cambridge, 2015.

44 Suraiya Faroqhi, *A Cultural History of the Ottomans*, I.B. Tauris, London—New York, 2016, p. 2.

Acknowledgments

Hasan Umut was awarded the McGill-Ingenium Fellowship in the History of Science, Medicine, and Technology in 2016, which made it possible for him to conduct research during the summer of 2016 in the collections of Ingenium: Canada's Museums of Science and Innovation in Ottawa. He is grateful to McGill University and Ingenium: Canada's Museums of Science and Innovation for providing him with this opportunity. Special thanks go to Assia Kaab with whom the authors exchanged ideas and insights into the Petrovic Collection. They thank Anna Adamek, Jelena Bogdanovic, Ziya Erdinç, Sandra Judge, Sajjad Nikfahm-Khubravan, Abdullah Taha Orhan, F. Jamil Ragep, Sally P. Ragep, and Hasan Fehmi Topal for their help in various ways during the research on the collection. The authors also thank the National Gallery of Canada, the Montreal Museum of Fine Arts, the Art Gallery of Ontario, and the Serbian Heritage Museum for providing information about George Petrovic's artistic work.

CHAPTER 12

Instruments of Knowledge and Power in a Colonial Context

Scientific Instruments during the French Occupation of Egypt, 1798–1801

Patrice Bret

1 Introduction

> Learned Europe could not be indifferent to the power of science used in a country to which it has been brought back by armed wisdom and the love of mankind after being exiled for a long period by barbarity and religious fervour.[1]

Boldly claiming the wisdom and humanity of Bonaparte's army, the very first issue of *La Décade égyptienne*—the learned journal published in Cairo by the French during their occupation of Egypt (1798–1801)[2]—foresaw social and intellectual progress from the power of the sciences used in the country where they were supposed to have been born. Scientific instruments were the means to gain those expected positive outcomes in an Ottoman province actually ruled by the Mameluke Beys. The Egyptian astronomer and chronicler Abd al-Rahmān al-Jabartī (1754–1825) expressed strong criticism of the occupying army. Nonetheless he wrote most appreciatively of the French instruments displayed in Cairo, deplored their looting by the rioters on 21 October 1798, and recounted how the French tried to retrieve them (see *Section 4* below, first paragraph). In 1809, Jacques-Marie Le Père (1763–1841)—the former chief civil engineer of Bonaparte's expedition—also complained briefly about the looting in the narrative of his topographical survey of the Suez Isthmus for the *Description de l'Egypte*:

1 "Travaux de l'Institut", *La Décade égyptienne, Journal littéraire et d'économie politique* 1 (1799), p. 15.

2 Among the huge bibliography on the French occupation, see Henry Laurens *et al.*, *L'expédition d'Égypte, 1798–1801*, Armand Colin, Paris, 1989; Yves Laissus, *L'Égypte, une aventure savante 1798–1801*, Fayard, Paris, 1998; Patrice Bret, *L'Égypte au temps de l'expédition de Bonaparte (1798–1801)*, Hachette Littératures, Paris, 1998; André Raymond, *Égyptiens et Français au Caire, 1798–1801*, IFAO, Cairo, 1998; Juan Cole, *Napoleon's Egypt: Invading the Middle East*, Palgrave Macmillan, New York—Basingstoke, 2007.

 | DOI:10.1163/9789004412842_013

> We had brought from France very good bubble levels with telescopes but they were plundered from General Caffarelli's house during the Cairo riot, along with many valuable objects, the lack of which was deeply regretted in the latter part of our stay in Egypt.[3]

Those instruments of knowledge and power have not yet been studied. Even Anne Godlewska hardly mentioned them in her outstanding work on the Napoleonic mapping of Egypt.[4] This chapter, based on mostly unpublished French public and private archival sources and Oriental printed primary sources, investigates the actual scientific instruments the French worked with in Egypt. Lack of space restricts the study to instruments for astronomy, geodesy and topographical surveying.[5]

The first section describes the instruments collected in Paris in early spring 1798, along with the instrument-makers enrolled for the expedition. The second section is an attempt to point out and illustrate the actual use of those instruments, as well as others brought privately by scientists and officers. The third and last section explores the repairing of instruments and the designing and making of new ones in the mechanical workshops set up in Cairo by the painter and chemical and mechanical engineer Nicolas-Jacques Conté (1755–1805).[6] Throughout the chapter, al-Jabartī's descriptions will reflect the view of an Oriental savant, who owned the best private collection of astronomical instruments in Cairo,[7] on that unprecedented display of Western European science and technology in the East.

---

3 Jacques-Marie Le Père, *Mémoire sur la communication de la mer des Indes à la Méditerranée par la mer Rouge et l'isthme de Soueys*, Imprimerie Royale, Paris, 1815, p. 24, n. 1.

4 Anne Godlewska, "The Napoleonic Survey of Egypt: A Masterpiece of Cartographic Compilation and Early Nineteenth-Century Fieldwork", *Cartographica* 25, 1–2 (1988), pp. 17–18.

5 On the instruments of the chemical and pharmaceutical laboratories, the cabinet of physics and the surgical service, see P. Bret, "Les premiers laboratoires coloniaux: monstration, production et recherches chimiques pendant l'expédition d'Egypte (1798–1801)", *Revue d'histoire des sciences*, forthcoming.

6 P. Bret, "Conté et 'les arts de la France au milieu des déserts': vicissitudes d'un transfert de technologie pendant l'occupation française", in M. Dewachter, A. Fouchard (eds.), *L'Égyptologie et les Champollion*, Presses universitaires de Grenoble, Grenoble, 1994, pp. 331–345; P. Bret, "Nicolas-Jacques Conté, ou l'intelligence technique", *Musée des arts et métiers. La revue* 25 (1998), pp. 4–13.

7 Especially instruments made in 1700 by Riḍwān al-Falakī, including golden bronze armillary spheres shown on his portraits by André Dutertre during the French expedition and by Pascal Coste in 1822. See André Raymond, "À propos de deux portraits de la *Description de l'Égypte*: 'l'astronome' et 'le poète'", *Annales islamologiques* 35 (2001), pp. 385–392: 387. Al-Jabartī had inherited them from his father Hasan (1698–1774), the last Sheikh who taught astronomy and mechanics at al-Azhar; see Abd al-Rahmān al-Jabartī, *'Ajā'ib al-ātār*, Bulaq,

2 Collecting the Necessary Scientific Instruments in Paris

The acquisition of scientific instruments for Bonaparte's expedition to Egypt was funded by the Ministry of War through the Corps of Military Engineers, and was organised by General Louis-Marie-Joseph-Maximilien Caffarelli du Falga (1756–1799). A military engineer by profession, Caffarelli was also a member of the *Institut national des sciences et arts* founded in 1795 to replace the old *Académies royales* suppressed in 1793, and he was helped by a few scientists who knew the aim of the expedition, especially Claude-Louis Berthollet (1748–1822), his colleague at the *Institut national* and a professor of chemistry at the *École polytechnique*.[8] Several state institutions lent scientific instruments and other material, mainly the *Bureau des longitudes*—closely linked to the Paris Observatory—the cartographic services (*Dépôt général de la Guerre, Dépôt général de la Marine* and *Bureau du Cadastre*), and the *École nationale aérostatique* (National School of Ballooning) at Meudon.

2.1 *The Inventory of the Expedition Instruments*

Caffarelli's *Toisé des dépenses* fortunately provides a detailed list of the instruments purchased from Paris merchants and instrument-makers (*Tab. 12.1*).[9]

Obviously, cartography was a main concern for Bonaparte and Caffarelli, as it was for many members of the *Institut national* in Paris and the *Commission des sciences et arts* attached to the military expedition under Caffarelli's command. Topography accounted for a large part of the scientific expenses of the expedition, especially for the scientific instruments. Although far behind the library (which also contained many astronomical and geographical books) the cost of the topographical equipment was nearly the same as the cost of the printing material, another well-resourced section supplemented by Gaspard Monge with what he collected at the printing press of the Sacred Congregation for the Propagation of the Faith in Rome. The total amount for astronomy and topography reached 15,042 francs, which is more than the cost of all the other

Cairo, 1297 H/1879 CE, 4 vols.: v. 1, pp. 74, 114 and 397; English translation: Thomas Philipp, Moshe Perlmann (eds.), *Abd al-Rahman al-Jabartī's History of Egypt*, Franz Steiner Verlag, Stuttgart, 1994, 2 vols.: v. 1, pp. 122, 186 and 664.

8 Berthollet was a member of the section of chemistry in the division of "sciences mathématiques et physiques", Caffarelli a non-resident member of the section of political economy in that of "sciences morales et politiques".

9 "Armée d'Angleterre/Génie/Comptabilité/An 6e/Toisé des dépenses faites par le Cen Caffarelli Général de brigade du Génie, d'après les ordres du Général en chef Buonaparte sur les fonds ordonnés par le ministère de la Guerre pour l'expédition de la Méditerranée", Vincennes, Service historique de la Défense (now SHD, GR), no shelfmark. Unless otherwise stated in footnotes, all details about instruments are from this document.

TABLE 12.1 Instruments, tools and other objects purchased for scientific purposes by General Caffarelli

1. Astronomical Instruments	5,534 francs	6.75%
2. Topographical Equipment	9,508	11.59%
3. Cabinet of Physics	3,515	4.28%
4. Chemical Laboratory	1,909.05	2.33%
5. Ballooning Material	4,001.40	4.88%
6. Cabinet of Natural History	1,257	1.53%
7. Surgical Instruments	2,804	3.42%
8. Pharmacy	1,233.55	1.50%
9. Machines and tools	1,450	1.77%
10. Printing Material	10,161.88	12.39%
Purchase of books for the Library	25,329.15	30.87%
Transport and shipping costs	6,812.01	8.30%
Office and drawing materials	8,524.66	10.39%
TOTAL	82,039.70	100.00%

instruments (14,720 francs for sections 3 to 8), or only a little less if we also include machines and tools (to make a total of 16,170 francs). Like the cabinet of physics[10] and also surgical instruments, the astronomical and topographical equipment consisted mainly of very sophisticated and expensive instruments, whereas other sections contained only a few instruments among various other kinds of equipment.

The astronomical instruments section included mainly horological and optical items. In horology, Louis Berthoud (1754–1813), one of the best clockmakers in Paris, provided the most expensive instrument on Caffarelli's list: his marine watch No. 34, which costs 2,000 francs. A certain E. Lamy—so far unidentified—not only sold "a portable astronomical clock, and a compound pendulum with a metal thermometer, the whole golden brown, contained in two boxes, one for the clock and the other for the pendulum" for 1,200 francs, but also a copy of the *Atlas Cœlestis* by Flamsteed, in its most recent third edition (London, 1781). Most of the optical instruments were made by Elisabeth-Théodore de La Poix de Fréminville (1756–1818), a civil engineer who had been director of the national workshop of Gros-Caillou, in charge of making achromatic telescopes

10 Physical instruments were supplied through a well-known scientist (Dolomieu) and instrument makers (Mossy, Dumotiez brothers, Jean-François Richer), along with untraced other providers (Rochette and Gaudette).

for the navy and the telegraph since 1795.[11] Fréminville used excellent British optics. He fitted a 12-franc micrometer to an "astronomical telescope mounted on a brass stand with azimuth and elevation movements, two celestial and one terrestrial eyepieces, the said telescope having a Dollen [Dollond] lens 32 lines [72 mm] in diameter, all enclosed in a box of mahogany" costing 600 francs. His other sales were one 5-foot telescope, one 3.5-foot telescope with a 59-mm lens, one 2-foot telescope, one astronomical eyepiece, and two spare 68-mm lenses.[12] Of the total of 5,776 francs paid to Fréminville, the instruments for the sole use of the *Commission* cost only 1,832 francs, plus 30 francs for shipping, the major part being for the Army and listed with the topographical equipment. Lastly, a "reflecting circle and the English tables"[13] were provided for 400 francs, not by an instrument maker or a merchant, but through a young *ingénieur-géographe*, Edme-François Jomard (1777–1862). Trained at the *École polytechnique* then at the *École des Géographes* of the *Cadastre*, Jomard belonged to that cartographic and fiscal establishment.[14] Where he bought those instruments and plane tables is not known, but the number of his invoice indicates that Jomard probably advanced the money in order to save time at the last moment prior to joining his colleagues and the fleet at Toulon.

Two of the merchants or instrument makers mentioned above also contributed to the topographical equipment. Rather strangely, Fréminville is listed again here for an unspecified number of "military telescopes", only a few of which were used for surveying. Lamy provided one plane table, one alidade with telescope, and a box of mathematical instruments for the *Commission* for a total of only 120 francs.[15] For a total of 5,474 francs, Étienne Lenoir (1744–1832)[16] was the one who provided the greatest number of items: one reflecting circle, one 13-inch azimuthal quadrant fitted with two telescopes, one meridian telescope, two plane tables, nine alidades (five with telescopes),

11 Maurice Daumas, *Les instruments scientifiques aux XVIIe et XVIIIe siècles*, Presses universitaires de France, Paris, 1953, pp. 141 and 373; Maurice Daumas, *Scientific Instruments of the Seventeenth and Eighteenth Centuries and their Makers* (trans. by Mary Holbrook), Batsford, London, 1972, pp. 104 and 283.

12 Note that the dimensions provided here do not agree perfectly with those given by Nouet.

13 The 'English tables' probably refer to the most recent edition of the *Nautical Almamac* published annually by the Commissioners of Longitude, London, from 1766.

14 See Y. Laissus, *Jomard, le dernier Égyptien*, Fayard, Paris, 2004.

15 Other mathematical instruments (dozens of rules, dividers and squares) provided by the natural history artists Jean-Henri Redouté and Jacques-Gérard Milbert are listed in the 'Office and drawing material' section.

16 Anthony J. Turner, *From Pleasure and Profit to Science and Security. Etienne Lenoir and the Transformation of Precision Instrument-making in France 1760–1830*, Whipple Museum of the History of Science, Cambridge, 1989.

six surveying compasses, one dip compass, four graphometers (three with telescopes), three bubble levels with telescopes, one pantograph, "which has already been used, but as good as a new one", two protractors (6-inch and 7.5-inch) and four brass metre bars.

2.2 *Instrument Makers and "Artists" of the Expedition*

Most of Caffarelli's expenses for the *Commission des sciences et arts* were purchases of instruments plus their packing and sometimes their setting up (for example, the micrometer fitted by Fréminville to the Dollond telescope), and for repairs (to several plane tables from the *Bureau de la Guerre*, a pinnule graphometer by Lenoir, and two scales by Gaudette). Obviously, such a quantity of valuable and fragile precision instruments also required instrument makers or at least specialized "artists" and artisans for their maintenance, as well as other mechanics.[17] Several were commissioned for General Bonaparte's secret expedition.

Three instrument makers in Paris—J.-F. Richer himself, Breguet Jr., and Paul-Étienne-Marie Lenoir (1776–1827)—were even asked to participate as members of the *Commission des sciences et arts*.[18] Only Lenoir Jr. eventually went to Egypt, with two unknown clockmakers, Favre and Lemaire, also nominated for the *Commission*.[19]

Although two artisans are better documented because they had worked previously at the *École polytechnique*, they are not known as instrument makers. Calixte-Victor (or Calixte-Charlemagne) Cirot (d. 1801) was probably the "artist for mathematical and physical instruments", who had to "explain the processes of his art to the students" in the early months of the *École centrale des travaux publics*, which became the *École polytechnique*.[20] There, in June 1796, he also replaced Savart as deputy curator of the gallery of models, drawings and prints. Son of a carpenter, François-Sébastien Aimé (1762–1843) joined a company of military workers (*compagnie d'ouvriers*) in the Artillery as a

17 On those "artists" cited in the following paragraphs, see P. Bret, "Les oubliés de Polytechnique en Égypte: les 'artistes' mécaniciens de la Commission des Sciences et des Arts", in *Scientifiques et sociétés pendant la Révolution et l'Empire*, CTHS, Paris, 1990, pp. 497–514.

18 His father Étienne Lenoir had previously made an astrolabe for the Sultan of Morocco in 1789; Turner, op. cit. (n. 16), p. 50. The plates of his reflecting circle, printed in Borda's book, were adapted into Turkish by Ishak Effendi, *Aksü'l-merâyâ fi ahzi'z-zevâyâ*, Matbaa-i Amire, Istanbul, 1834–1835.

19 Favre, "clockmaker to the army", auctioned a clock made in Paris which displayed the day of the month; *Courrier de l'Égypte* 16 (14 Nov. 1798), p. 4.

20 Ambroise Fourcy, *Histoire de l'École polytechnique*, Chez l'auteur, Paris, 1828 (reprinted and ed. by Jean Dhombres, Belin, Paris, 1987), p. 55.

carpenter in 1783. Ten years later he managed the manufacturing of ordnance at the Paris arsenal, and soon became workshop manager at the *Atelier de perfectionnement des armes portatives* where new ways of manufacturing were being tried out under the supervision of Vandermonde and Hassenfratz.[21] In January 1795, Aimé became *artiste-instituteur* at the *École centrale des travaux publics*. In the course of descriptive geometry he taught under Gaspard Monge's supervision, he gave practical lessons on the *trait de charpente* (the procedures used to find the exact dimensions and angles needed for cutting the joints of timbers for buildings). Aimé also made wooden models and looked after their collection. In October 1796, he was requisitioned by the Ministry of the Navy for a technical commission to be sent to Istanbul. When the project was cancelled Aimé became a commissioned officer in the Military Engineers, and in December 1797 he was attached to the training staff at the *École du Génie* at Metz. Soon afterwards he was recruited by Caffarelli for the expedition.

Besides those specialists who had previously worked for and with the scientists, either in their own workshops or in state technical schools and experimental establishments, other workers from the navy and the army were also required (see *Section 4* below). Some belonged to the company of military balloonists (*aérostiers militaires*) previously trained by Conté at the *École nationale aérostatique* in Meudon, but also to companies of military workers attached to the Artillery and the Military Engineers.

As well as the instruments obtained by Caffarelli for the *Commission*, the scientists, engineers and members of the medical staff who joined the expedition brought their own instruments for use in Egypt.

3 Scientists and Engineers, and Their Instruments in Egypt

3.1 *The Instrument Users: Astronomers and Engineers*

There were two senior astronomers in the *Commission des sciences et arts*, and a third joined them in Cairo. All of them were elected to the *Institut d'Égypte* founded by Bonaparte in August 1798, the meeting room of which was furnished with "one of the biggest and most beautiful clocks by Berthoud" seized from a Mamluk house.[22]

21 P. Bret, *L'État, l'armée, la science. L'invention de la recherche publique en France, 1763–1830*, Presses universitaires de Rennes (coll. Carnot), Rennes, 2002, passim.

22 Most probably by Ferdinand Berthoud (1727–1807). See Geoffroy Saint-Hilaire to Georges Cuvier, 23 August 1798, in E.T. Hamy (ed.), *Lettres écrites d'Égypte* [...] *par Étienne Geoffroy Saint-Hilaire ...*, Hachette, Paris, 1901, p. 74.

The chief astronomer was Nicolas-Antoine Nouet (1740–1811), who has been attached to the *Bureau du Cadastre* under the mathematician and engineer Gaspard Clair François Marie Riche de Prony (1755–1839) for two years. Formerly a monk, Nouet had learnt astronomy at the Paris Observatory where he became an assistant to Cassini in the 1780s. He contributed to the annual French nautical almanac, *Connoissance des temps*, and became the most senior professor after the 1793 reform and Cassini's resignation. He joined the *Dépôt général de la Guerre* in 1795, then the *Cadastre* in 1797 where he was in charge of mapping the newly annexed departments of Palatinate and Savoy.[23] Nouet replaced André Méchain, who had recently returned from his lengthy measurement of the Paris Meridian in Spain. Méchain refused to participate in the expedition but let his son Jérôme-Isaac Méchain (1778–1851) accompany Nouet as a student (*élève astronome*). The Knight of Malta Jean-Auguste d'Angos (1744–1833), formerly director of the Malta Observatory, also refused to join. In Egypt, Nouet used the Berthoud marine watch No. 34 mentioned above, made in 1797. He also brought a *compteur astronomique* (an attachment to a telescope for measuring time intervals) and a micrometer, both lent by the Paris Observatory through the *Bureau des longitudes*. Later, in Cairo, he was given Beauchamp's Berthoud marine watch No. 29, although it was in a very bad state.

François-Marie Quénot (1761–1803?) who had been in the navy since 1781, was an officer with expertise in astronomical observation. He had worked for Jérôme Lalande, who probably arranged for him to be assigned to the *Commission des sciences et arts* when he was asked to borrow instruments through the *Bureau des longitudes*. Quénot brought his own Berthoud marine watch No. 19 to Egypt—the model for the first gimbaled marine watches of tobacco-box shape—, that he had bought on 6 May 1792.[24] Having sailed back to France after only a few months because he did not feel comfortable in the climate, he wrote an account of his work in Alexandria, one of the first to be published in Paris by a member of the expedition.[25]

Also belonging to Lalande's network, the third astronomer was Pierre-Joseph de Beauchamp (1752–1801), a non-resident member of the *Institut national des*

23 P. Bret, "L'astronome Nicolas-Antoine Nouet (1740–1811), membre de l'Institut d'Égypte, directeur de la Carte de Savoie," in *Les scientifiques et la montagne: actes du 116e Congrès national des sociétés savantes, Chambéry, 29 avril-4 mai 1991, Section d'histoire des sciences et des techniques*, CTHS, Paris, 1993, pp. 119–147.

24 Private collection; Jean-Claude Sabrier, *La Longitude en mer à l'heure de Louis Berthoud et Henri Motel*, Antiquorum, Geneva, 1993. Quénot first bought a marine watch, No. 16, in March 1791.

25 François-Marie Quénot, "Extrait des observations astronomiques", *Magasin encyclopédique des sciences, des lettres et des arts* 26 (1799), pp. 145–152.

sciences et arts.[26] From 1782 to 1789, Beauchamp was Vicar General in Baghdad, and sent meteorological data to Lavoisier and astronomical data to Lalande.[27] At the Missionary Fathers' convent in Baghdad he set up an exceptionally well equipped observatory, with a *télescope paralactique* (for measuring very small angles such as the parallax of stars), a half-a-second clock, a quadrant moving under a central turning roof, and a meridian telescope.[28] Elected a corresponding member of the Academy of Sciences in 1785, Beauchamp also corrected the longitude of the Caspian Sea in 1788, although he was not able to actually reach the sea shore as there was a local rebellion nearby. Back in France, he was named consul in Muscat in March 1795. The next month General Calon, director of the *Dépôt général de la Guerre*, purchased Berthoud marine watch No. 29 specially for Beauchamp's new journey, and in June he asked him to correct the map of Asia Minor on the Black Sea coast. Beauchamp was later provided with a complete set of scientific equipment: a quadrant with two telescopes, a 4-foot achromatic telescope (with a lens of 136-mm diameter), a night telescope and an astronomical counter, a 4-inch (108-mm) surveying compass, a variation compass, a pocket compass, an "excellent" Borda-Lenoir reflecting circle, an achromatic surveying telescope, a smaller silvery telescope, a set of *poids de marc* (old system weights), an ivory *pied du roi* trimmed with silver and an 'academy foot' (two standards of length: the academy foot was slightly too long!), along with "Borda's book", astronomical tables and an *Ephemerides*.[29] On his way to the Sultanate of Oman Beauchamp spent several months in Istanbul in 1796, until he obtained the authorization to sail to Trebizond (present day Trabzon, Turkey) to observe its actual position, which he did the following year. While he was in Aleppo he was urged by the government to join Bonaparte in Cairo instead of going directly to Muscat. Beauchamp's long expertise in and knowledge of the Ottoman Empire led the General-in-Chief to send him on a diplomatic mission to Istanbul on Idris Bey's ship in

26 For a biography of Beauchamp see Marie-Liesse Pierre, *Pierre-Joseph Beauchamp, 1752–1801, missionnaire, astronome, géographe, diplomate*, Lycée Claude Mathy, Luxeuil-les-Bains, 1990.

27 P. Bret, "L'expérience préalable de l'Empire ottoman dans la Commission des sciences et arts de l'expédition de Bonaparte en Égypte (1798–1801)", in Ekmeleddin İhsanoğlu, Ahmed Djebbar, Feza Günergun (eds.), *Science, Technology and Industry in the Ottoman World*, Brepols, Turnhout (Belgium), 2000, pp. 101–114.

28 "Plan of my observatory in Baghdad, 1786", in Pierre, op. cit. (n. 26), pp. 23–25.

29 "Borda's book" refers to Jean-Charles Borda, *Description et usage du cercle de réflexion*, Didot L'aîné and Jombert, Paris, 1787. It is not specified which 'astronomical tables' Beauchamp provided, probably the English *Nautical Almanac*. The *Ephemerides* refers to Jérôme de Lalande, *Ephémérides des mouvemens célestes, pour le méridien de Paris*, Veuve Hérissant, Paris, 1792, v. 9.

December 1798. Unfortunately he was arrested and kept in jail in Istanbul until 1801. Beauchamp had left most of his instruments in Aleppo. He left the others to Nouet before leaving Cairo, notably his marine watch No. 29. Thus with Quénot's own marine watch No. 19, three high quality Berthoud chronometers were used in Egypt, even though Caffarelli's *Toisé des dépenses* listed only No. 34 used by Nouet.

Besides those experienced astronomers, a fourth young scientist played a crucial role for the astronomical and geodetic work in Egypt. Jean-Baptiste Corabœuf (1777–1859), another *ingénieur-géographe* of the *Cadastre* trained at the *École polytechnique* and the *École des Géographes*, served as Nouet's most able assistant.[30] Many other *ingénieurs-géographes*, military engineers, mining engineers and civil engineers had a lot of experience with topographical instruments. Some forty of the youngest had been regular students at the *École polytechnique*, where half a dozen of the others had also extended and updated their own qualifications.

3.2 *The Instruments in the Field*

Nouet himself described or at least mentioned the instruments he actually used in Egypt, mainly a repeating circle of 25-cm diameter with decimal divisions, Louis Berthoud's marine watch No. 34, and Dollond's achromatic telescope. He sometimes gave further information: for instance, he had had the first two with him since leaving Paris, keeping the watch in his pocket in a vertical position when it was not suspended in its box, and verifying it and Quénot's No. 19 against the astronomical clock of the Naval Observatory at Marseilles, then at the observatory in Malta.[31] The telescope, which had a 63-mm aperture and was mounted on a brass stand with a smooth motion, was shipped out by the Military Engineers but it is unclear whether it was that one that General Kléber allowed Nouet to 'choose' at their headquarters in Alexandria on 31 August—which is highly probable since they were responsible for the

30 P. Bret, "Le Dépôt général de la Guerre et la formation scientifique des ingénieurs-géographes militaires en France (1789–1830)", *Annals of Science* 48, 2 (1991), pp. 113–157; P. Bret, "Du concours de l'an II à la suppression de l'École des géographes: La quête identitaire des ingénieurs géographes du Cadastre de Prony, 1794–1802", in Ana Cardoso de Matos, André Grelon, Irina Gouzévitch, Maria Paula Diogo (eds.), *Jogos de identidade profissional: Os engenheiros, a formação e a acção—Les enjeux identitaires des ingénieurs: Entre la formation et l'action—The Quest for a Professional Identity: Engineers between Training and Action*, Edições Colibri, Lisbon, 2009, pp. 121–154.

31 [Nicolas-Antoine] Nouet, "Rapport sur les Observations faites pour déterminer la position géographique d'Alexandrie et la direction de l'Aiguille aimantée", *La Décade égyptienne, Journal littéraire et d'économie politique* 1 (1799), pp. 165–181: 165–166.

instruments of the *Commission des sciences et arts*—or whether that authorization was for another achromatic telescope.[32]

In Alexandria, helped by Coraboeuf, Nouet measured the baseline for the triangulation of the town and made ten astronomical observations from the terrace of the Batavian consulate with his 13-inch Lenoir quadrant in order to determine the latitude of the town accurately.[33] It is interesting to note an addition made to adapt the instrument to the local context:

> We had attached a metal mirror to the movable telescope so as to reflect the images of the objects perpendicularly to the axis of the telescope. That adaptation is necessary in climates where the Sun approaches the zenith.[34]

That was the first step for mapping Alexandria after fixing the geodetic markers for its triangulation network. That dangerous collaborative work—one soldier of the escort was killed—required twenty-five people (astronomers, topographers, military engineers, civil engineers and naval officers). Nouet also used a surveying compass and a dip needle.[35] Quénot made fifteen observations with a reflecting circle, seven of them using the horizon of the sea and the other eight with an artificial horizon, along with eighty series of determinations of the distance between the Moon and the Sun, totalling 480 distances measured.[36]

On 11 September 1798, Bonaparte ordered that all the objects of the *Commission des sciences et arts*, notably those used for chemistry, were to be sent from Alexandria to Cairo, but "the tools, however, are what must be sent first [along with] one or two plane tables, bubble levels and one graphometer".[37] In March 1799, prior to going to Upper Egypt with Coraboeuf, Nouet asked for the astronomical instruments he had left in Damietta.

3.3 *The Failure of Cairo Observatory*

Nouet left Alexandria for Cairo in early September 1798, but Berthoud marine watch No. 34 was dropped on the journey in Rosetta. Because of this when, in the same month, he had to calculate the longitude of Cairo, he had to do it

32 Henry Laurens (ed.), *Kléber en Egypte, 1798–1800: Kléber et Bonaparte, 1798–1800*, Institut français d'archéologie orientale, Cairo, 1988, 2 vols.: v. 1, p. 282.

33 Nouet, op. cit. (n. 31), p. 169.

34 *Ibid.*, p. 169.

35 He gives their details and sizes, *ibid.*, pp. 171–172.

36 Quénot, op. cit. (n. 25), p. 147.

37 Caffarelli to Liédot, 11 September 1798: SHD, GR, B6* 95.

by observing eclipses of the satellites of Jupiter. After that, little triangulation work was done in Egypt, and the 1/100,000 topographical map of the country—the work of 27 engineers and officers, including fourteen *polytechniciens*—was mostly prepared by surveying with plane tables and compasses adapted to the geodetic canvas based on 36 astronomical markers observed by Nouet and Coraboeuf.

On 2 August 1798, only ten days after occupying Cairo, Bonaparte asked Berthollet, Monge and Caffarelli to choose a building on Esbekieh square or nearby to house a printing press, a chemical laboratory, a cabinet of physics, an observatory and the *Institut d'Égypte*, which was to be founded three weeks later. On 12 October, at the tenth meeting of the Institute, Bonaparte "stressed the usefulness of an observatory and suggested ways of speeding up the establishment of one".[38] The experts put in charge of choosing its location were the architect Charles Norry, the astronomers Beauchamp and Nouet, and Caffarelli himself. Five days later Nouet gave an account of the investigation of several locations proposed for the observatory and gave the expert's opinion that "a separate building especially devoted to astronomical and meteorological observations" was needed.[39] After the riot of 21 October the new building was replaced by more modest facilities within the *Institut* buildings. There is no evidence of the installation of an observatory similar to Beauchamp's in Baghdad, and the only geodetic reference mark for Cairo was the top of the Great Pyramid in Giza. In the first version of his chronicle, probably written after a visit to the observatory, al-Jabartī gave a brief depiction of the astronomical instruments:

> [The French] possess extraordinary astronomical instruments of perfect construction and instruments for measuring altitude of wondrous, amazing, and precious construction. And they have telescopes for looking at the stars, and measuring their scopes, sizes, heights, conjunctions, and oppositions, and the clepsydras and clocks with gradings and minutes and seconds, all of wondrous form and very precious, and the like.[40]

38 "Institut", *La Décade égyptienne, Journal littéraire et d'économie politique* 1 (1799), pp. 116–122: 118.

39 *Ibid.*, p. 121.

40 'Abd al-Raḥmān Jabartī, *Napoleon in Egypt. Al-Jabartî's Chronicle of the French Occupation, 1798* (translation of *Târîkh muddat al-Faransîs bi-Misr* by Shmuel Moreh), Markus Wiener Publishing, Princeton—New York, 1993, p. 110.

In the third and last version of his chronicle al-Jabartī also described the sundials of various kinds built by Nouet in the *Institut d'Egypte* complex. The largest one was horizontal:

> The astronomer Tût [Nouet] occupies the house of Hasan Kâshif Jarkis. In the upper courtyard of the house, he has drawn regular lines on the entire paved surface, in order to mark the hourly graduation until noon. Instead of a style he has placed on the pediment of the house a circular plate, pierced by numerous holes, so that the Sun's rays pass through them down to the lines and their divisions. Thus one can know the time up to noon, and which is the current sign of the Zodiac, the images of which show the exact position of the Sun every month.

A mural dial was fixed in a more visible place in 1798 or 1799. It indicated hours from 6.30 a.m. to 6 p.m., as a painting by Conté also shows:[41]

> He also built, on the highest wall of the ground floor courtyard between the two houses, a dial with a style to indicate the hours before and after the middle of the day. But that dial is not like our dials which indicate the time of the ʿ*asr* (noon), the hourly graduations until sunset, the twilight and dawn curves, the direction (*samt*) of the *qibla*, with the divisions in degrees, etc., in order to determine times of prayer. Since the French are not affected by those indications, they do not care about them.

Lastly, in the garden of the *Institut*, there was a third, horizontal sundial—the only one still existing:[42]

> The same astronomer has also drawn many lines on the surface of a square plate of brass. He has put that plate on a column slightly shorter than a man in the middle of the garden. The style is replaced by an iron triangle, the shadow of which is projected down to the lines which are divided into degrees. It is quite a nice piece, bearing all the required indications

41 Engraved in *Description de l'Égypte, ou recueil des observations et de recherches qui ont été faites en Égypte pendant l'expédition de l'armée française*, L'Imprimerie impériale, Paris, 1809, 20 vols.: v. 7, "État moderne", pl. 56.

42 It was sold at Christie's in London in April 17, 2007; see www.christies.com/lotfinder/lot/a-napoleonic-brass-sundial-made-by-nicholas-4892277-details.aspx (accessed 17 April 2019). I am grateful to Denis Schneider for this information.

on its border. The maker's name is written in Arabic in fine lettering, engraved on the copper and decorated with silver inlay (*tanāzil*).[43]

3.4 *Measuring Levels on the Isthmus of Suez and Measuring the Great Pyramid*

In a hostile environment, the taking of levels on the Isthmus of Suez had to be divided into four successive operations, one in February 1799, and then three from October to December. The surveyors not only measured levels between Suez on the Red Sea and Peluse on the Mediterranean Sea, but also between Mouqfar, a central station on the Isthmus, and the Nilometer (*Mekyas*, a structure for measuring the depth of the river) on the Nile island of Rhoda, near Cairo. For the levelling along the 22 leagues between the Isthmus and the *Mekyas* the young civil engineer Jean-Baptiste Simon Fèvre (1775–1850) had a short water level, plus a bubble level between the 472nd and 478th stations only.[44]

There were many sources of error caused either by the instruments or by human error in observing, operating and reading the sights, calculating the heights and transferring the results onto the map.[45] To minimise the errors four engineers performed the levelling between each pair of stations. There was one at each end with a surveyor's pole and one in the middle using a bubble level with two telescopes pointing in opposite directions, who checked his readings by reversing the telescopes. The zero point of the height measurements was spring tide high water level at Suez. The fourth engineer made a straightforward topographical survey with a graphometer. To minimise the distortions caused by refraction and by water vapour in the atmosphere, the stations were mostly only 300 to 400 metres apart, but up to 600 metres apart during the first three and last two hours of the day, which were the best times for observations.[46]

As observations from balloons required lightweight instruments, Conté invented three hypsometric barometers for measuring heights more accurately than the common method that relied on reading the height of an ordinary

---

43 The three quotations on the French sundials are translated from 'Abd-al-Rahmān al-Jabartī, *Journal d'un notable du Caire durant l'expédition française, 1798–1801* (ed. by Joseph M. Cuoq), Albin Michel, Paris, 1979, pp. 134–135. In fact, the names of the astronomer Nouet and the carver Fouquet are written in French. Only the date and the explanation of how to read the time has been translated into Arabic.

44 Le Père, op. cit. (n. 3), p. 43. Lastly, they also surveyed from the *Mekyas* to the northeastern angle of the Great Pyramid.

45 *Ibid.*, pp. 52–53.

46 *Ibid.*

barometer by eye. First he tried to measure the time necessary to fill an evacuated vessel with any liquid at two different altitudes. Then he built the first aneroid barometer, based on what was later called 'Conté's principle'.[47] Not satisfied with its accuracy, he decided to weigh the quantity of mercury that moved as the barometer was taken to a different altitude, in an instrument later known as '*baromètre-balance*'. He presented it to the *Institut national* on 31 March 1798 and participated in a test the next month, just before leaving Paris for Toulon. After further tests, on 30 May, while Conté was sailing to Malta, Louis Lefèvre-Gineau, a professor at the *Collège de France,* and the reporter of a commission of the *Institut national des sciences et arts,* gave a positive assessment of the barometer to the Institute.[48]

Sailing back to France in the Autumn of 1801, Conté listed the inventions to be presented to the *Institut national.* About his barometer, he said "With the instrument that I had shown to this assembly I have measured the height of the Great Pyramid of Giza".[49] His result was slightly more accurate than the trigonometric measurements made by Nouet and the straightforward additions of the heights of the pyramid steps made in December 1799 by artillery officers, and again in January 1801 by the directors of the Giza excavations, Colonel Jean-Marie-Joseph Coutelle (1748–1835) and the architect Jean-Baptiste Lepère (1761–1844), who had designed a special apparatus and had it made at the Cairo mechanical workshops, where scientific instruments were repaired and made.[50]

---

47 L.C. [Louis Costaz], "Mémoire sur un nouveau Baromètre au moyen duquel on mesure immédiatement les changemens de densité de l'air, par le poids de mercure, par le citoyen Conté", *Bulletin des sciences par la Société philomathique* 2, 14 (Apr.–May 1798), pp. 106–107 plus one pl.

48 Other experts at the Panthéon and Notre-Dame cathedral were Guyton de Morveau, Alexandre-Henri Tessier, Jean-Baptiste Leroy and Charles.

49 Baron Thenard family archives, CIb/57, no. 3. Conté actually climbed the three pyramids there and found the height of the Pyramid of Cheops to be 427 feet; Conté to his colleagues at the Conservatoire, Alexandria, 5 May 1800, Musée des arts et métiers, Paris, CNAM 10.38. Jomard gives wrong and contradictory information about Conté's result.

50 On Coutelle, see P. Bret, "Le physicien, la pyramide et l'obélisque: problèmes d'archéologie monumentale selon Coutelle", in P. Bret (ed.), *L'expédition d'Égypte, une entreprise des Lumières (1798–1801)*, Tec & Doc Lavoisier, Paris, 1999, pp. 129–156. On Lepère, see Françoise Labrique, Uwe Westfehling, *Mit Napoleon in Ägypten. Die Zeichnungen des Jean-Baptiste Lepère,* Philipp von Zabern, Mayenz; Wallraf-Richardtz-Museum & Fondation Corboud, Cologne, 2009; Bret, "L'expérience préalable …", cit. (n. 27).

4 Repairing, Making and Designing Scientific Instruments: Conté's Mechanical Workshops in Cairo

The accident to Berthoud watch No. 34 in Rosetta was not the first damage to the instruments. In Alexandria, "after a couple of weeks, [the] instruments were spoiled by rust"[51] and Conté devised a process to protect them (and the muskets) with a varnish.[52] In late October he had far more to do after the plundering of the instruments recently stored at Caffarelli's house. Al-Jabartī wrote:

> In that house there was a large number of precision instruments, wondrous telescopes, astronomical devices, measuring devices for the engineering and mathematical sciences, etc. It goes beyond what we can imagine! As to the value of the items, only the ones who built them and knew their use were able to estimate it. Alas! It was dispersed and broken by the crowd. The French were quite affected by this disaster. They spent lot of time looking for scattered pieces, and highly praised those who brought one back.[53]

As Bonaparte asked worriedly: "What shall we do now? We do not even have tools". Conté replied: "Well, we'll make the tools".[54] In fact, he had already started as soon he arrived in Cairo in September: "I have done the experiments and the designs for all that myself", he wrote. "Moreover, I have been obliged to design most of those devices in a novel way, since I did not have the materials needed to make them in the European manner".[55] Claude-Pierre Molard (1759–1837) had provided the *Commission des sciences et arts* with more than 200 machines and tools which were probably transferred to the mechanical workshops before the riot, thus escaping the crowd who were not able to enter. In early October Beauchamp told Lenoir Jr. that he wanted to sell various tools to the mechanical workshops that Conté was setting up.[56] From then onwards, despite the isolation of the colony after the defeat of the fleet off Aboukir in August 1798 and further continuing difficulties created by the war, those

51 Quénot, op. cit. (n. 25), p. 151.

52 N.-J. Conté, "Procédé pour empêcher les effets de la rouille et l'oxidation des métaux", *Bulletin de la Société d'encouragement pour l'industrie nationale* 1, 9 (Apr.–May 1803), pp. 74–75.

53 Translated from 'Abd-al-Raḥmān al-Jabartī, *Journal* ..., cit. (n. 43), p. 84.

54 Quoted by E.-F. Jomard, *Nicolas-Jacques Conté*, Imp. Thunot, Paris, 1849, p. 34.

55 Conté to his wife, Cairo, 22 September 1798; Baron Thenard family archives, CIb/20.

56 Caffarelli to Conté, Cairo, 9 October 1798: SHD, GR, B6* 95, p. 42.

workshops made tools, machines and sophisticated instruments. Once more, al-Jabartī provides a depiction of them:

> Buildings were also fitted out for carpenters, for craftsmen making instruments, windmills and carriages, for carpentry and everything required for major works and structures, and for all other craftsmen. There was another building exclusively for the smiths. Enormous forges had been set up there, with impressive bellows from which a continuous powerful draught was drawn by a slight movement of the hand. They had made anvils and large hammers to make iron instruments, as well as lathes. They also had large lathes for turning metal blocks: workers turned large wheels while the master turner worked with hardened cutting tools, and water dropped from a box with a hole in it onto the part worked by the tool so as to reduce the heating produced by friction. A little further away were the precision mechanics specializing in compasses, horology or measuring tools for engineers, etc.[57]

4.1 *The Organisation of the Workshops*[58]

In a report to the *Institut d'Egypte* Le Père praised "the orderliness and competence of Conté's workshops".[59]

The two largest workshops dealt with heavy machinery for all the French industrial plants and mills. Under Pierre-Onézime Adnès (1760–1819), the first dealt with ironwork. Wooden devices were built under Aimé, in a workshop equipped for carpentry and woodwork. Jean Collin—a young workshop manager from the *École nationale aérostatique*—was in charge of woodturning and machines for that work.

Other workshops were more specialized. Three others were devoted mainly to the needs of the scientists and engineers of the *Commission des sciences et arts*: those of Lenoir Jr. for astronomical instruments, Cirot for geographical and topographical instruments and François-Michel Hérault (d. 1800), for "precision, geometry, horology and precious metal work". A fourth workshop supplied "special weapons and other objects of that kind" under a man called Couvreur, who had been a chief instructor armourer at the national weapons factory in Versailles and later became the armourer of the Mamluks of the Guard (1802–1804). Lastly, the former artist of the *École polytechnique*

57 Translated from 'Abd-al-Raḥmān al-Jabartī, *Journal* ..., cit. (n. 43), p. 94.

58 Bret, "Les oubliés ...", cit. (n. 17); and Bret, "Conté et 'les arts de la France ...", cit. (n. 6).

59 "Institut", *La Décade égyptienne, Journal littéraire et d'économie politique* 1 (1799), pp. 219–222: 221.

Pierre-Denis Fouquet (1767–1847) was responsible for carving, and the military balloonist Hochu for line engraving.[60]

4.2 *Making and Repairing Instruments*

Although by September 1798 Le Père had already designed a special apparatus for measuring the diameter at different heights of Pompey's Pillar in Alexandria,[61] most of the common instruments were made in the mechanical workshops in Cairo.

Before leaving for Syria in early February Caffarelli asked Conté to provide the civil engineer Girard with a surveying compass, a plane table and a level for his journey to Upper Egypt, along with twenty dividers, rules and squares for the French school, and to make and send as soon as possible compasses, standard metres and measuring rods, and dividers.[62] As well as many kinds of industrial machinery, weapons and other products for the army and the colony (windmills, a gunpowder mill, swords, trumpets, syringes, cloth), the workshops had to make not only such instruments as a precision balance for the Cairo Mint and a probe for the Giza archaeological digs, but also the tools required to make these things (vices, files, chisels, planes, hammers, punches) and sometimes the raw materials too (steel, tinplate, common glass and flint-glass, cardboard, varnished canvas).

4.3 *Conté's Inventions in Egypt*

The French military ballooning equipment was developed and perfected by 1796, mainly by Conté, in continuous consultation with the chemist and balloonist Guyton de Morveau about political questions, and the chemist and military balloonist Coutelle in the field. Conté invented several devices related to his ballooning system, two of which were taken to Egypt. He improved the handling of his barometer in Egypt after using it at Giza, with the aim of presenting it at the *Institut national* again at a later date.[63] On the other hand, his "ingenious application of the expansion of metals to temperature measurement"[64]

60 "Notice sur l'attelier de méchanique établi au Kaire sur la direction du citoyen Conté, chef de brigade des Aérostiers et membre de l'Institut d'Egypte", *Courrier de l'Égypte* 40 (12 Oct. 1799), pp. 2–3: 2.

61 Charles Norry, *Relation de l'Expédition d'Egypte, suivie de la description de plusieurs monumens de cette contrée*, Pougens, Paris, 1799, pp. 60–64.

62 Caffarelli to Liédot, 11 September 1798: Paris, Service historique de la défense. Département de l'armée de terre (SHD/DAT) B6* 95.

63 Conté to his colleagues at the Conservatoire, Alexandria, 5 May 1800: Paris, Musée des arts et métiers, CNAM 10.38.

64 Guyton de Morveau, "Essai de pyrométrie …", *Mémoires de la classe des sciences mathématiques et physiques de l'Institut de France*, Baudouin, Imprimeur de l'Institut, Paris, 2nd semester 1808 (published 1809), pp. 1–35: 26–27.

in the first metallic high-temperature pyrometer (with a 50-cm dial), instead of the Wedgwood clay pyrometer which was not suited to the production of hydrogen, was never used in Egypt. Actually, the furnaces could not be built because many of the fire bricks had been broken while being unloaded from the transport ship *Le Patriote*, and fuel was lacking too. And the special varnish that Conté had used successfully in Meudon to keep hydrogen inside the silk envelope of the balloon was not suitable for the Egyptian climate.

In Cairo Conté invented an apparatus to measure time in milliseconds, in order to estimate the quality of gunpowder from the speed that it burned. As with his *baromètre-balance*, he decided to weigh the mercury flowing from an upper vessel to a lower vessel while a line of gunpowder was burning. As two French pounds or 18,432 grains (979 g) of mercury flowed down in one second the theoretical accuracy was 1/18,432 of a second. Conté read his memoir about this at the *Institut d'Égypte* on 12 November 1799, and *La Décade égyptienne* briefly described the apparatus:

> By measuring the weight of the mercury that has flowed out [of the upper vessel], the duration of the flow will be determined with a precision that one could hardly have been expected and cannot be obtained today by any other means. Citizen *Conté* applied this ingenious invention to measuring the flammability of [French] gunpowder [made in Cairo]. The experiment was performed in the public session of the Institute [in Cairo].[65]

Conté later increased the theoretical precision up to 1/36,864 of a second, as he stated in the list of his inventions that he made on the ship back to France.[66]

None of the numerous other instruments on that list had definitely been invented in Egypt and most of the inventions made or improved there were machines (especially hydraulic machines) and processes rather than scientific instruments, although some were useful for instrument making, such as his

65 "Précis des séances et de travaux de l'Institut d'Egypte …", *La Décade égyptienne, Journal littéraire et d'économie politique* 3 (1800), pp. 292–306: 299. The memoir has been published by P. Bret, "Contrôle de qualité, expertise et recherche expérimentale à la fin du XVIIIe siècle: de l'éprouvette d'ordonnance à la mesure de l'inflammabilité", in *Instrumentation, expérimentation et expertise des matériaux énergétiques (poudres, explosifs et pyrotechnie) du XVIe siècle à nos jours* (*Proceedings of the Troisièmes Journées Paul Vieille*, Paris, October 2000), CRHST-A3P-CEA/DAM, Paris, 2001, pp. 29–59, and P. Bret, "Gunpowder Manufacturing in Cairo from Bonaparte to Muhammad 'Alî: Adaptation, Innovation and Transfer of Technology (1798–1820)", in Brenda J. Buchanan (ed.), *Gunpowder, Explosives and the State: A Technological History*, Ashgate, Aldershot, 2006, pp. 206–238.

66 Baron Thenard family archives, CIb/57.

machine for "making the surfaces of the hardest materials plane" and processes for making tinplate and flint glass, unfortunately not described.[67]

4.4 *Telegraphs and Telescopes*

Conté also made telegraphs in Egypt. Despite what many historians have said he did not use Chappe's system, used in France since 1794, but developed several original systems.[68] In 1796 he designed and tried out a new kind of aerial telegraph after joint experiments by the *École polytechnique* and the *École nationale aérostatique* at Meudon.[69] On 24 November, he also participated as an expert with Prony at trials of Breguet and Betancourt's first telegraphic system. Immediately after he landed in Alexandria in July 1798 Conté planned a telegraphic line from Alexandria to Aboukir, Rosetta and Damanhour, and later to Cairo.[70] After a first trial General Kléber ordered the building of the first five stations. Because of bureaucratic delays they were still being built when Nelson's fleet defeated the French fleet in Aboukir Bay on 1 August.[71]

When General Menou decided to set up a colony, Conté started to build the crucial line between Alexandria and Cairo with a new type of telegraphic instrument, similar to but simpler than the second system of Breguet and Betancourt. In July 1800 Desgenettes commented:

> One could hardly imagine how optical instruments might have been supplied in a country where their manufacture and the necessary materials are unknown if we did not say that those telegraphs were made in the mechanical workshops managed by Colonel Conté, whose genius never fights against the greatest difficulties without overcoming them.[72]

---

67 *Ibid.*, nos.

68 P. Bret, "L'origine du télégraphe aérien en Afrique: Conté et l'établissement des télégraphes en Égypte durant l'expédition française (1798–1801)", unpublished paper given at the 11th International Conference on the History of Posts and Telecommunications (FNARH), Saverne, 28–30 May 1998.

69 P. Bret, "Les épreuves aérostatiques de l'École polytechnique en l'an IV: de la géométrie descriptive à l'origine de l'École des Géographes", *Sciences et Techniques en Perspective* 19 (1990–1991), pp. 119–142; "Aerostatic Institute" (trans. from Friedrich Johann Lorenz Meyer's German account), *Monthly Magazine* 6 (Nov. 1798), pp. 336–339.

70 Kléber to Bonaparte, Alexandria, 10 July 1798; in Laurens, *Kléber en Egypte* ..., cit. (n. 32) v. 1, p. 114.

71 Kléber to Bonaparte, Alexandria, 17 July 1798; *ibid.*, p. 133; Conté to Kléber, Alexandria, 21 July 1798, sold at auction by Arcole, Drouot, Paris, 12–13 February 1990, no. 227.

72 "Etablissment des Télégraphes en Egypte", *Courrier de l'Égypte* 74 (16 Jul. 1800), p. 4.

One month later the telegraphs were tested at the Citadel and five weeks later Menou wrote that Conté and his balloonists "had made sixty excellent telescopes".[73] Telegraphic instruments were made mainly of wood. Optical instruments were more of a challenge. A long time afterwards the chemist Chaptal recalled a crucial problem: "Telescopes were lacking. He [Conté] produced flint glass and made excellent ones".[74] Nothing is known about that flint glass but in November 1800 Conté wrote that he had made "quite a large quantity of astronomical telescopes, and others for the telegraph".[75] The only remaining workbook of the mechanical workshops includes the statement "five telescopes started" in February and March 1801.[76] On 20 January 1801 Conté announced that he had "a telegraphic line built to connect Alexandria and Cairo. 17 telegraphs are nearly complete and will be put in place soon".[77] It is doubtful that was ever done because the British forces landed at Aboukir only five weeks later, and besieged Alexandria in March. Although no archival evidence has yet been found to show that Conté's telegraph actually functioned, his work impressed the Ottomans, and Seyyid Abdürrahim Muhibb Efendi wrote in his short account of his embassy to Paris (1806–1811):

> The telegraph is used in the countries of England and France, and it does not exist in other states. When the French went to Egypt they brought some of the smallest of them, and it is said that it is how they communicated with Alexandria.[78]

5 Conclusion

The unpreceded—both in quantity and quality—range of scientific instruments that accompanied Bonaparte's conquest of Egypt in 1798 and the three following years of military occupation and novel colonial experiment must be put in the broader context of the Franco-Ottoman scientific relationship

73 Conté to Sanson, Cairo, 14 August 1800: SHD, GR, B6 50; Menou to the Minister of War, Cairo, 23 September 1800.

74 Jean-Antoine Chaptal, *De l'industrie française* (ed. by Louis Bergeron), Imprimerie nationale Editions, Paris, 1993, pp. 300–302; originally published as J.-A. Chaptal, *De l'industrie françoise*, chez Antoine-Augustin Renouard, de l'imprimerie de Crapelet, Paris, 1819.

75 Baron Thenard family archives, CIb/36, 10 November 1800.

76 "Armée d'Orient—Atteliers de mécanique. Etat des ouvrages du mois confectionnés aux atteliers de mécanique pendant le mois de ventose an 9e", Private collection.

77 Baron Thenard family archives, CIb/39, 20 January 1801.

78 Translated from Stéphane Yerasimos (ed.), *Deux Ottomans à Paris sous le Directoire et l'Empire*, Sindbad /Actes Sud, Arles, 1998, p. 195.

and compared with the European scientific technologies introduced into the Ottoman Empire during the last two decades of the eighteenth century.

On the one hand obviously—above all when we consider the aggressive and unexpected French invasion of an Ottoman province—the huge amount and colonial use of Napoleonic scientific instruments must be seen as a break with previous practice which developed in the context of military cooperation.

On the other hand—above all if, despite the failure of the planned diplomatic initiative,[79] we accept both Bonaparte's official explanation that he went to Egypt in order to expel the Mamluks and "protect" the country on behalf of the Sultan—the instruments can be seen as an aspect of the continuation of the previous military missions sent to Istanbul for the modernisation of the Ottoman Empire, so as to help it against Russian expansion towards the Black and Mediterranean seas.[80] The main result of those friendly missions in the context of a long-lasting alliance had been the creation of engineering schools in the period between the Russo-Ottoman treaty of 1774 and the French conquest of Egypt of 1798.[81] Besides teaching, at that time the scientific instruments were mainly used for mapping and fortification planning, plus shipbuilding.[82]

Finally, it was both a continuation of and a break with local smaller collections of instruments in French consulates and other venues with more scientific purposes, such as Beauchamp's observatory in Baghdad. In that case

79 The French minister of Foreign Affairs Talleyrand was expected to sail to Istanbul and inform Sultan Selim III, but he stayed in Paris.

80 Frédéric Hitzel, "La France et la modernisation de l'Empire ottoman à la fin du XVIIIe siècle", in P. Bret (ed.), *L'expédition d'Égypte, une entreprise des Lumières, 1798–1801*, Académie des sciences—Tec & Doc Lavoisier, Paris, 1999, pp. 9–19.

81 F. Hitzel, "Les écoles de mathématiques turques et l'aide française (1775–1798)", in Daniel Panzac (ed.), *Histoire économique et sociale de l'Empire ottoman et de la Turquie (1326–1960)*, « *Collection Turcica* 8 », Peeters, Paris, 1995, pp. 813–825; Mustafa Kaçar, "Osmanlı Devleti'nde Mühendishanelerin Kuruluşu ve Bilim ve Eğitim Hayatındaki Değişmeler [The foundation of Engineering Schools and the changes of educational and scientific systems in the Ottoman Empire]", PhD thesis, Institute of Social Sciences, University of Istanbul, 1996; F. Hitzel, "L'introduction des techniques occidentales et la naissance des premiers ingénieurs turcs", in Irina Gouzévitch, André Grelon, Anousheh Karvar (eds.), *La formation des ingénieurs en perspective. Modèles de référence et réseaux de médiation, XVIIIe–XXe siècles*, Presses universitaires de Rennes, Rennes, 2004, pp. 83–93.

82 On the cartographic work, see Jean-François Pérouse, Feza Günergun (eds.), *Entre Trois Mers—Cartographie ottomane et française des Dardanelles et du Bosphore/Üç Denizin Arasında—Osmanlı ve Fransız Boğaz Haritaları*, Arkas Sanat Merkezi, Izmir, 2016; F. Hitzel, "François Kauffer (1751?–1801), ingénieur-cartographe français au service de Selim III", in E. İhsanoğlu, F. Günergun (eds.), *Science in Islamic Civilisation*, Research Centre for Islamic History, Art and Culture, Istanbul, 2000, pp. 233–243.

sources are lacking to know better how far Turkish and Arab notables might have been invited to visit, watch and use them. In Cairo on the contrary, despite the October 1798 uprising, the scientific instruments were duly shown to the notables, and experiments were even demonstrated. From autumn 1798 to winter 1801 the French had close contacts with Egyptian savants, especially two of them. One was the previously-mentioned astronomer and chronicler Sheikh al-Jabartī, who was also a member of the last Divan (*Dīwān*) installed by General Menou (October 1800–June 1801). The other was his friend Sheikh Hasan al-ʿAttār al-Misrī (1766–1835), whose links with the French led him to run away into Upper Egypt during the second Cairo uprising in March 1800, and not return until years after the French capitulation. Later, as a professor at the University of al-Azhar, he was a strong influence on his students in promoting the renewal of science in Egypt. He chose one of them, Rifāʿa al-Tahtāwī (1801–1873), to be the imam of the mission of Egyptian scholars sent to Paris from 1826 to 1831, under Jomard's direction. Al-ʿAttār's teaching had prepared his student well to understand French science, technology and society better, so that al-Tahtāwī carefully described and commented on it in Arabic in his *Takhlīs al-ibrīz fi talkhīs Bārīz* (The fineness of Gold: a synopsis of Paris; Bulaq, 1834) which he translated into Turkish (*Sefaretname-yi Rifa'at Bey*) in 1839, the very first year of the *Tanzimat* which actually began the process of social and scientific change in the Ottoman Empire.[83]

Thus the Western scientific instruments brought by Napoleon's scientists—as well as other technological items—indirectly assisted in increasing the consciousness of their importance in society, thus contributing to the rise of the Egyptian *Nahda* (Renaissance) and a modernizing spirit in the Near East, which had first slowly begun around the centre of power in Istanbul.

83 French translation by Anouar Louca, *L'Or de Paris*, Sindbad—Actes Sud, Paris, 1988; English translation by Daniel A. Newman, *An Imam in Paris. Al-Tahtawi's Visit to France (1826–1831)*, Saqi, London, 2002; re-edited as *An Imam in Paris. Account of a Stay in France by an Egyptian Cleric (1826–1831)*, Saqi, London, 2011.

CHAPTER 13

The Magic Lantern as an Ambassador between Cultures and Religions

Imrich Emanuel Roth and the First Dissolving View Shows in the Ottoman Empire, 1845–1846

Bernd Scholze

1 Magic Lanterns, Limelight, and the First Dissolving View Shows

In the mid-eighteenth century travelling showmen with scientific equipment, in particular for demonstrating electricity and magnetism, created widespread interest in natural science amongst urban populations. The magic lantern was more a part of their exhibition of philosophical instruments for sale than for use in real demonstrations.[1] Until the start of the Seven Years' War (1756–1763) communities in the countryside might see a magic lantern brought by a Savoyard, followed at the beginning of the nineteenth century by travelling opticians, usually members of the many Jewish communities, selling magic lanterns in small villages in Southern Germany as the only toy amongst their spectacles.[2] In comparison with other philosophical instruments like the microscope or the telescope, towards the middle of the nineteenth century the magic lantern was a familiar instrument in all parts of the country amongst people of all social classes. The poor light of the lantern did not allow demonstrations to large audiences.

The limelight was invented in England in the 1820s. Starting from the end of the year 1832 it was adopted by Cooper and Cary in their London showroom at No. 287 Strand for projecting microscope images.[3] This marked the start of a new era for the magic lantern. From now on a projected image was bright enough to be shown at a staggering size on a screen to entertain hundreds if

1 E.g. philosophical instrument directories by Peter Joseph Reuter, *Das Verzeichnis von den vollkommensten Erfindungen* …, [no printer and place], ca. 1760; Jacob v. Bianchy, *Auserlesenes Cabinet Physicalischer und Mathematischer Instrumenten, zum Besten der Liebhaber dieser Wissenschaften zusammen getragen*, Johann Thomas Trattnern, Vienna, 1763, pp. 59–60; Johann Conrad Gütle, *Verzeichnis eines Physikalisch und Mathematischen Maschinen Instrumenten und Experimenten Kabinets* …, [no printer and place], 1780, p. 52.

2 Heinrich Buurman, *Optiker und Oculisten in Ostfriesland*, Verlag Sollermann, Leer, 1994.

3 *The Tourist*, London, No. 22, 8 Jan. 1833.

 | DOI:10.1163/9789004412842_014

FIGURE 13.1 Typical professional dissolving view lanterns such as Roth could have used
AUTHOR'S COLLECTION

not thousands of spectators at the same time. A new public entertainment appeared on the already oversupplied market of public shows in London, the dissolving view show. Two magic lanterns, powered by the oxy-hydrogen limelight, stood side by side next to each other, slightly angled so that the two slides mounted into the lanterns would be shown in the same place on the screen. By opening one shutter and closing the other in front of the lenses, one image faded in as the other image faded out on the screen (*Fig. 13.1*). The images dissolved into each other.

The heyday of dissolving views started at Easter 1841 when the famous Royal Polytechnic Institute in London presented them for the first time, in a special room within the building.[4] Despite the unbelievable beauty of projected slides, accompanied by music, and the great success with audiences which lasted for

4 Jeremy Brooker, *The Temple of Minerva. Magic and the Magic Lantern at the Royal Polytechnic Institution, London 1837–1901*, The Magic Lantern Society, London, 2013, pp. 52–59.

decades, it took nearly three years for the first dissolving view shows to cross the Channel to mainland Europe.

2 The First Dissolving View Shows in Europe

At this time, the Viennese Ludwig Döbler (1801–1864) was one of the most celebrated magicians in Europe. The owner of the St James's Theatre in London hired Döbler for a number of performances in the summer season of 1842. When Döbler arrived in London one of his first visits was to the Royal Polytechnic Institute to see a dissolving view show.[5] Döbler, a curious man always looking for new attractions for his own shows who had already introduced limelight microscope projection into Germany in the Spring of 1836,[6] immediately understood the value of the new attraction for his own business. As a man of action, with a generous budget, Döbler commissioned a dissolving view device and slides from the optician of the Royal Polytechnic Institute to be collected at the end of his contract some weeks later. When he travelled home to Vienna Döbler carried a dissolving view device with him, but no slides as the slide painter did not finish the job in time.[7]

Back in Vienna, Döbler was in desperate need of a glass-painter, a profession which had almost become extinct in his home town, but he found one in Mr. Havliczek.[8] He had been taught glass painting in a style which already looked old-fashioned in the early 1840s.[9] However, Döbler had little choice and did not pay too much attention to this, but the newspaper journalists of the time did. Döbler gave the first dissolving view show on mainland Europe on 18 March 1843 in his Josephstädter theatre,[10] with additional shows until 4 May. The slides were accompanied by music specially composed for the theme of the slides by Anton Emil Titl (1809–1882).[11] According to contemporaries, the quality of the painting of the slides was average and did not match

5 Werner H.A. Debler, *Leopold Ludwig Döbler 1801–1864*, Einhorn-Verlag, Schwäbisch Gmünd, 2001, p. 248.

6 *Die Biene. Fränkisches Unterhaltungs-Blatt*, Bamberg, No. 94, 26 Nov. 1835.

7 Debler, op. cit. (n. 5), p. 252.

8 *Allgemeine Musik-Zeitung*, Vienna, No. 141, 25 Nov. 1845.

9 Walter Spiegl, *Die Fenster in Laxenburg und am Brandhof*, www.glas-forschung.info/pageone/pdf/lax.pdf (accessed 22 March 2019).

10 *Der Humorist*, Vienna, No. 56, 20 Mar. 1843.

11 *Allgemeine Musik-Zeitung*, op. cit. (n. 8).

the expectations of an experienced audience.[12] The criticism did not stop Döbler undertaking an extensive tour in the following months.

Döbler travelled to Budapest at the end of October 1843,[13] and gave fifteen shows in the National Theatre. The theatre provided enough space for 1,000 spectators and usually the shows were sold out, so about 15,000 people saw Döbler's shows out of a total population of 100,000 people at this time. Why did the dissolving view shows immediately have such great success? The images—various country and city scenes—dissolved one into another without any connecting storyline. Each image stayed on the screen for a few minutes and was accompanied by a musical piece from the theatre orchestra. The combination of a projected image and classical music was a beautiful effect that nobody could resist. The lack of a storyline gave spectators the unconscious freedom for their own fantasies.

3 Imrich Emanuel Roth

Imrich Emanuel Roth (1814–1885) from Košice (Slovakia), at this time part of the Austro-Hungarian Empire, was almost certainly one of the spectators at Döbler's shows in Budapest. Roth was an academic painter who had studied art in Vienna, Düsseldorf and Munich.[14] His mother operated a kosher restaurant in Košice and, according to an 1831 inventory of the residents, they were one of the few Jewish families in the town.[15] Imrich Emanuel had three brothers, one of whom may be known to historians of scientific instruments. David Roth was the inventor of a pin-wheel calculating machine, for which he was granted a French patent in 1840. The Musée des Arts et Métiers in Paris owns one of the few surviving machines.[16]

Roth moved to Budapest in 1840. Probably with the help of his brother David in Paris, Imrich Emanuel experimented with the daguerreotype at the end of 1843,[17] the same time as Döbler was giving his dissolving view shows

12 *Der Adler*, Vienna, No. 82, 10 Apr. 1843.

13 *Der Ungar*, Budapest, No. 247, 24 Oct. 1843.

14 I. Chalupecký, "Roth, Imrich Emanuel (1814–1885), Maler und Photograph", in *Österreichisches Biographisches Lexikon 1815–1950*, Verlag der Österreichischen Akademie der Wissenschaften, Vienna, 1986, v. 9, p. 277.

15 Yehoshua R. Bucher, Ruth Shachak (eds.), *Encyclopedia of Jewish Communities, Slovakia* (translation of *Pinkas hakehillot Slovakia*), Yad Vashem, Jerusalem, 2003, s.v. "Košice", pp. 495–505.

16 Judith Brody, "An émigré physician: Dr. David (Didier) Roth, Homeopath, Art Collector, and Inventor of Calculating Machines", *Journal of Medical Biography* 8 (2000), pp. 215–219.

17 *Der Ungar*, Budapest, No. 215, 16 Sep. 1843.

in the city. With his expert eye as a painter Roth must have been very disillusioned by the quality of the painted slides. This must have been the moment when Roth had the idea of presenting dissolving views of much higher quality. He returned to his home town only to come back a few months later with a complete device to show dissolving views. He obviously painted the slides himself,[18] including the following scenes: *The stock exchange in Paris by day and by night, A chalet in Appenzell canton in winter and summer* and *The great London bridge*;[19] *Holyrood chapel in Scotland, by day and by night, The Munich Isartor in moonlight* and *The harbour of la Rochelle*;[20] as well as a slide of *The Two Towers in Bologna* and the *Hall of the Palais de Justice in Paris* (*Figs. 13.2*).[21] The majority of the scenes are not known as standard images of London dissolving view makers. The origin of his lantern is not known. His first trials in front of connoisseurs of the art took place in his home town Košice, followed by the first public shows there.[22] After that his intention seems to have been to travel to Budapest and then to his final destination, Constantinople. At the beginning of the nineteenth century, the Orient was a desired travel destination among painters. Roth followed the lure of the East.

4 Roth's Travels in Europe

On his way he stopped in Prešov (Slovakia),[23] Bratislava[24] and Tata (Hungary)[25] for shows before he arrived in Budapest in May 1844. His debut there was on 20 May 1844 at the Royal Theatre.[26] The local press referred very enthusiastically to his shows: "During the third performance, the audience clapped their hands continuously and expressed their very obvious astonishment and satisfaction with the show".[27] According to the press, Roth showed sophisticated slides, outstanding in their pictorial language and their paint quality. In particular, the light and colour of the night and snow slides showed an enchanting effect.[28]

18 *Ibid.*, No. 122, 28 May 1844.
19 *Der Siebenbürger Bote*, Sibiu, No. 70, 3 Sep. 1844.
20 *Ibid.*, No. 73, 13 Sep. 1844.
21 *Der Ungar*, Budapest, No. 146, 26 Jun. 1844.
22 *Der Humorist*, Vienna, No. 25, 29 Jan. 1844.
23 Chalupecký, op. cit. (n. 14), p. 277.
24 *Der Ungar*, Budapest, No. 83, 10 Apr. 1844.
25 *Gemeinnützige Blätter zur Belehrung und Unterhaltung*, Budapest, No. 36, 5 May 1844.
26 L. Wolff (ed.), *Almanach für Freunde der Schauspielkunst auf das Jahr 1844*, Julius Gittenfeld, Berlin, 1845, p. 451.
27 *Der Ungar*, Budapest, No. 122, 28 May 1844.
28 *Ibid.*

FIGURE 13.2 Example of a fine hand-painted dissolving view set, *Spectre of the Brocken*, produced by the Royal Polytechnic Institution, London, 1858
AUTHOR'S COLLECTION

The minor role of the magic lantern as a physical instrument was replaced by its ability to unite all classes of society during a performance.

Budapest was only one stop in his travel plans to a series of places where dissolving views were unknown. Constantinople was his ultimate destination. His success so far opened doors to the high-profile and profitable theatres in the larger towns. He stopped in Szeged,[29] Timișoara,[30] Mehadia and Sibiu.[31] In Sibiu, Roth needed to wait ten days until the theatre was available to him, so for 10 Kreutzer he exhibited the calculating machine made by his brother, which was apparently part of his luggage, in his guesthouse, Zum römischen Kaiser.[32] Roth left Sibiu early in September 1844 for Brasov, and then gave dissolving view shows in Bucharest over the winter. He set off in early spring 1845 for Iași, the capital city of the Principality of Moldavia.[33] After shows there in March 1845 he travelled further east and entered the Russian Empire for his next destination, Odessa.[34] Not much is known about his stay in Odessa. Most likely he gave dissolving view shows there, which would have been the first dissolving views in the Russian Empire. He took a passage on a ship bound for Constantinople where he arrived at the end of April or beginning of May 1845.

29 *Der Humorist*, Vienna, No. 88, 12 Apr. 1845.
30 *Der Ungar*, Budapest, No. 146, 26 Jun. 1844.
31 *Der Siebenbürger Bote*, Sibiu, No. 66, 20 Aug. 1844.
32 *Ibid.*, No. 68, 27 Aug. 1844.
33 *Der Humorist*, Vienna, No. 88, 12 Apr. 1845.
34 *Illustrirte Theaterzeitung*, Vienna, No. 147, 20 Jun. 1845.

5 Roth in Constantinople

Roth's arrival in the metropolis of the Ottoman Empire coincided with the wedding preparations of Adilé, the youngest sister of the Sultan Abdülmecid I (1823–1861), to Mehmed Ali Pasha, the Mushir of Tophane, who was the Marshal of the Imperial Arsenal of Ordnance and Artillery.[35] For the occasion of the wedding, the broad-minded and musically-knowledgeable Sultan built a small theatre within the walls of the Topkapı Palace[36] to watch an Italian opera for the first time in his life.[37]

News of the sensational success of Roth's dissolving views had preceded him and Count Bartholomäus von Stürmer, a well-connected man in Constantinople who acted as an unofficial ambassador for the Habsburg Monarchy, helped his fellow-countryman Roth to take part in the wedding ceremonies with a dissolving view presentation in the theatre for the Sultan. Roth had the privilege and honour of giving the first presentation in the Ottoman Empire, which took place on 20 May 1845.[38] The French-language paper *Journal de Constantinople* summarized the evening:

> Roth had shown the dissolving views in presence of the Sultan and high officials. The Sultan was so pleased that he requested another show on the next evening, at which the Valide Sultan [the Sultan's mother] together with the sultanesses [concubines and wives of the Sultan] of the Imperial Harem [part of the Topkapı Palace] should be present. Roth received warm acknowledgement from the audience and the Sultan.[39]

The invitation to the Imperial Harem was not a matter of course: there was no trespassing beyond the gates of the Harem for foreigners. The Constantinople-based newspaper correspondent Alexander Jacobi wrote about this evening, that the dissolving view production in front of the Harem was an exceptional rarity.[40] It is not known whether Roth later painted from memory scenes from the Harem (*Fig. 13.3*).

What made this inconspicuous physical instrument, the magic lantern, so attractive? It was the beauty of the hand-painted slides and the dissolving effect on the screen which simply touched the senses of the spectators.

35 *Deutsche Allgemeine Zeitung*, Leipzig, No. 111, 21 Apr. 1845.
36 *Illustrirte Theaterzeitung*, Vienna, No. 133, 4 Jun. 1845.
37 *Frankfurter Oberpostamts-Zeitung*, Frankfurt, No. 188, 10 Jul. 1845.
38 *Beilage zur Frankfurter Oberpostamts-Zeitung*, Frankfurt, No. 188, 10 Jul. 1845.
39 *Illustrirte Theaterzeitung*, Vienna, No. 140, 12 Jun. 1845.
40 *Ibid.*, No. 147, 20 Jun. 1845.

FIGURE 13.3 Hand-painted slide *The sweet waters of Asia*, showing a scene in Istanbul, Turkey. English, 1850s
AUTHOR'S COLLECTION

Independently of religion, social class or ethnic background, when two images dissolved into one another on the screen, the audience became united, whether in Budapest, Bucharest, Odessa or Constantinople.

The favourable reception by the Sultan opened the door for future generations of dissolving view showmen to travel, with the permission of the public authorities, through the Ottoman Empire, such as the Hungarian magician Joseph Vanek (1818–1889) who in 1859 gave "Phantasmagoria and astronomic-geological presentations with oxy-hydrogen gas" in the theatre of the Dolmabahçe Palace in Constantinople,[41] or a Mr. Döring in 1885, also in Constantinople.[42]

Roth was rewarded for this magnificent presentation with splendid gifts from the Sultan, and as the news of his success spread in Constantinople many

41 *Der Sammler*, Vienna, No. 120, 1 Nov. 1859.

42 David Robinson, Stephen Herbert, Richard Crangle (eds.), *Encyclopaedia of the Magic Lantern*, The Magic Lantern Society, London, 2001, p. 93.

persons of rank approached him wanting to see his presentations.[43] Not much is known about Roth's activities for the following months. He fulfilled the requests of many rich residents. As a painter, he may also have used the time for making sketches. He tried to obtain commissions for painting portraits, but he felt he was held back because as a presenter of dissolving views he was generally speaking seen as an entertainer and therefore not a suitable person for the 'serious' art of painting.[44]

6 Roth in Alexandria and Cairo

Early in 1846 Roth left Constantinople for the Mediterranean harbour town of Alexandria, in Egypt, which had formerly been part of the Ottoman Empire. The foreign minister of the Ottoman Empire, Mustafa Reşid Pasha (1800–1858), gave Roth a letter of recommendation to his Egyptian colleague Artin Bey Shoukry.[45] Thus in 1846 Roth set foot on new ground in North Africa taking with him the dissolving view apparatus and his painting utensils.

The letter of recommendation opened doors for Roth into the circle of political power in Egypt. Traditionally the governor of Egypt spent the summer months away from the hot and dusty capital city of Cairo, in Alexandria on the Mediterranean Sea with its fresh breeze. Soon after Roth arrived, Artin Bey Shoukry introduced him to Prince Muhammad Saʿid Pasha (1822–1863), the son of Muhammad Ali Pasha of Egypt (1769–1849) who had been the Viceroy of Egypt when it was part of the Ottoman Empire, and who is regarded as the 'Father of Modern Egypt'. Roth spent a lot of time with the Prince in the newly completed palace of Ras El Tin; he was invited to the wedding of the first valet Saulfarar Efendi and worked as a portrait-painter. Within five weeks he portrayed the Prince and members of his family as well as members of the staff. In his rare moments of spare time he prepared one of the largest rooms in the palace for showing dissolving views. As a result of his hard work under the scorching sun, Roth suffered from the heat and sore eyes.[46] He showed his paintings to Muhammad Ali Pasha of Egypt for comment. After half an hour during the second audience Roth took the opportunity to ask him to be allowed to finish his preparations for the dissolving views and to show them later that day. After sunset on 17 May 1846, Roth projected his slides before "all the great and the

43 *Illustrirte Theaterzeitung*, Vienna, No. 147, 20 Jun. 1845.
44 *Kunstblatt XVI. Beilage zu den Sonntagsblättern*, Vienna, No. 31, 2 Aug. 1846.
45 *Kunstblatt XIII. Beilage zu den Sonntagsblättern*, Vienna, No. 25, 21 Jun. 1846.
46 *Ibid.*

FIGURE 13.4 Hand-painted slide with a winter scene. English, 1850s
AUTHOR'S COLLECTION

small of the court", with Muhammad Ali Pasha of Egypt in the centre. Roth was asked to explain every slide in detail and was sweating behind the lantern. The winter scenes made a particularly good impression in the hot climate (*Fig. 13.4*), as did images of Paris, because Muhammad Ali's son was staying in Paris at this time. When the last of thirty slides passed the slot in the dissolving view lantern, the Viceroy asked Roth whether the presentation had finished. It was received with exceptional acclaim. That night's dissolving view presentation by Roth is believed to have been the first on the African continent.[47]

Another presentation was given to Prince Muhammad Saʿid Pasha on 20 June 1846, when Roth asked the music ensemble of the navy to accompany the projected images. Two public presentations were given by command of the young Prince Mehmed Ali Bey, who printed programme sheets for this occasion on fine wove paper.[48] In the meantime, the Viceroy departed with his entourage for Constantinople.

---

47 *Kunstblatt XVI. Beilage zu den Sonntagsblättern*, Vienna, No. 31, 2 Aug. 1846.
48 *Kunstblatt XIX. Beilage zu den Sonntagsblättern*, Vienna, No. 37, 13 Sep. 1846.

A few days later Roth left Alexandria for Cairo. He hired a ship to sail on the Nile and painted sketches of the landscape. When he arrived in Cairo his health was not at its best, and probably for this reason he soon abandoned his plan to travel further to India.[49] Instead he decided to go back to his home town of Košice. It is not known what he did with his dissolving view equipment but it can be assumed that he left it behind in Cairo and sold it to a Mr. Kurth, who gave public dissolving view shows there at the end of the year 1846.[50]

Back in Košice, Roth founded the first photographic studio in his hometown.[51] He died in 1885 and was buried at the Kalvarienberg cemetery. The local press described him in an obituary as a fine, educated aesthete and the best-known personality in the town.[52] As far as is known neither his dissolving view lantern nor his slides are still in existence.

7 Conclusion

Roth was an early pioneer spreading dissolving views across the East and the West. Along with other presenters such as Julius Laschott or Johann Bosch, he laid the foundations of a new era for the magic lantern in society. However, an array of superb slides is only interesting for so long and soon started to be boring for the spectators. Interest in dissolving views developed quickly in society but was soon set aside in favour of a new interest, popular science. The showmen with their dissolving views turned into serious scientific lecturers, when talking about the history of the formation of the Earth or about astronomy, themes which superseded the collections of scenic slides and had no connection with them. The success story of the spread of dissolving views across Europe, Asia and Africa lasted until the end of the nineteenth century, when cinematography arrived.

---

49 *Ibid.*

50 *Die Gegenwart. Politisch-literarisches Tagsblatt*, Vienna, No. 16, 20 Jan. 1847.

51 Paul Tischler, "Geschichte der Zipser Fotografie: III. Gründler Fotografen", *Karpatenblatt* 18, 4 (Apr. 2009), p. 9.

52 *Kaschauer Zeitung. Kassa-Eperjesi Értesitö*, Košice, No. 112, 29 Sep. 1885.

CHAPTER 14

Scientific Instrument Retailers in Istanbul in the Nineteenth Century, and Verdoux's Optical Shop

Meltem Kocaman

1 Introduction

It is only recently that the material culture of science in the Ottoman Empire has become a subject of research. New studies concentrate predominantly on the history of scientific instruments. Their principal motivation, it can be argued, is to disentangle the history of Ottoman science from an intellectual history that is founded on bookish knowledge only. The history of scientific instruments can be studied from various perspectives. This chapter treats them as commodities and tries to understand their commercial circulation in the late nineteenth-century Ottoman context. Historical studies so far assert that, for the most part, the scientific instruments used by the Ottomans came from Europe. Coming from a workshop or a factory in Europe, these instruments ended up in the Ottoman lands for two main reasons: either they were diplomatic gifts from Europeans to the Ottoman Court, to the Sultan himself or to highly-placed bureaucrats (in the pre-modern era especially, watches and clocks and some other marvels of science were much-admired gifts),[1] or they could be used to meet institutional requirements such as building up a collection of instruments for military or civilian educational institutions or for engineering projects. Acquiring scientific instruments to satisfy individual curiosity, though it was not prevalent, was another reason.[2] Obtaining scientific instruments required contact with the manufacturer or with a retailer in Europe, and without the help of many intermediaries these transactions could not be completed. However, these intermediaries were not specialists

1 For examples see Otto Kurz: *European Clocks and Watches in the Near East,* The Warburg Institute, University of London, London, 1975; Michael Talbot, "Gift of Time: Watches and Clocks in Ottoman British Diplomacy, 1693–1803", *Jahrbuch für Europäische Geschichte* 17 (2016), pp. 55–79; Jerry Brotton, *This Orient Isle: Elizabethan England and the Islamic World,* Allen Lane, London, 2016.

2 See Feza Günergun, "The Ottoman Ambassador's Curiosity Coffer: Eclipse Prediction with de La Hire's 'Machine' Crafted by Bion of Paris", in F. Günergun, D. Raina (eds.), *Science Between Europe and Asia,* Springer, New York, 2011, pp. 103–123.

 | DOI:10.1163/9789004412842_015

in scientific instruments.[3] In some cases Ottoman ambassadors in Europe or translators were used.

In the second half of the nineteenth century a new route was opened: some European-style shops selling a variety of scientific instruments sprang up in Istanbul. An international directory of the metal industry published in 1874 in Vienna recorded four Istanbul-based firms which imported scientific instruments.[4] Two of them sold surgical instruments while the other two sold optical instruments. Clearly the compilers of the directory saw these firms as scientific instrument retailers. In fact, as we shall see below, the number was higher.

The situation had been rather different in the first half of the nineteenth century according to Austrian geologist Ami Boué (1794–1881). Boué visited the Ottoman lands three times and his comment is telling:

> There is no optician in Turkey, and it is not easy even to buy good spectacles. We are obliged to go to certain cities to make these purchases. Thus, more than once in the interior of the Serbia, even in Turkey, we have seen our spectacles coveted by some doctor or writer. There are even Turkish cantons, especially in Southern Bosnia, where spectacles, especially those of colour, appeared to be objects unknown until our passage. Telescopes, like certain instruments of physics such as bar magnets, etc., are possessed only by some great Turkish lords.[5]

What is intriguing in this paragraph is not that the Ottomans were deprived of the latest fashionable eyewear, which would change towards the end of the century as we will see later. More remarkable is Boué's observation about who could possess scientific instruments in the Ottoman Empire. His sociological observation agrees with Fatma Müge Göçek's findings about the possession of

---

3 One of these intermediaries was Gabriel Defrens, a special envoy charged with buying clocks and instruments for Sultan Murad III (r. 1574–1595). Through Defrens the Sultan ordered a terrestrial and a celestial globe which were probably made in Mercator's (1512–1594) workshop. See Feza Günergun, "Science in the Ottoman World", in George N. Vlahakis *et al.*, *Imperialism and Science: Social Impact and Interaction*, Santa Barbara, California, ABC-CLIO, 2006, pp. 71–118: 80.

4 J. Reiner, F. Pfeiffer (eds.), *Internationales Adressbuch der gesammten Metall-Industrie*, J. Plaut II, Vienna, 1874, p. 476.

5 Ami Boué, *La Turquie d'Europe ou observations sur la géographie, la géologie, l'histoire naturelle, la statistique, les moeurs, les coutumes, l'archéologie, l'agriculture, l'industrie, le commerce, les gouvernements divers, le clergé, l'histoire et l'état politique de cet empire*, Libraire de la Société de Géographie de Paris, Paris, 1840, 4 vols.: v. 3, p. 109. Author's translation.

Western goods in the eighteenth-century Ottoman Empire, which she derived from empirical study of inheritance registers.[6] Göçek compares the possessions of three social groups, namely the "top-level administrators", "the military", and "the populace", and concludes that only the first two groups owned binoculars and telescopes. The populace did not own any, and the administrators' instruments were more valuable than those of the military group.

"Great Turkish lords", thanks to their strong connections, could have obtained scientific instruments in one of the ways mentioned above, but what about those who did not have such opportunities, such as doctors, technically-trained men, academics and writers? How could they buy scientific instruments when there was not even an optician's shop?

The fact that some shops started to trade in scientific instruments, and some of them even developed into businesses that could be called scientific instrument retailers, was very much connected with the emergence of a new consumer group in addition to the "great Turkish lords" cited in the quotation above. The social, cultural and economic transformations of the Ottoman Empire in the second half of the nineteenth century created new social groups among non-Muslims and Muslims, such as state bureaucrats, learned men, engineers, doctors, etc. In addition, the development of well-organised commercial networks with European countries from which the Ottomans imported scientific instruments was very significant. Although demand increased gradually from the second half of the century onwards, it did not reach a level that necessitated the establishment of new commercial routes. The quantity of imported scientific products was very low in comparison to other Western goods. It was only through certain organizations that it was possible to find ways of connecting producers and consumers of scientific instruments, and this chapter argues that the retailers in Istanbul were the critical link in this commercial enterprise.

2 Scientific Instrument Retailers in Istanbul

Antoine Urgliavich (sometimes written as Anton Urglievich) was the owner of one of the optical instrument firms in the directory mentioned above. He was a member of the Austrian community in Istanbul in the second half of

6 Fatma Müge Göçek, *Rise of the Bourgeoisie, Demise of Empire: Ottoman Westernization and Social Change*, Oxford University Press, New York and Oxford, 1996, p. 107.

the nineteenth century.[7] His occupation was *opticien* according to *L'indicateur ottoman* and *Annuaire oriental* (a Constantinople directory) from 1881 to 1900.[8] He had a shop located near Galata tower. After his death his wife Madame Urgliavich took over the shop and ran it for a while by herself, but in about 1900 she started collaborating with A. Marinovich (*Tab. 14.1*). Information from directories alone does not reveal much about Urgliavich or his shop, but he becomes a highly important figure for this chapter because of an Austrian publication dated 1873. This was an economic study of Istanbul and the adjoining territory using the data gathered by consulates and the Austro-Hungarian Chamber of Commerce in Istanbul.[9] The part of the study about the trade in scientific instruments in Istanbul was written by vice-consul Ritter von Bosizio with contributions from Antoine Urgliavich.[10] This portrayal of the scientific instrument sector in the Ottoman Empire is more reliable than Boué's observations because it is written with the insight of an experienced local tradesman—Urgliavich—who specialized in the subject. According to this study the scientific instruments that were used in the Ottoman Empire were not produced domestically, they were imported from foreign countries. This observation was true not only for sophisticated scientific instruments but also for the most ordinary ones. Even the repair of optical instruments was problematic in the Ottoman Empire. In theory, from the perspective of a scientific instrument retailer this could have been a good commercial opportunity. To meet the local demand he could have made wholesale purchases from Europe. In practice the market did not favour retailers. For example, Urgliavich, who owned "the only large establishment" in Istanbul, sold large quantities of

7 His presence in Istanbul is dated to 1869 by the *Hof- und Staats-Handbuch der Österreichisch-Ungarischen Monarchie für 1876*, Druck und Verlag der K. K. Hof- und Staatsdruckerei, Vienna, (1876), p. 130. According to the 25 June 2011 edition of the popular Turkish daily newspaper *Sabah*, Urgliavich had constructed an apartment in Galata in 1867; www.sabah.com.tr/ekonomi/2011/06/25/ferit-sahenke-komsu-olmak-85-milyon- (accessed 23 March 2019). Binoculars and a combined barometer-thermometer signed "Urgliavich—Constantinopoli" have been offered for sale by online antique dealers. A reflecting circle bearing his signature is on display in the Rahmi M. Koç Museum, Istanbul.

8 *L'indicateur ottoman. Annuaire almanach du commerce de l'industrie, de l'administration et de la magistrature*, Constantinople: 1881, 1883, 1885 and 1888; and *Annuaire oriental (ancien indicateur oriental) du commerce, de l'industrie, de l'administration et de la magistrature,* Constantinople: 1889–1890, 1892–1893, 1893–1894, 1894, 1895, 1896, 1898 and 1900.

9 Ritter von Schwegel (ed.), *Volkswirthschaftliche Studien über Constantinopel und das anliegende Gebiet*, Druck der kaiserlich-königlichen Hof- und Staatsdruckerei, Vienna, 1873, pp. III–IV.

10 Ritter von Bosizio, A. Urglievich, J. Dollinger, "Wissenschaftliche Instrumente und Uhren," in Schwegel, op. cit. (n. 9), pp. 272–281.

nautical charts to recoup the capital outlay on the purchase of purely scientific instruments. A firm which wanted to sell scientific instruments had to sell other items as well for its economic survival. The explanation for their small share of the market was because the main users of scientific instruments, such as local institutions including the growing number of telegraph stations, factories and even individual engineers in some cases, met their requirements by direct contact with manufacturers in Europe, without using any intermediaries.[11]

The study also informs us about the most common categories of instruments that were sold in the Istanbul market, though without giving any details of the specific products. The main purpose of the report was to assess the place of Austria among the other European countries from which the Ottoman Empire imported scientific instruments. So the study was restricted to finding which instruments were bought from which country. According to this account the Ottomans obtained physical and optical instruments, as well as those for geodesy, exclusively from France. Unsurprisingly, among the optical instruments the great variety of ordinary spectacles and other eyeglasses was most prominent. Approximately 1,000 dozen spectacles were imported every year. They were obtained from Paris, mostly from the company *Société des lunetiers Muneaux, Videpied & Cie*. All the scientific instruments imported from France, mostly from Paris, were handed over to the shipping company *Messageries Maritimes*, and were transported by rail to Marseilles from where they were shipped by the *Messageries*' steamers.[12] Mathematical instruments were imported from both France and England, but England was the only source of naval instruments. They were sent by English and Russian steamships directly to Istanbul. Most of the surgical instruments came from Paris, but Vienna and London were also among the suppliers. The preferred companies were *Charrière* for surgical instruments and *Blaux* for dental instruments, both located in Paris, but for instruments for use in eye surgery *Berliner-Platz* was favoured. The capital city was the main consumer of surgical instruments, some were sent to Anatolia but in almost insignificant quantities.[13]

Likewise, the number of chemical instruments imported was very low, and they were brought only on request. The study claims that the total quantity imported was worth barely 1,000 francs per year.[14] The majority came from

11 *Ibid.*, p. 272.

12 *Ibid.*, p. 273.

13 *Ibid.*, p. 274.

14 1,000 francs in 1870 would be worth about 10,000 euros in 2018 (calculated by comparing the price of gold at the two dates.)

France, from a trading house in Paris called *Galante*. Only a few chemical instruments, namely retorts, flasks, and Woulfesche distillation apparatus, came from Vienna. A London based company *Rorweger* was the source of imported extract presses. It was noted, however, that the Ottomans used local distillation apparatus made of copper, and this was the only scientific instrument produced in the country.[15]

The study's claim that few scientific instrument retailers existed in Istanbul except Urgliavich might be valid for the 1870s but there was a change after 1880. Although the character of the firms was similar to the earlier ones in that they did not specialize in the scientific instrument trade alone, the number of shops offering such commodities increased. *Table 14.1* lists scientific instrument retailers in Istanbul in this period, but it does not claim to be comprehensive.[16]

TABLE 14.1 Retailers of scientific instruments in late nineteenth-century Istanbul

Company	Address
H. Adjarian and A. Ghazarossian	Grande Rue de Péra [İstiklâl Cad.] No. 257, Next to Galata Saray High School (1881)
P.J. Armao	Yüksek Kaldırım No. 100, Kevork Bey Han, Galata (1895–1900)
L.P. Caracas	Bahçe Kapı No. 6, next to Hacı Bekir Confectionery (1891–1894)
J.B. Copello	Yüksek Kaldırım No. 67, Galata (1881–1883)
J. Schapira His widow I. Schapira, after 1889	Next to Hotel Missirie (1874) Grande Rue de Galata [Büyük Galata Cad.], No. 9, Karaköy Square (1881–1888) Yüksek Kaldırım No. 90, Kevork Bey Han, Galata. Home: Tulumbacı St. No. 8, Pera (1888) Yüksek Kaldırım No. 100, Galata (1889–1894)

15 *Ibid.*, p. 275.

16 The data in *Table 14.1* is taken from: *L'indicateur ottoman*, cit. (n. 8); *Annuaire oriental*, cit. (n. 8); *Annonces-Journal de Constantinople* No. 76, 23–30 Jul. 1874, p. 2; No. 80, 21–27 Aug. 1874, p. 2; and No. 81, 27 Aug.–3 Sep. 1874, p. 2.

TABLE 14.1 Retailers of scientific instruments in late nineteenth-century Istanbul (*cont.*)

Company	Address
A. Urgliavich His widow (1885–1900) His widow and A. Marinovich (1900)	Galata Tower Square. Home: Küçük Hendek Sokak No. 1, Pera (1881) Galata Tower Square. Home: Küçük Hendek Sokak No. 2, Pera (1883–1888) Yorgancılar Cad., No. 26, Galata (1895–1898) Fermeneciler Cad., No. 173, Galata (1900)
E. Vitalis et Cie.	Grande Rue de Péra [İstiklâl Cad.] No. 494, near Teke [Pera Dervish Lodge] (1881) Grande Rue de Péra [İstiklâl Cad.], opposite the Russian Consulate, No. 491 (1883–1885) Home: Chancellerie St. [Şehbender Sokak] No. 4, Pera (1885)
J. Verdoux	Tünel Square and Grande Rue de Péra [İstiklâl Cad.], No. 482

In the above list two partners, H. Adjarian and A. Ghazarossian, were registered as *opticiens* in *L'indicateur ottoman* catalogues.[17] In their shop, however, customers could find surgical instruments, watches and some stationery materials in addition to spectacles and optical instruments. Their small company also had an electrical business, selling telephones, bells and lightning conductors, as well as installing and repairing all kinds of electrical machines. Most likely Adjarian did the electrical work: he was a member of the *Société internationale des électriciens* from 1884 onwards and his membership details show his profession as electrician.[18] After 1895 he and his brother K. Adjarian started a new business, this time specializing only in electricity.[19]

P.J. Armao owned one of the scientific instrument shops in Galata. His trading card indicates that he founded his company in 1867.[20] It was active up to 1900. Like the others he sold spectacles and optical instruments but

17 *L'indicateur ottoman*, cit. (n. 8), 1881, p. 347.

18 *Bulletin de la Société internationale des electriciens* 1 (1884), p. 33; and 7, (1890), p. 176.

19 *Annuaire oriental*, cit. (n. 8), 1895, p. 548; 1896, v. 1, p. 69; 1898, v. 1, p. 681; 1900, v. 1, p. 692.

20 An image of Armao's trade card is in the writer's possession. The source is unknown.

also instruments for physics, mathematics, meteorology, geodesy, etc. The Zappeion Greek High School in Istanbul has a maximum and minimum thermometer bearing Armao's label. He also imported electrical devices for medical use from Vienna.[21]

The variety of commodities sold by L.P. Caracas was similar to the others: eyewear, compasses, barometers, thermometers, mathematical instruments (levels, geodetic and surveying instruments) and surgical instruments for doctors, midwives, dentists and veterinarians. He supplied schools, hospitals, pharmacists, doctors and engineers. He advertised that he offered a better price because he imported goods directly from the factories in Europe. He could also provide a repair service. From 1894 he added the installation of electric bells and the supply of accessories to his services.[22]

J.B. Copello listed himself in the 1881 and 1883 directories as an optician and as a manufacturer of surgical instruments. He also provided a repair service for the products he sold.[23] Unfortunately no further information is available about the intriguing statement that he manufactured instruments.

J. Schapira had a shop in Istanbul from 1874 to 1894. Given that he was listed as *opticien* and *électricien* and it was stated that he sold precision instruments,[24] he might have traded in other scientific instruments as well, like the optician-electricians mentioned above.

E. Vitalis ran a stationery and printing business but also sold eyewear as well as some scientific instruments such as barometers, thermometers, binoculars, and maps.[25]

3 A Closer Look: Verdoux's Optical Shop

The optical shop located at number 482 on Tünel Square in Istanbul, named *à la Méridienne*, but often called *Verdoux's optical shop*, was once a very well-known place to inhabitants of Istanbul because it occupied the same corner

21 *Başbakanlık Osmanlı Arşivi* (Ottoman Archives of the Prime Minister's Office; hereafter 'Ottoman Archives'), Istanbul, BEO, 1551/116308, 19 Eylül 1900 (19 Sep. 1900).

22 *Annuaire oriental*, cit. (n. 8), 1891, p. 112; 1892–1893, p. 112; 1893–1894, p. 1370; and 1894, p. 1370.

23 *L'indicateur ottoman*, cit. (n. 8), 1881, p. 347; and 1883, p. 432. His surname resembles that of the Levantine family Copello (or Copella or Capella).

24 *Annonces-Journal de Constantinople*, No. 80, 21–27 Aug. 1874, p. 2; and No. 81, 27 Aug.–3 Sep. 1874, p. 2; *L'indicateur ottoman*, cit. (n. 8), 1881, p. 347; 1883, p. 432; 1885, p. 262; 1888, p. 303; *Annuaire oriental*, cit. (n. 8), 1889–1890, p. 487; 1892–1893, p. 555; and 1893–1894, p. 585.

25 *L'indicateur ottoman*, cit. (n. 8), 1883, p. 360.

for more than seventy years from the last decades of the nineteenth century. The location of the shop was strategic. Galata and Pera, two popular neighbourhoods of Istanbul, cross at Tünel Square. Thousands of people walked in the streets of Pera in the late nineteenth century, some to wander around and enjoy urban life, others simply to go to work. They all passed this optical shop. The window display attracted the attention of these passers-by. E. Giraud reported in 1897 that there was a huge thermometer in the shop window. It was seen every day by thousands, and "they regarded what this instrument measured as true as law".[26] Unsurprisingly, the optical shop and its owner Jean Verdoux (1841–1928) left their mark on the memories of some inhabitants of Istanbul, among them writers such as Sermet Muhtar Alus and Ahmet Rasim.[27]

Jean Verdoux was born in Labassère (Hautes-Pyrénées) in France.[28] His first profession had nothing to do with optical instruments. He was apprenticed to a wine producer in France and worked there for some time, at least for a sufficiently long period to become a master. In 1865 he moved to Istanbul with a small amount of capital in his pocket to look for a job.[29] What kind of job he dreamed of in Istanbul is unknown. Istanbul, being the capital city of the Muslim-ruled Ottoman Empire was not a centre of attraction for ambitious wine producers to make a career despite a considerable number of non-Muslim wine consumers (and some Muslims as well). Verdoux was not one of those French experts that the Ottoman state hired frequently such as engineers or technicians: he did not have a higher education or industrial experience. Notwithstanding these drawbacks, there was a solid French community in Istanbul and well-established commercial networks with France which would encourage a French entrepreneur. So, by using his fellow-countrymen, he began a new career in Istanbul from scratch. First, he worked as an apprentice to an elderly French optician named L. Poirier. In his unpretentious shop located in a back-alley of Pera (in Yüksek Kaldırım) Verdoux learned the business of an optician. Within a couple of years (probably in 1869), he took over

26 Erol Üyepazarcı, "Bir Zamanlar İstiklal Caddesi", *Atlas Tarih* (Apr.–May 2013), pp. 74–81: 75.

27 Sermet Muhtar Alus (1887–1952) was a prolific columnist. He wrote a newspaper article based on childhood memories of his first visit to Verdoux's shop in 1898 and on an interview with Horace Louis Guerillotile who directed the shop after Verdoux died. See Sermet Muhtar Alus, "Gördüklerim Duyduklarım: Eski Verdoux Mağazası", *Akşam* (4 Nov. 1943), p. 5, released online by Şehir Üniversitesi, Taha Toros Arşivi, Dosya No. 123, http://earsiv.sehir.edu.tr:8080/xmlui/handle/11498/19490 (accessed 8 May 2018). Another source that mentions Verdoux's shop is Ahmet Rasim, *Şehir Mektupları* (ed. by Figen Berk), Karanfil Yayınları, Istanbul, 2005, p. 151.

28 He was the son of Chérèze Arrodère and Baptiste Verdoux. See Paris, Archives nationales, *Légion d'honneur*, LH/2688/89.

29 *Ibid.*; Alus, op. cit. (n. 27).

the business and moved the shop to a better and certainly more visible location, to Tünel Square, number 482.[30]

Monocles or spectacles were primarily medical devices for correcting visual impairment. In time they became fashionable accessories for gentlemen in a similar manner to pocket watches. Nahide Işık Demirakın in her study of the life-style of late nineteenth-century Ottoman urbanites observed that fine-quality spectacles and monocles became a mark of the social status of the wealthy Ottoman men who adopted European fashion in their dress, and they became desirable possessions for those who followed upper-class fashion.[31] Sermet Muhtar Alus, for instance, confessed how much in his childhood he had envied the appearance of young gentlemen who wore spectacles. His parents tried to counter his fixation on spectacles with golden frames or with slim golden chains by pointing out the snobbishness of wearing them in spite of having no visual defect.[32] This enjoyment of eyewear together with the growing culture of consumption in Pera-Galata was nothing but a good business opportunity for an optician with a French name such as Verdoux. He took advantage of this very successfully. Ahmet Rasim, in 1896, compared the prices of spectacles in Istanbul and pointed to Verdoux's shop as the most expensive.[33] His "fine eyewear from Paris" was for the people who lived well. His reputation even reached the Ottoman Palace in the 1880s. Sultan Abdulhamid II bought two pairs of spectacles with blue-tinted extra-thin lenses from Verdoux.[34] The Sultan and his household subsequently became regular clients, as bills kept in the Ottoman Archives confirm (*Fig. 14.1*).

Verdoux's shop was not only selling goods but also selling a European way of life, as were other shops in Pera. Furthermore, Verdoux made a good impression on people. Sermet Muhtar Alus's description shows him to be an extrovert personality, friendly and generous.[35] Having such social skills would have contributed to him adapting smoothly to the cosmopolitan social life of Pera, which would have been crucial to enhancing his business connections. The

30 *Ibid.*

31 Nahide Işık Demirakın, *The City as a Reflecting Mirror: Being an Urbanite in the 19th Century Ottoman Empire*, Unpublished PhD Thesis, İhsan Doğramacı Bilkent University, Ankara, 2015, pp. 73–74.

32 Meral Demiryürek, "Paşazade Bir Muharrir: Sermet Muhtar Alus Hayatı ve Eserleri Hakkında bir Bibliyografya Denemesi", *Kebikeç* 22 (2006), pp. 218–283: 222.

33 Demirakın, op. cit. (n. 31), pp. 131–132.

34 Nuran Yıldırım, Bülent Özaltay, "Sultan II. Abdülhamid'in Sağlığı ve Sağlık Hizmetleri", in Coşkun Yılmaz (ed.), *Sultan II. Abdülhamid ve Dönemi*, Sultanbeyli Belediyesi, Istanbul, 2012, pp. 123–146: 131.

35 Alus, op. cit. (n. 27).

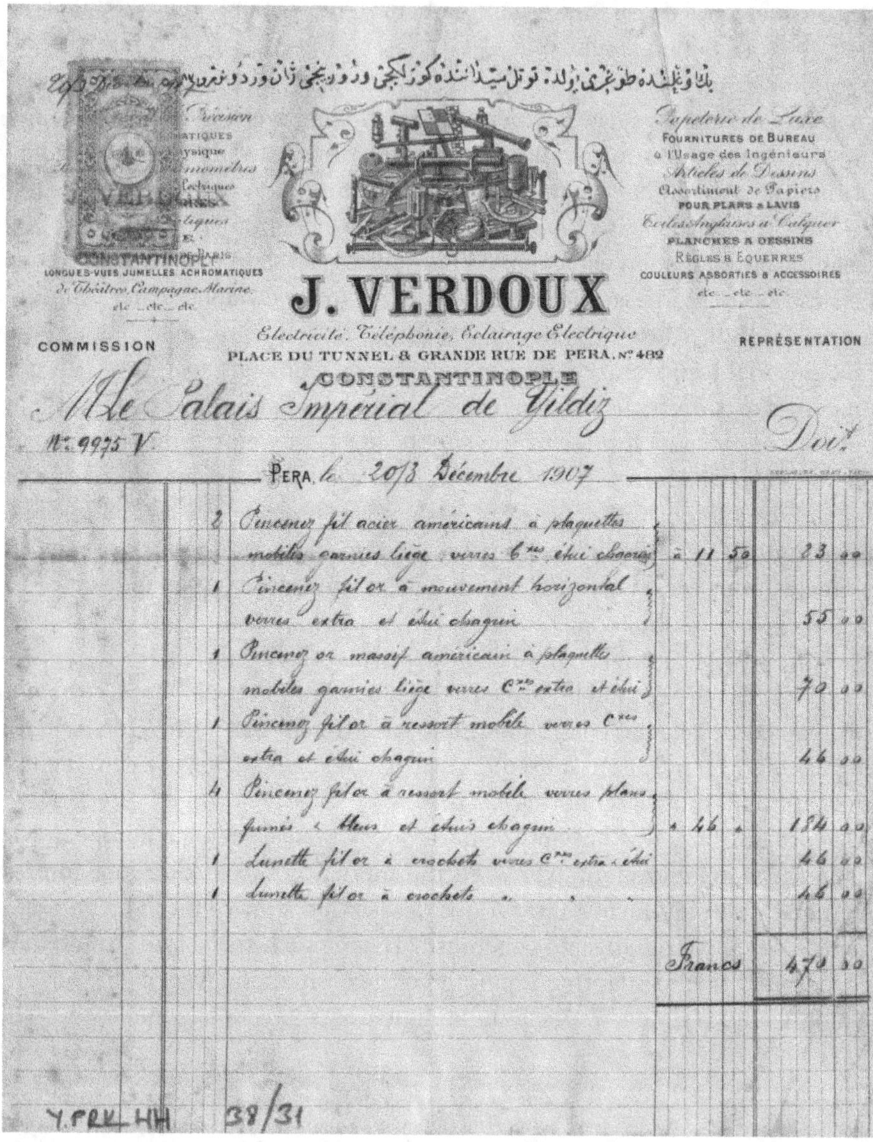

Précision
…ATIQUES
…ysique
…nomètres
…lectriques
CONSTANTINOPLE
LONGUES-VUES JUMELLES ACHROMATIQUES
de Théâtre, Campagne, Marine
etc. etc. etc.

Papeterie de Luxe
FOURNITURES DE BUREAU
à l'Usage des Ingénieurs
Articles de Dessins
Assortiment de Papiers
POUR PLANS & LAVIS
Toiles Anglaises à Calquer
PLANCHES A DESSINS
RÈGLES & EQUERRES
COULEURS ASSORTIES & ACCESSOIRES
etc. etc. etc.

J. VERDOUX
Electricité, Téléphonie, Eclairage Electrique
PLACE DU TUNNEL & GRANDE RUE DE PERA, N° 482
CONSTANTINOPLE

COMMISSION — REPRÉSENTATION

M. Le Palais Impérial de Yildiz — Doit
N° 9975 V.

Péra, le 20/3 Décembre 1907

Qté	Désignation	Prix	Total
2	Pincenez fil acier américains à plaquettes mobiles garnies liège verres C… étui chagrin	à 11 50	23 00
1	Pincenez fil or à mouvement horizontal verres extra et étui chagrin		55 00
1	Pincenez or massif américain à plaquettes mobiles garnies liège verres C… extra et étui		70 00
1	Pincenez fil or à ressort mobile verres C… extra et étui chagrin		46 00
4	Pincenez fil or à ressort mobile verres plans fumés & bleus et étuis chagrin	à 46	184 00
1	Lunette fil or à crochets verres C… extra & étui		46 00
1	Lunette fil or à crochets		46 00
	Francs		470 00

Y.PRK.HH 38/31

FIGURE 14.1 Jean Verdoux's bill for the Yıldız Palace, December 1907
ISTANBUL, *BAŞBAKANLIK OSMANLI ARŞIVI* [OTTOMAN ARCHIVES], Y.PRK.HH, 38/31/1

following example which involves Verdoux demonstrates how social and business networks were entangled in late nineteenth-century Istanbul.

The father of Sermet Muhtar Alus, Ahmet Muhtar Pasha (1861–1926), after having started to work as a teacher at the Ottoman Military Academy in the early

1880s, was invited to dinner by his former teacher and new colleague, Joseph Lecoq Pasha, at Morand's café (*Café Louvre*). Ahmet Muhtar met Verdoux at this café for the first time. Morand, the owner of the café, was a French veteran. After having completed his military service in Algeria he moved to Istanbul and opened a café in Pera.[36] Serving French dishes in a congenial social environment made Morand's café popular particularly among the French, including Verdoux and Lecoq, but also among some of the Ottoman elite and other Europeans. For example, the Constantinople Lodge of the Independent Order of Odd Fellows organized their weekly gatherings in Morand's café. Members of the Lodge were chiefly British artisans employed in the public works of the Ottoman state.[37] So this space attracted the technicians, the tradesmen and the educated men of science who shared similar tastes and class culture, and paved the way for making new acquaintances and strengthening professional bonds.

Verdoux had brought such a café-like atmosphere to his optical shop where he could build friendly social connections with the Ottoman bureaucratic elite, as this quotation from Sermet Muhtar Alus's writing illustrates very well:

> [Verdoux] welcomed Pashas and Effendis to his shop every Friday, for instance Hasan Rüştü Pasha, the director of the accounts department at the Ministry of War; Hacı Hüseyin Pasha, a chief artillery officer; Rami Bey a high official from the accounts department; my grandfather; Hüsnü Pasha of Kadıköy, a director at the department of fortification and construction; Çerkes İzzet Pasha, director at the cavalry department. Conversation was accompanied by coffee and cigarettes in the winter and by iced sherbet and ice cream in the summer. Friendship thrived as if it were not an optical shop but Arif's coffee house on Divanyolu Street.[38]

For an optician who sold spectacles and binoculars only, even if his products were luxury items targeted at elite consumers, such extravagant efforts to be linked with the Ottoman elite seem somehow inconsistent but Verdoux's range of products was beyond that of a typical Ottoman optical shop, at least after the 1880s. Having good relations with the state—the biggest client in the Ottoman market—would have been of utmost necessity for his business.

36 *Ibid.*

37 Henry Ratcliffe, *A List of the Lodges Composing the Independent Order of Oddfellows, Manchester Unity, Friendly Society, with a Table of the Lodge Nights for 1876–7*, Manchester, 1876, p. 214; *The Quarterly Magazine of the Independent Order of Odd-Fellows, Manchester Unity Friendly Society*, 3 (1862), pp. 193–194.

38 Alus, *op. cit.* (n. 27).

4 Verdoux's Commodities and Clients

Verdoux's published advertisements in the *Annuaire oriental* show the difference between him and the other retailers included in *Table 14.1*. From 1885 he published mostly full-page advertisements. By itself that was a sign of good financial standing, but the real advantage of a full-page advertisement was that, because of the larger space, Verdoux could bring a variety of products to the attention of buyers, including those that could not be exhibited physically in the shop window or—more importantly—those that were not expected to be found in a conventional optical shop in Istanbul. Verdoux's advertisements listed scientific instruments under general categories. These were precision instruments, geodetic instruments, mathematical instruments, physics apparatus and optical instruments. Among them were barometers, thermometers, telescopes, opera glasses and achromatic binoculars for the army and the navy. Rulers and set squares were mentioned specifically, most likely because they were the best sellers. We do not now know specific details of the products that Verdoux sold because no sales catalogue survives, if there ever was one, but a few instruments bearing Verdoux's label, such as a brass barometer and a theodolite, can be found in online auction catalogues.

Who bought scientific instruments from Verdoux? Ministries and educational institutions rank top of the list. In Verdoux's own words from the entry in his *Légion d'honneur* file, he sold physics and chemistry apparatus to the Ottoman Ministry of Public Instruction, and the Ministry of Public Works bought all the instruments it needed for geological studies from him.[39] Having made contact with the Minister of the Navy (Bozcaadalı Hasan Paşa), he sold mariners' compasses, barometers, telescopes, mathematical instrument sets and engineering instruments to the Imperial Naval School and to some naval vessels.[40] The Zografeion High School, a Greek community school in Istanbul, was one of his customers. Engineers and doctors were among the clients he courted. He sold almost all the equipment necessary to set up an engineering business ranging from furniture to engineering instruments and documents.

We also know that he imported electrical devices for medical purposes at the request of doctor clients.[41] Like some other opticians Verdoux sold electrical instruments probably at least from the 1880s. Europe gradually incorporated Ottoman lands into the electricity market, and even though there were restrictions on the use of electricity during the reign of Sultan Abdulhamid II,

39 Archives nationales, *Légion d'honneur*, cit. (n. 28).

40 Alus, op. cit. (n. 27).

41 Ottoman Archives, BEO, 3373/252943, 17 Ocak 1908 (17 Jan. 1908).

and strict regulations on the importation of electrical devices, the irresistible diffusion of this new technology into industry, warfare, medicine and domestic use increased the number of people who wanted to benefit from it, the state and the Palace included. Verdoux was very aware of this change. In this regard, his marriage to Maria Lacoine in 1884 was useful.[42] She was the daughter of Émile Lacoine (1835–1899), the *conseiller électricien de gouvernement ottoman*, the technical consultant for the Ottoman Ministry of Telegraphs, and a teacher of courses on electricity among many other duties.[43] He had powerful connections with the state as a well-respected electrical engineer. Being son-in-law of Lacoine would have strengthened Verdoux's business network with various institutions. The sale of optical and other military telegraph equipment to the Ottoman army, and cables and other materials for constructing telegraph networks to the Ministry of Posts and Telegraphs,[44] was probably helped by his family connections. *Table 14.2* lists some of the contracts Verdoux made or bills signed by him. The data has been gathered from the Ottoman Archives. The quantity of electrical commodities is striking, as well as the status of his clients.

TABLE 14.2 Some of Verdoux's contracts and bills

Date	Imported articles	Importing company/ country	Client	Sale price
18 Sep. 1880[a]	Various stationery	–	Lycée Impérial Ottoman de Galata-Sérai	1,600 francs
22 Jul. 1882[b]	Eyeglass	–	La Majesté Imperiale le Sultan	37.50 (francs)

42 Paris, Archives nationales, *Minutes et répertoires du notaire Ernest LEGAY, 25 février 1875–14 mai 1902 (étude XXIII)*, MC/ET/XXIII/1277. A *Répertoire numérique détaillé* (Detailed digital directory) of these papers available on the Archives nationales website lists the marriage contract on p. 46. See www.siv.archives-nationales.culture.gouv.fr/siv/rechercheconsultation/consultation/ir/pdfIR.action?irId=FRAN_IR_041698 (accessed 7 April 2019).

43 For biographical notes on Émile Henri Lacoine (1835–1899) see Feza Günergun, "Salih Zeki ve Astronomi: Rasathane-i Amire Müdürlüğü'nden 1914 Tam Güneş Tutulmasına", *Osmanlı Bilimi Araştırmaları* 7, 1 (2005), pp. 97–122: 117.

44 Archives nationales, *Légion d'honneur*, cit. (n. 28).

TABLE 14.2 Some of Verdoux's contracts and bills (*cont.*)

Date	Imported articles	Importing company/ country	Client	Sale price
20 Nov. 1890[c]	75 large accumulators for electric lighting and 150 brackets for fixing them	–	Yıldız Palace	8,649 francs
8 Apr. 1893[d]	Telephone	France	–	–
13 Apr. 1896[e]	Phonograph, clock	–	Yıldız Palace	230 lira
24 Apr. 1901[f]	Electrical equipment	Paris	Electrical factory of the Imperial Arsenal	–
21 May 1903[g]	Lamp glasses, electrical wires and cables, voltmeters, etc. (6 crates)	–	*İdare-i Mahsusa* steamships (Government fleet)	–
12 Jul. 1905[h]	Four *Mangin* electrical projectors (75 cm) for land armies	Sautter, Harlé & Cie (Paris)	The Ministry of Military Supplies	170,000 francs, (without 5% discount)
Dec. 1907[i]	Various eyeglasses	–	Yıldız Palace	470 francs
17 Jan. 1908[j]	An induction-machine for therapeutic use	Paris	–	–
17 Aug. 1908[k]	Copper wires for the telegraph line from Yedikule to Çorlu	–	The Ministry of Posts and Telegraphs	24,900 francs

a Ottoman Archives, MF.MKT, 65/171, 18 Eylül 1880 (18 Sep. 1880).
b Yıldırım, Özaltay, op. cit. (n. 34), p. 131
c Ottoman Archives, Y.MTV, 47/145, 20 Kasım 1890 (20 Nov. 1890).
d Ottoman Archives, Y.PRK.ASK, 89/37, 8 Nisan 1893 (8 Apr. 1893).
e Ottoman Archives, Y.MTV, 139/105, 13 Nisan 1896 (13 Apr. 1896).
f Ottoman Archives, Y.MTV, 143/37, 24 Haziran 1896 (24 Jun. 1896).
g Ottoman Archives, Y.MTV, 244/66, 21 Mayıs 1903 (21 May 1903).
h Ottoman Archives, Y.MTV, 276/46, 12 Temmuz 1905 (12 Jul. 1905).
i Ottoman Archives, Y.PRH.HH, 38/31, Aralık 1907 (Dec. 1907).
j Ottoman Archives, BEO, 3373/252943, 17 Ocak 1908 (17 Jan. 1908).
k Ottoman Archives, BEO, 3385/253843, 17 Ağustos 1908 (17 Aug. 1908).

Less is known about the manufacturers that produced Verdoux's commodities than about his clients. Thankfully, he took generous advantage of the full-page advertisements to show the companies he represented. Verdoux was the representative of the *Société générale des téléphones et des usines Rattier* which manufactured telephones and cables, and of Lazare Weiller's company that specialized in forges and foundry work, wire drawing and rolling-mill products. He was also a representative of Gustave Trouvé's company[45] and, as shown in *Table 14.2*, of *Sautter, Harlé & Cie*.

He bought everything from France. He described himself proudly as a tradesman who struggled against foreign competition—German competition in particular—from 1869 onwards to sell in the Ottoman market. For "his most noticeable services to the French industry" he was awarded *Chevalier de la Légion d'honneur* in 1907.[46]

5 Conclusion

The manufacture of scientific instruments became industrialized in the nineteenth century. Market conditions, the distribution of goods, and marketing became significant factors in the circulation of scientific instruments. In the Ottoman context, France had a more advantageous position than other European countries not only because of the superiority of its scientific instruments but also because of its advantage in commerce in general, with its better trading network and greater long-term investment in trade with the Ottoman Empire. This favoured a French retailer against other retailers in Istanbul, but Istanbul shops were not just representatives of European retailers, they evolved in the Istanbul market and were shaped by local needs. Verdoux had no experience in the field at first, but he took over a small optical shop and turned it into a scientific instrument retailer, arguably the most advanced in Istanbul, and made a career for himself in the Ottoman Empire.

The progressive character of his business was shaped by two factors. The most important was his entrepreneurship in reaching out to influential clients. The consumers of scientific instruments were increasing in number towards the end of the nineteenth century; there were more engineers, more doctors and more schools than in the earlier period. But the state was the most important client, and it was Verdoux who supplied the Sultan's spectacles. His efforts

45 *L'indicateur ottoman*, cit. (n. 8), 1885, p. 36 and 1888, pp. 20–21; *Annuaire oriental*, cit. (n. 8), 1889–1890, p. 22.

46 Archives nationales, *Légion d'honneur*, cit. (n. 28).

to grow his business were outstanding. He was affiliated to various organizations and societies: he was the founder of the *Chambre de commerce française* in Istanbul, a member of the *Alliance française*, the *Ligue maritime française* and the *Société de Géographie*.[47] He was also elected a member of *Société internationale des électriciens* in 1884.[48]

Acknowledgments

I am indebted to Professor Patrice Bret for directing my attention to J. Verdoux's *Légion d'honneur* files in the online catalogues of the *Archives nationales*, and to Panagiotis Lazos who gave me information about the scientific instrument collections in the Greek community schools of Istanbul that were supplied by some of the retailers mentioned in this chapter. I am grateful to Nihal Uzan and Yılmaz Aksoy for support with French and German texts, and to Feride Akın for improving the English of the manuscript. The writing of this chapter was facilitated by a three-month research visit to Buenos Aires made possible by the WORLDBRIDGES scheme (EU, 7th Framework, Marie Curie-IRSES, GA: 612644).

47 *Ibid.*
48 *Bulletin*, cit. (n. 18), 1884, p. 193.

Index

Notes:

- 'al' at the beginning of an Arabic name or title has been ignored for positioning in the index. Thus al-Ashraf is listed between 'ascension' and Asia.
- European names are listed surname first, followed by given names. Arabic names are not inverted, except in cases where a person is usually known by a later part of his name, which are cross-indexed.
- Instruments are all listed under 'instrument', sub-divided by type. Thus optical instruments are listed under 'instrument, optical'.

Abbas Eqbal Ashtiani 150(n20)
Abbas Mirza 150
Abd al-A'imma 118, 119
'Abd al-'Aziz ibn Muhammad al-Wafā'ī al-Miqātī 113
'Abd al-laṭīf b. Aḥmad al-Dimashqī 24, 29
Abd al-Rahmān al-Jabartī, *see* al-Jabartī, Abd al-Rahmān
'Abd al-Razzāq ibn Muḥammad, *also known as* Mu'īn al-Munajjim Kāshānī 124
Abdolrahman al-Jami 142
Abdul Wahid ibn Mas'ud 116
Abdulhamid II, Sultan 60, 249, 252
Abdülmecid I, Sultan 235, 236
abjad (decimal numeral system) 8, 10
Aboukir 221, 225, 226
Abū al-Ḥasan al-Marrākushi 23
Abu Taleb ibn Hassan Hosseini Safavi 147
Abu Taleb Kalim of Kashan 118
academic 232, 242
Académie royale, France 208
Academy of Architecture, Berlin 171
Academy of Sciences
  of Romania 166
  Paris 214
  Russian 157, 158
acoustics 172, 180, 186
Adilé (sister of the Sultan Abdülmecid I) 235
Adjarian, H. 245, 246
Adjarian, K. 246
admiral 1, 3, 96
Adnès, Pierre-Onézime 222
advertisement, advertise 160, 163, 164, 165, 247, 252, 255
aerial balance, *havai terazi* (a form of level) 24, 200, 201, 202
aerial *kūniyā*, (a form of level, see *al-kūniyā al-hawā'iya*) 21, 23, 24, 25, 37
aérostiers militaires (military balloonists) 212
Africa 237, 238, 239
Agia Triada (Holy Trinity) monastery, Istanbul 174, 175
Ahmad al-Mansur 115
Ibn al-Amin Ahmed b. Yahya 10
Ahmed Pasha the Bombardier, *also called* Bonneval, Claude Alexandre de 43
Ahmed Ziya Akbulut, Major 56
Ahmet Muhtar Pasha 250, 251
Ahmet Rasim 248, 249
Aimé, François-Sébastien 211, 212, 222
air pump, *see also* vacuum pump 180, 183, 184
Aktien-Gesellschaft Institut für angewandte Geophysik, Göttingen 166
al-āla 'alā ṣūrat jāmāt al-ḥammām (instrument resembling bowls for the public bath) 33
al-āla al-kulliya yumkin bihā ramī al-khumbara (the universal instrument with which one can throw a bomb) 31, 33
'Alā' ad-Din 'Alī bin Muhammad al-Qawshajī (*or* al-Qūshjī), *see* Ali Qushji
'Alā' al-Dīn b. al-Shātir 113
Albania 196
Albrecht of Prussia, Duke 81(n3)
Alemdar Mustafa Pasha 76
Aleppo, Syria 3, 101, 214, 215
Alexandria, Egypt 197, 213, 215, 216, 221, 223, 225, 226, 237, 239
Alfani, Guido 164
Alfonsine tables 86(n11)
Algeria 251

Ali ibn Ahmad ibn Muhammad al-Sharfi 6
ʿAlī ibn Ibrahīm ibn al-Shātir 112
Ali Qushji, ʿAlī Qūshjī' 3, 74(n53), 110, 114
alidade 7, 84, 87(n13), 95, 96, 98(n39), 127, 210
Aligarh Muslim University, India 147
alla franca instruments (European style instruments) 55, 57
Alliance française 256
Almagest (Ptolemy's) 3, 56(n6), 130
almanac, Ottoman 78
  nautical almanac, *see also* ephemerides 210(n13), 213, 214(n29)
almucantar (parallel of altitude) 55, 87
altitude (of a celestrial body) 7, 12, 55, 59, 60, 67, 68, 71, 73, 84, 91, 217, 220
  altitude of the Sun 13, 14, 55, 57, 65, 68, 69, 70, 77, 78, 90
Ambassadors, The (by Holbein, Hans, Junior) 81, 83
ambassador 102, 110, 142, 229, 235, 241
  English ambassador 115
  French ambassador 57(n7), 115
  Moroccan ambassador 116
  Ottoman ambassador 52, 52(n18), 241
America, American, *see also* United States of America 150, 184, 189, 195
Amir-Arjomand, Kamran 144, 145
Anatolia, Anatolian 1, 11, 101, 102, 244
Andréossy, Antoine François 24
Anfangs-Gründe aller Mathematischen Wissenschaften (by Wolff, Christian Freiherr von) 19, 22, 25, 26, 29
Angos, Jean-Auguste d' 213
Annuaire oriental (a Constantinople directory) 243, 252
anomaly (in astronomy) 132, 132(n32), 133, 139
Apianus, Peter 81, 85
apogee 127, 129, 130, 132–138
apsidal line 130, 132, 136, 137
Aq Qoyunlu Turkomans 142
Arab, Arabian, Arabic 36, 80, 87, 93, 95, 96, 98, 99, 102, 103, 109, 113, 115, 121
Arabic (language) 3, 6, 8, 11, 14, 16, 18, 24, 41, 43, 46, 49, 51, 61, 72, 73, 87, 94, 95, 96, 98, 99, 102, 103, 114, 116, 117, 121, 122, 123, 129, 197, 198, 199, 219, 228
architect 173, 176, 179, 192, 197, 202, 217, 220
architecture, architectural 171, 191, 192, 193, 194, 195, 197, 202, 203, 204
archive, archival 63, 67, 76, 77, 170, 177, 190, 204, 207, 226
  Ottoman Archives 67(n38), 247(n21), 249, 253
Argentina 165
Aries, first point of 133(n33), 134, 136, 137
Aristotle, Aristotelian 3, 196
arithmetic, arithmetical 19, 24, 55
Armao, P. J. 245, 246, 247
armillary sphere 7, 84, 92, 94, 102, 115, 183, 207(n7)
army 40, 56, 58, 60(n21), 61, 74, 76, 149, 206, 210, 212, 223, 252, 253
Arsakeion Girls' School, Athens 174
arsenal
  Imperial Arsenal, Istanbul 235, 254
  Maritime *or* Naval Arsenal, Istanbul 1, 57, 59
  Paris arsenal 212
  School of Geometry, Maritime Arsenal, Istanbul 57
  artificial horizon 67, 69, 216
arshin (a Serbian ruler or unit of length) 191, 196, 197, 198
artillery 76, 211, 212, 220, 235, 251
Artin Bey Shoukry 237
artisan 52, 211, 251
artist (*see also* painter) 34, 61, 65, 67, 193, 194, 222
Asakir-i Mansure-i Muhammediyye (a new Ottoman army) 60(n21)
ascendant 87, 91
ascension (in astronomy) 91, 128
al-Ashraf, (Yemenite sultan) 112
Asia, Asia Minor 146, 214, 236, 239
assassination 76
Association for Women's Education, Istanbul 173
Association of Cultural Executives, Montreal 190
astrolabe 2, 3, 4, 6, 7, 12, 13, 14, 55, 81, 87, 89, 92, 103, 109, 114, 116, 118, 120, 122(n1), 126, 127, 131, 143, 146
  simple astrolabe (*al-usṭurlāb al-basīṭ, see also* half circle) 29

astrologer 3(n6) 6, 11, 47(n14), 56, 78(n69), 117, 122, 131
astrology, astrological 6, 88, 91, 115, 117
astronomer 6, 11, 47(n14), 56(n6), 71, 73(n51), 74, 80, 91, 92, 108, 110, 112, 122, 123, 124, 129, 130, 131, 141, 143, 144, 149, 150, 164, 206, 212, 213, 215–219, 228
astronomical 89, 97, 100, 102, 109, 110, 131, 142, 209, 210, 213, 215, 216, 217
  book, literature, manuscript, text 7, 14, 63, 71, 98, 102
  calculations, data, operations 7, 10, 71, 73, 74, 77, 214
  instrument, apparatus, device 1, 4, 5, 6, 7, 12, 13, 14, 15, 43, 55, 80, 93, 94, 95, 98, 99, 102, 108, 120, 142, 143, 149, 207, 208, 209, 210, 214, 216, 217, 221, 222, 226
  parameters 13, 65, 69, 71, 74, 77, 129, 130, 132, 135, 136, 139, 140, 141
  refraction, *see also* refraction, atmospheric 70, 74
  tables *see also zij and* ephemerides 10, 59, 64, 69, 88, 97, 126, 130, 214
Astronomicum Caesareum (by Peter Apianus) 84, 85
astronomy 1–6, 3(n6), 12, 13, 43, 47, 56, 64, 84, 88, 92, 93, 109, 110, 111, 115, 116, 117, 120, 132, 133, 142, 143, 146, 147, 149, 150, 175, 207, 208, 213, 239
astrophysics 157
Atelier de perfectionnement des armes portatives, Paris 212
Athens 171, 174
Atlas Cœlestis (by John Flamsteed) 209
Atwood machine 180
Australia 163
Austria, Austrian 17, 23, 241, 242, 243, 244
  Austrian instrument maker 181, 182, 242
Austro-Hungarian Chamber of Commerce, Istanbul 243
Austro-Hungarian Empire 232
authentic, authenticity 124, 190, 191, 192
automaton clocks 115, 117
al-Azhar, University of 228
azimuth 7, 12, 71, 84, 87, 97, 120, 210

Babo device 184
Baghdad 101, 214, 217, 227
Bahāʾ al-dīn al-ʿāmilī 24
Bailly, Jean 93
Baily, Richard, *also known as* İngiliz Selim 73(n51)
balance 24, 35, 160, 184, 200, 201
  aerial, *see also* aerial balance 24, 200, 201, 202
Baldric of Bourgueil 102
balloon, ballooning, balloonist 209, 212, 219, 223, 224, 226
  National School of Ballooning, *see École nationale aérostatique*
Barbarossa, *see* Hayreddin Pasha 1
barometer 172, 176, 180, 181, 219, 220, 223, 247, 252
  aneroid barometer 220
  baromètre-balance 220, 224
Baronian, Petros, *see* Bārūn
Bartholomäus von Stürmer, Count 235
Bārūn (Petros Baronian) 121, 121(n51)
Bassorah (Basra), Iraq 1
Bausch & Lomb (scientific instrument maker) 186
Bayezid II, Sultan 122
Bayram b. Ilyās 113
Beauchamp, Pierre-Joseph de 213, 214, 215, 217, 221, 227
Beaujour, Félix de 111
Bedini, Silvio A. 190
Belgrade 16, 17, 18, 190
  University of Belgrade 191, 195
Bélidor, Bernard Forest de 18, 19, 20, 30, 32, 33, 38, 63
Benghazi, Libya 198, 199, 202
Berge, Matthew (scientific instrument maker) 68
Berlin, University of 171
Berliner-Platz (supplier of instruments for eye surgery) 244
Bernard de Verdun (Bernardus de Virduno) 88, 90, 99, 100, 102, 103, 106
Bernhere, Thomas 115, 116, 117, 121
Berthollet, Claude-Louis 208, 217
Berthoud, Louis 209, 212–216, 221
Betancourt, Agustín de 225
Bézout, Étienne 64
Bicaj, Albania 196

bimetallic coil 159
binoculars 242, 247, 251, 252
Bion, Jean-Baptiste (son of Nicolas Bion) 52
Bion, Johann Theobald (translator) 18
Bion, Nicolas 22, 26, 29, 40, 43, 45(n9), 47, 48, 49, 49(n17), 51, 52, 54
Birkenmajer, Aleksander 87
Black Sea 11, 214
Blaux, Paris (surgical instrument maker) 244
Blondel, François 19, 20, 30, 31, 32, 38
Bombardiers 43, 52, 56
  Corps of Bombardiers (*Ulufeli Humbaracı Ocağı*) 43, 52
bombs, throwing of 19, 29, 30, 31, 33, 38
Bonaparte, *see also* Napoleon 206, 208, 211, 212, 214, 216, 217, 221, 226, 227
Bonneval, Claude Alexandre de, *also called* Ahmed Pasha the Bombardier 43
Bonnycastle, John 73
Borda, Jean-Charles de 214
Borri, Cristoforo 144, 146
Bosch (scientific instrument maker) 162
Bosch, Johann (magic lantern presenter) 239
Bosizio, Ritter von 243
Bosnia 241
Bossut, Charles 64
botany 169
Boué, Ami 241, 243
Bouguer, Pierre 63
boussole, see also compass, magnetic 37
Boyle, Robert 35
Bozcaadalı Hasan Pasa 252
Bradley, James 63, 64, 69, 71
Brahe, Tycho 80, 93
  Tychonic system, Tychonic cosmology 74, 143
brass 29, 32, 35, 36, 65, 68, 80, 84, 87, 103, 113, 116, 159, 210, 211, 215, 218, 252
Brasov, Romania 234
Bratislava 233
Breguet, Abraham-Louis 225
Breguet Jr. (Antoine-Louis, son of Breguet, Abraham-Louis) 211
Brentjes, Sonja 143
Breslau 17
Breton Frères (scientific instrument makers) 180
Brice, William 96
Britain, 149, 162
British 60, 148, 210, 251
  British army, British forces 76, 226
  British fleet 77
Brockmann, Hubert J. 165
bronze 199, 202
Brussels, Belgium 182
Bucharest, Romania 234, 236
  University of Bucharest 166
Budapest, Hungary 232, 233, 234, 236
build, builder, building trade 188, 192, 196, 197, 202, 203
Bulgaria 189, 202
bureaucrat, bureaucratic 52, 225, 240, 242, 251
Bureau de la Guerre, France 211
Bureau des longitudes, Paris 208, 213
Bureau du Cadastre, France 208, 210, 212, 213, 215
Bursa, Turkey 102
Busbecq, Ghislain de 111
al-būṣula, (*see also* magnetic compass) 35, 36, 37
Bylica, Martin 81(n3)
Byzantines 101
  Byzantine Empire 168

Caffarelli du Falga, General Louis-Marie-Joseph-Maximilien 207, 208, 209, 211, 212, 215, 217, 221, 223
Cagnoli, Antonio 64
Cairo, 112, 113, 196, 206, 207, 212–217, 219, 220, 221, 223–226, 228, 237, 239
  Cairo uprising *or* riot 207, 221, 228
cabinet (of instruments) 175, 180
  geological cabinet 169
  natural history cabinet 209
  physics cabinet 172, 180, 209, 217
calendar 13, 70, 71, 74, 100, 114, 198
  Gregorian 70, 71
  Hijra, Hijri 70, 113, 198
Callet, Jean François 72
Calon, General 214
Cambridge and Paul Instrument Company 161

Canada 187, 189, 195
Canada Science and Technology Museum (*see also* Ingenium) 187, 190, 202
Canadian Conservation Institute 202
cannon 32, 142
Caracas, L. P. 245, 247
Carlowitz & Co., Hamburg (scientific instrument supplier) 164
carpenter, carpentry 211, 212, 222
cartography, cartographer, cartographic 6, 78, 120, 208, 210
Cartuja Seismological Station, Granada 165
Caspian Sea 100, 214
Cassini, Jacques 63, 64, 69, 71, 72, 74
Cassini, Jean-Dominique (Cassini IV) 213
catalogue
  catalogue of the Petrovic collection 190, 192, 200
  trade catalogue 163, 167, 180, 182, 184, 246, 252
Catalogue of Incipits of Mediaeval Scientific Writings (by Lynn Thorndike and Pearl Kibre) 104
cathetometer 180
cathode ray tubes 181
Çelebi, Evliya 11(n40), 111(n15)
Çelebi, Katib 7, 37
Celtis, Konrad 91(n24)
cellar 157, 159
  gunpowder cellar (Tartu) 158
CENCO 184, 186
centrifugal machine 180
Çerkes İzzet Pasha 251
Chaldeans 80, 93
Chaptal, Jean-Antoine 226
Chardin, John 143
Charrière, Paris (surgical instrument supplier) 244
chart, nautical 244
chemist, chemistry, chemical 18, 160, 169, 171, 172, 173, 175, 177, 186, 208, 216, 223, 226
  chemistry *or* chemical laboratory 173, 174, 177, 178, 184, 209, 217
China 162, 164
Christian, Christianity 20, 102, 117, 147, 169, 170
chronometer 215
Cihannüma, (Cosmorama by Kātib Çelebī) 7, 11(n40), 37
cinematography 239
Cirot, Calixte-Victor, *or* Calixte-Charlemagne 211, 222
Citadel (in Cairo) 225
Clairaut, Alexis Claude 63
classical studies, classical education 170, 171,172
clepsydra 217
clergy 169, 170, 175
climatology 175
clock 56, 97, 115, 117, 212, 214, 217, 240, 254
  alarm clock 58
  astronomical clock 209, 215
  automaton clock 115, 117
  organ clock 115
  public clock, tower clock 111, 111(n13, n15)
  striking clock 114
clockmaker 114, 209, 211
coat of arms, Ottoman 61
Collège de France 220
Collège de Sorbonne, Paris 92
Collin, Jean 222
colony (French, in Egypt), colonial 206, 221, 223, 225, 226, 227
comet 92, 147
Comité International pour la Métrologie Historique 193
Commentarii Bellici 53
commercial, commerce 114, 120, 168, 173, 240, 242, 243, 248, 255
Chamber of Commerce, Austro-Hungarian, in Istanbul 243
  Chambre de commerce française, Istanbul 256
Commercial School of Halki 171
Commission des sciences et arts 208, 210, 211, 212, 213, 216, 221, 222
compagnie d'ouvriers (company of military workers) 211
compas de proportion 42, 45
compass, magnetic 11, 13, 14, 35, 36, 37, 47, 112, 113, 114, 117, 118, 119, 120, 214, 217, 222, 247
  compass, dry 117
  compass, wet 112

compass, magnetic (cont.)
dip compass, dip needle 211, 216
mariner's compass 252
surveying compass 211, 214, 216, 223
variation (declination) compass 214
compass needle 10, 35, 36, 37, 95, 112, 113, 117, 118, 119, 120
compasses (mathematical instrument) 21, 37, 44, 188, 195, 223
compendium 1, 4, 5, 40, 53, 114
compteur astronomique (a telescope attachment) 213
conjunction, astronomical 97, 123, 217
Connaissance des temps, *see also* almanac, nautical *and* Ephemerides 64, 70, 213
Conseiller électricien de gouvernement ottoman, *see* also Lacoine, Émile 253
Constantinople *see also* Istanbul 24, 76, 100, 169, 170, 172, 233, 234, 235, 236, 237, 238, 243, 251
Construction and Principal Uses of Mathematical Instruments (by Nicolas Bion, translated by Edmund Stone) 40(n2), 49(n17)
consul, consulate 214, 216, 227, 243
consultant 191, 253
consumer, consumption 114, 242, 244, 248, 249, 255
Conté, Nicolas-Jacques 207, 212, 218–226
Conté's inventions 220, 223–226
Cooper and Cary 229
coordinates (altitude-azimuth, ecliptic, equatorial, geographic) 84, 97, 120
Copello, J. B. 245, 247
Copernicus, Copernican system 74, 149
copper 10, 21, 30, 219, 245
Corabœuf, Jean-Baptiste 215, 216, 217
Çorlu, Turkey 254
cosmographe de Girod (a form of tellurium) 182, 183
cosmography 84, 143, 171, 175, 176, 177
Coulomb torsion balance 181
Coutelle, Colonel Jean-Marie-Joseph 220, 223
Couvreur 222
craftsmen, *also* 'artists' 211, 222
crista 84, 88
Croatia 189, 198, 199, 200
crusade, crusader 100, 102
cubit 20, 21, 22, 31, 33, 126, 126(n12), 198
Culpeper, London 121(n49)
culture 26, 109, 115, 168, 169, 187, 188, 195, 203, 204, 229, 242, 249, 251
material culture 23, 80, 192, 240
Cuno, Jakob 81(n3)
curriculum, in schools 169, 170, 171, 173, 177
Cyprus 189

d'Alembert, Jean Le Rond 63
Daguerreotype 232
Dallam, Thomas 115
Damanhour, Egypt 225
Damascus 101
Damietta, Egypt 216
Dār al Funūn (first modern higher education institution in Iran) 142, 153
Dardanelles 59, 77
De torqueto (by Regiomontanus) 89
declination
magnetic 10, 11, 11(n40), 14, 15, 95, 117
solar 13, 70, 71, 74
deferent 122, 126, 130, 132, 133, 134, 135, 137, 139
Dei gesta per Francos (by Guibert of Nogent) 102
Denmark 162
dentist 244, 247
Dépôt général de la Guerre, France 208, 213, 214
Dépôt général de la Marine, France 208
Descotes, Pierre Marie 165
Description de l'Egypte (by Le Père, Jacques-Marie) 206
Description et usage du cercle de reflexion (by Jean-Charles Borda) 214(n29)
Desgenettes, René-Nicolas Dufriche 225
dial, *see* sundial
dialling 108, 111, 112, 114, 120, 121
Dictionnaire de l'ancienne langue française (by Frédéric Godefroy) 99
Diderot, Denis 63
Die Kunst Bomben zu werffen (The Art of Throwing Bombs, by François Blondel) 19
Dimyat, Egypt 120

diplomat, diplomatic 17, 52, 58, 115, 150, 168, 214, 227, 240
dissolving view (magic lantern show) 229–239
distillation apparatus 245
Divān (of Abu Taleb Kalim of Kashan), *qasīda* from 118
dividers (mathematical instrument) 69, 210(n15), 223
Dīwān, Divan (Council) 228
Diwan-i-farsī (by Mir David) 118
Döbler, Ludwig (magician) 231, 232
doctor 173, 241, 242, 247, 252, 255
Dollond, John 210, 211, 215
Dolmabahçe Palace, Istanbul 236
Dolomieu 209(n10)
Doppelmayr, Johann Gabriel 23, 26, 29
Döring, Mr. 236
Dorn, Hans 81(n3)
drawing (instruments, materials) 54, 209
Dreux, Robert de 111
Drosten, Robert (scientific instrument maker) 182
Dubrovnik, Croatia 198
Ducretet, Eugène Adrien (scientific instrument maker) 182, 183
du Mans, Raphael 146
Dumotiez Frères (scientific instrument maker) 183, 209(n10)
Düsseldorf, Germany 232
Duterme, Ch. (scientific instrument retailer) 183

Earth, the 7, 35, 74, 126, 128, 130, 132, 133(n33), 139, 158, 239
  centre of the Earth 130, 132, 137, 139
earthquake 157
East, Eastern 3, 16, 19, 37, 72, 80, 95, 96, 100, 187, 189, 192, 203, 207, 228, 233, 239
East India Company 117, 147
Ebubekir Râtib Efendi 54
eccentricity, eccentric centre 130, 134 137, 140
eclipsarium 52
eclipse 12, 13, 52, 126, 130, 217
ecliptic 8, 70, 71. 74, 84, 94, 96, 128, 130, 132, 133(n33), 139
  obliquity of the ecliptic 13, 71, 74, 84, 87, 88
École centrale des travaux publics, France, later the *École Polytechnique* 211, 212
École des Géographes, Paris 210, 215
École du Génie, Metz 212
École nationale aérostatique (National School of Ballooning), Meudon, France 208, 212, 222, 225
École polytechnique, France 208, 210, 211, 215, 217, 222, 225
economy 78, 162, 168, 169, 173, 178, 195, 242, 243, 244
Edison Lamp Works of the General Electric Company 161
education, educational 43, 102, 150, 160, 169, 170, 171, 173
  education of girls 169, 173
  educational institution 102, 153, 170, 240, 252
  higher education 142, 153, 178, 248
  scientific education 168
  secular education 169
  theological education 175
  Western European education 173
Egypt, Egyptian 1, 96, 102, 108, 113, 120, 189, 196, 197, 206–228, 237, 238
  Upper Egypt 216, 223, 228
electric egg 184
electric lighter, Volta's 184
electric whirl 184, 185
electrical
  apparatus, device, equipment 247, 252, 253, 254
  bell 246, 247
  Bonetti electrical machine 182
  business 246
  Carré electrical machine 180, 183
  lightning conductor 246
  machine 181, 184, 246
  medical electrical device 252, 254
  Ramsden machine 183
electrician, *électricien* 246, 247, 253, 256
electricity 172, 179, 180, 184, 229, 246, 252, 253
electronics 160
electroscope, Zeleny 186
Elements (Euclid's) 3
Elements of Geometry (by John Bonnycastle) 73
elevation, angle of 4, 30, 31, 32, 210

Elizabeth I, Queen (of England) 116
el-Muhit, see Kitabü'l-Muhit
el-Muwaqqit, *see* Mustafa bin Ali el-Muwaqqit
Encyclopédie (of Diderot and d'Alembert) 63
Enderun (a section of the Imperial Palace, Istanbul) 58, 59, 60, 62, 63
engineer 35, 54, 55, 56, 58, 59, 60, 64, 66, 73, 74, 76, 77, 78, 93, 146, 177, 212, 213, 217, 219, 222, 242, 244, 247, 248, 252, 253, 255
  chemical engineer 177, 207
  civil engineer 206, 209, 215, 216, 219, 223
  electrical engineer 253
  mechanical engineer 177, 207
  military engineer 208, 212, 215, 216
  mining engineer 215
  Ottoman engineer 73
engineering 53, 64, 76, 221, 240
  engineering business 252
  engineering documents 252
  engineering instrument 221, 252
  engineering school 227
Engineering School of the Maritime Arsenal, Istanbul 57, 77
England, English 67, 68, 115, 117, 120, 147, 149, 157, 182, 210, 226, 229, 236, 238, 244
English instrument maker 182
English (language) 26, 39, 51, 96, 99, 123, 163, 165
engraving 119, 223
entertainer, entertainment 99, 229, 230, 237
Enlightenment 63
ephemerides, *see also* almanac, nautical 64, 74
Ephémérides des mouvemens célestes, pour le méridien de Paris (by Jérôme de Lalande) 214(n29)
epicycle, epicyclic (in the Ptolemaic system) 130, 132, 133, 134, 137, 138, 139, 140
Epirus, Greece 173
Epperies, Johann de 91, 92, 103, 107
equant 130, 132, 133, 134, 137, 139
equation of time 126, 128
equator, equatorial 84, 92, 94, 96, 97, 134
  equatorial circle (instrument) 2, 3, 4, 7, 8, 10, 11, 12, 13, 13(n42), 14, 95, 96
equatorium 122–133, 135, 139, 141
  alidade of 124, 125, 127, 129, 133, 134, 137, 138, 139
  ring of 124–27, 129, 133–137, 139
  ruler of 124, 125, 133, 134, 137, 138
  plate of 122–129, 132–141
ERDA, Aktien-Gesellschaft Institut für angewandte Geophysik, Göttingen 166
Eşkalü'l-Alatü'l-Hendesiyye (On Geometrical Instruments) 53
Estonia, Estonian 154, 158, 159, 160, 161, 162, 163, 167
  Estonian (language) 159
Euclid, Euclidean geometry 3, 19
equinox 90, 136, 133(n33), 137
Europe, European countries 11, 22, 39, 40, 45, 54, 57, 71, 78, 100, 102, 103, 112–117, 119, 120, 142, 146, 149, 163, 169, 181, 206, 231, 233, 239, 240–244, 247, 252, 255
European (*see also* Western) 10, 14, 29, 37, 53, 54, 58, 61, 62, 63, 64, 71, 74, 84, 93, 99, 100, 101, 102, 114–117, 119, 120, 121, 142, 143, 146, 147, 149, 158, 173, 198, 206, 207, 221, 240, 241, 244, 249, 251, 252, 255
  books 18, 37, 63, 73
  instruments 14, 23, 26, 29, 57, 59, 74, 116, 119, 121
  science, technology 115, 143, 149, 207, 227
  sundials 114, 117, 118, 120
  symbols 62, 63
  tradition 10, 23
Expedition to Egypt, Bonaparte's 206–228
expenses, for Bonaparte's expedition to Egypt 208, 209, 211
experiment, experimental 35, 169, 171, 172, 173, 176, 179, 184, 186, 195, 212, 221, 224, 225, 228, 232
Experimentator (shop name) 159
Exposition Universelle, Paris, 1900 181
eyeglasses, eyewear, *see also* spectacles 142, 143, 241, 244, 247, 249, 253
eyepieces, celestial and terrestrial 210
Eyüb Pasha 60

factory 186, 222, 240, 244, 247, 254
Fanari (a district of Istanbul, modern Fener) 173
Faroqhi, Suraiya 204

al-Fathiyya (astronomical work by Ali Qushji) 3, 74(n53)
Fatih mosque and *kulliye* (place of higher learning) 110, 114
Favre (clockmaker) 211
ferrous metal, ferrous alloy (instrument made of) 196, 197
Fèvre, Jean-Baptiste Simon 219
Feyzi Bey 55, 58–71, 73, 74, 76, 77, 78
Feyzi Efendi 77
Finland 162
Flamsteed, John 209
fleet 254
  British fleet 77, 225
  French fleet 210, 221, 225
  Ottoman fleet 1, 3
Fleisch, Alfred 160, 167
Florence, Italy 164
Flores ex almagesti (by Geber the Spaniard) 81, 90
foot (unit of length) 20, 21
Fordham University, New York 164, 165
fortifications 59, 75, 77, 227
Fortunate Isles 139
Fotiadis, Pericles 173, 176
fountain in vacuum 180, 184
Fouquet, Pierre-Denis 223
France 23, 29, 52, 57, 93, 149, 157, 162, 169, 179, 207, 213, 214, 220, 224, 225, 226, 244, 245, 248, 254, 255
French 24, 29, 40, 43, 52, 63, 64, 76, 77, 93, 115, 120, 180, 206, 207, 213, 217, 218, 221, 222, 223, 224, 225, 226, 227, 228, 232, 248, 249, 251, 255
  Franco-Ottoman scientific relations 226
  French conquest *or* occupation of Egypt 206, 226, 227
  French scientific instruments 182, 183, 206, 207
  French instrument makers 180, 184
French (language) 18, 26, 37, 39, 45, 63, 64, 72, 74, 99, 150, 169, 179, 235
Franco de Colonia 87(n12)
Franco de Polonia, *also known as* Magister Franchone de polonia parisiensis 86, 86(n12), 87, 88, 90, 92, 93, 94, 96, 98, 100, 102, 103, 105
Franco of Lüttich (Liège) 86(n12)
Franks, the 102
Fréminville, Elisabeth-Théodore de La Poix de 209, 210, 211
Frisch, Karl, *also called* as Kirde, Kaarel 166
Fulcher of Chartres 102
Fünunü'l-Harb (manuscript on Warfare Techniques) 53

Gao *or* Gago (now in Mali) 117
Gaiffe, Adolphe (scientific instrument maker) 180, 181, 182, 183
Gaitanaki, Chatzi-Stefani 179
Galante, Paris 245
Galata, Istanbul 1, 243, 245, 246, 248, 249, 253
Galatali, see Seydi Ali bin Hüseyin
Galileo, Galilean, 19, 146
galvanometer, galvanometric 157, 163
  Deprez-d'Arsnoval galvanometer 159, 186
  Hartmann & Braun galvanometer 159
Gamzigrad, Serbia 189, 199, 200
Ganot, A. 169
gasometer (for measuring the quantity of a gas) 180
Gaudette 209(n10), 211
Geber the Spaniard *or* Jābir ibn Aflaḥ 80, 81, 90, 91, 93, 94, 95, 97, 98, 103
Gelenbevi Ismail Efendi 72
General-in-Chief, *see* Bonaparte
geocentric model of the universe 73, 74
geodesy, geodetic 43, 160, 207, 215, 216, 217, 244, 247, 252
geography, geographer, geographic 1, 3, 18, 54, 95, 96, 101, 120, 123, 139, 147, 195, 208, 222
Geography (by Pomponius Mela) 100
geological, geologist 169, 236, 241, 252
geometer 21, 25, 29
geometry, geometric (= engineering) 18, 19, 22, 39, 43, 46, 47, 49, 53, 57, 64, 65, 69, 73, 212, 222
geophysics 157, 165
George III, King (of Great Britain) 63
Georgetown University, Washington, D.C., Seismological Observatory 161
Gerard of Cremona 90
Germany, German 11, 14, 17, 25, 80, 93, 97, 100, 229, 231, 255
  German instrument makers 181, 182

German (language) 17, 18, 19, 20, 22, 25, 26, 39, 93, 99, 163, 169
Germanos IV, Patriarch 175
Ghazarossian A. 245, 246
Gheorghiu, Trajan D. 166
Ghīyāth al-Dīn Jamshīd ibn Mas'ūd ibn Maḥmūd ibn Muḥammad Kāshānī, *see* Kāshānī 122
gift, diplomatic 58, 115, 117, 240
Gilbert & Wright 68
Gilbert, John 121(n49)
Gilbert, William 116(n33)
Girard (civil engineer) 223
Giraud, E. 248
Giza, Egypt 217, 220, 223
glass 35, 36, 67, 172, 223
  flint glass 223, 225, 226
globe 89
  celestial globe 2, 4, 5, 7, 8, 9, 10, 12, 13, 14, 58, 81
  terrestrial globe 58
gnomon 113, 114, 119, 120
Goa, India 146
Göçek, Fatma Müge 241, 242
Godefroy, Frédéric 99
Godlewska, Anna 207
gold, golden, goldsmith 110, 228, 249
Golden Horn, Istanbul 57
Golitsyn, Prince Boris 154, 155–158, 160–164, 166, 167
Göttingen, Germany 158, 164, 166
graduates (of a school) 170, 174
Granada, Spain 165
  Great Mosque of Granada 108
graphometer, *graphomètre* 29, 211, 216, 219
Great Britain 149, 162
Great School of the Nation, Istanbul, *see* Phanar Greek Orthodox College 170
Greece, Greek 23, 29, 102, 168, 169, 170, 171, 173, 174, 179, 189, 196
  Greek community of Istanbul 168, 169, 173, 178, 186
  Greek Kingdom 173
  Greek Lyceum of Chatzihristou 171
  Greek Ministry of Education, Instruments Factory of 186
  Greek Orthodox 168
  Greek Philological Society of Constantinople 169, 171
  Greek revolution 170
  Greek school 180, 182, 185, 186
  Greek state 168, 169, 186
Greek (language) 98, 99, 169, 179
Greenaway, Frank 193
Gregorian dates 70, 71
Griffin, of London (scientific instrument maker) 182
Gros-Caillou, national workshop of, France 209
Grotzsch, Helmut 193
Grundriss Der Experimentalphysik (by E. Jochmann) 169
Guibert of Nogent 102
Gujarat, India 1, 14
gunner, guns 37, 38, 43, 56, 142, 221
gunpowder, gunpowder mill 223, 224
Gunther, Robert T. 86, 93
Gutenberg, Johannes 100(n44)
gymnasium, (high school, *see also* school) 169

Habermel, Erasmus 81(n3)
Habsburg 17, 235
Hachette (scientific instrument maker) 180
Hacı Hüseyin Pasha 251
Hadiyyat al-Muhtadī (The Gift of the Convert, by Osman Efendi bin Abdülmennan) 16, 17, 18, 19, 22, 24, 26, 27, 28, 34, 36, 37
Hadji Hasan mosque, Konya 110
Hadley, John 57
Hagia Sophia mosque, Istanbul 110
Haifa, Israel 197
Haldat apparatus for hydrostatics 181
half circle (simple astrolabe) 25, 28, 29, 30, 31, 32, 37, 84
Halifax, Canada 195
Halifezade Çınari Ismail Efendi 71, 72
Halki, Turkey (*see also* theological) 177
Halliwell, James Orchard 86(n12), 106(n53)
Hamburg, Germany 157, 164
Harborne, William 115
Harem, Imperial (part of the Topkapı Palace, Istanbul) 235
Hartl disc (optical apparatus) 186
Hartmann & Braun (scientific instrument makers) 159
Hartmann, Georg 114
Hartmann, J. 97, 98

Hartmann, Polykarp 88, 106(n54)
Has Oda (Imperial Chamber of the palace in Istanbul) 58, 59, 76
Hasan al-ʿAttār al-Misrī, Sheikh 228
Hasan al-Sā'igh 110
Hasan Kâshif Jarkis 218
Hasan Rüştü Pasha 251
Hassenfratz, Jean Henri 212
havai terazi, terazi 24, 200, 201, 202
Havliczek, Mr. (glass painter) 231
Hayreddin Pasha, *also called* Barbarossa 1
heat 35, 179
heliocentric model of the universe 73, 74
Helmholtz, double siren *and* resonators 180
Hérault, François-Michel 222
Herkov, Zlatko 193
Hero's fountain 180, 183
Herodotus 100
Herschel, William 149
  Herschel telescope 149
Hey'at-e Jadid-e Engriz (New English Astronomy by Mir Mohammad Hossein Isfahani) 147
Hezarfen Hüseyin Efendi 55(n1)
Hijra, Hijri dates 70, 113, 198
Histoire littéraire de la France 88
Historia Hierosolimytana
  by Baldric of Bourgueil 102
  by Fulcher of Chartres 102
historiography 80, 99, 161, 167, 195, 204
history 90, 93, 103, 142, 155, 170, 187, 190, 191, 192, 193, 195, 202, 203, 204, 240
Hoca İshak Efendi 53
Hoca Tahsin Efendi 48(n16)
Hochu (military balloonist) 223
Hocquet, Jean-Claude 193
Holbein, Hans, Junior 81, 83
horizon 8, 26, 30, 31, 32, 55, 66, 68, 71, 84, 97, 216
  artificial horizon 67, 69, 216
  dip of the horizon 69, 73
  horizon ring, horizon plate 8, 95, 97
horology, horological 209, 222
horoscope 128, 131, 141
hours, equal, Nuremberg *or* Babylonian, Italian 8, 113, 114
Hudadoroglou, S. 172
Hülasatü'l Hey'e (Epitome of Astronomy, by Seydi Ali bin Hüseyin) 2, 3
Hungary, Hungarian 91, 233, 236
Hunter, William 148
Hüseyin Hilmi Efendi 56, 78(n69)
Hüseyin Rıfkı Tamani 72, 73
Hüsnü Pasha of Kadiköy 251
Hüsrev Pasha, Mehmet *also known as* Koca Hüsrev Pasha 60
hydrogen 224, 230, 236

Iași, Moldavia 234
Iatropoulos, Alexandros 177
Ibn al-Khatīb 108
Ibn al-Sāffar 108(n2)
Ibn Baso 108
Ibn Simʿun 112
Ibrahim b. Sinan al-Harranī 109
İdare-i Mahsusa (steamship company) 254
Idris Bey 214
Imber, Colin 96
immigrant, immigration 102, 168, 187, 189
Imperial Palace, Istanbul 58, 59, 76, 77, 253
Imperial School of Engineering, Istanbul 39, 43, 53, 54, 56, 58, 59, 63, 64, 66, 67, 72, 74, 76, 77
imported, importing 11, 40, 68, 163, 169, 241–245, 247, 252–254
inch 20, 21
inclinometer 25
Independent Order of Odd Fellows, Constantinople Lodge of the 251
India, Indian 1, 3, 14, 118, 142, 146, 147, 149, 162, 239
  Indian circle (*Hind dairesi*, an astronomical instrument) 3
  Indian Ocean 1, 3, 4, 11
  Indo-Muslim civilization 118
L'indicateur ottoman 243, 246
industry 162, 241, 253, 255
ingénieur-géographe 210, 215
Ingenium: Canada's Museums of Science and Innovation 187, 199
İngiliz Selim (Richard Baily), *also called* Selim the Englishman 73
Inhāta (by Ibn al-Khatīb) 108
Institut d'Égypte 212, 217, 218, 222, 224
Institut national des sciences et arts, France 208, 213, 220, 223
Institute of Physics, of the Russian Imperial Academy of Sciences 160, 161

instrument maker 52, 67, 68, 108, 110, 132, 133, 135, 146, 154, 157, 178, 180, 181, 182, 183, 186, 207, 208, 210, 211, 219
instrument making 119, 127, 154, 161, 167, 206, 209, 207, 208, 211, 212, 221–223, 224, 240, 242
instruments, collections of 170, 171, 172, 177–186, 207, 227, 240
  Petrovic Collection 187–205
instruments
  astronomical 1, 4, 5, 6, 7, 12, 13, 14, 64, 80, 93, 98, 99, 102, 108, 120, 142, 143, 149, 153, 207, 209, 216 217, 221, 222, 226
  ceremonial or decorative 114, 188, 199, 200, 201, 202, 203
  chemical 160, 169, 184, 209, 244, 245, 252
  dental 144
  drawing
  electrical 247, 252
  engineering 221, 252
  geodetic 43, 160, 247, 252
  geographical 222
  geometrical 19, 47, 53
  geological 252
  mathematical 21, 39, 40, 56, 119, 160, 187, 210, 211, 221, 244, 247, 252
  measuring 21, 57, 188, 192, 193, 221, 222, 223
  meteorological 160, 247
  metrological 187, 192
  nautical, naval 78, 244
  observational 58, 59, 64, 71, 92
  optical 209, 225, 226, 241, 242, 243, 244, 246, 248, 252, 253
  Ottoman 195, 199, 201
  philosophical 229
  physics, physical 159, 169, 184, 209(n10), 241, 244, 247, 252
  precision 78, 159, 211, 221, 222, 247, 252
  scientific 37, 43, 52, 54, 61, 67, 142, 143, 155, 159, 170, 172, 176, 177, 183, 186, 190, 206, 207, 208, 220, 221, 224, 226, 227, 228, 232, 240–247, 252, 255
  seismological 157
  surgical 209, 241, 244, 246, 247
  surveying 21, 25–29, 43, 247
  topographical 208, 209, 210, 215, 222
Instruments (magazine) 163
Instruments of Science: An Historical Encyclopedia 99
Introductio geographica (by Peter Apianus) 84(n6)
inventor, invention 46, 64, 80, 86, 87, 88, 90, 92, 93, 95, 97, 98, 121, 127, 143
  Conté's invention's 220, 223–226
Ioannidis, Ioannis 174
Iran, Iranian 24, 40, 118, 122, 142, 143, 146, 147, 149, 150, 153
Iraq 102
iron 21, 218, 222
Isfahan 117, 118, 120, 146, 147
Iṣlāh al-Majisṭī 94
Islam 17, 96, 108, 112, 113, 114, 117
Islamic
  architecture 202, 204
  arts and science 204
  cartography 120
  cities 46
  countries 18, 23, 30
  globes 8
  instruments 37
  knowledge 18, 37
  life 119
  mathematics 29
  science 204
  society 109
  sources 14
  sundials 108, 113, 121
  texts 11
  tradition 10, 14, 120
  units 20
  world, lands 23, 74, 112, 113, 115, 120
Israel 189, 197
Istanbul, *see also* Constantinople 1, 3, 10, 14, 37, 39, 40, 41, 42, 52, 56, 57, 58, 59, 60, 68, 70, 72, 74, 77, 110, 114, 122, 146, 168, 169, 170, 173, 174, 175, 178, 180, 186, 198, 200, 201, 202, 212, 214, 215, 227, 227(n79), 228, 236, 240–256
Italian 37, 114, 144, 235
ivory 113, 214
Iyrcae (later Turcae or Tyrcae) 100
Iz al-din b. Muhammad al-Wafai 13(n42), 96
İznik, Turkey 102

al-Jabartī, Abd al-Rahmān 206, 207, 217, 217(n40), 218, 221, 222, 228
Jābir ibn Aflaḥ, *see also* Geber the Spaniard 90, 94, 95
Jacobi, Alexander (newspaper correspondent) 235
Jamshīd's Cup, *Jām-i Jamshīd* (a name for Kāshānī 's equatorium) 128
Jew, Jewish 229, 232
Jochmann, E. 169
John of Murs 92
John of Sicily 90
Jomard, Edme-François 210, 228
Josephstädter theatre, Vienna 231
Journal de Constantinople 235
Juchereau de Saint Denys, Antoine Baron de 59, 60, 64, 74, 75, 76, 77
judge 40, 58
Jupiter 123, 130, 133, 146, 147, 152, 217

Kaab, Assia 202
Kabakçı Mustafa insurrection 76
Kalvarienberg cemetery 239
Kapıdağlı, Konstantin 61
Kāshān, Iran, 118, 122
Kāshānī, (Ghīyāth al-Dīn Jamshīd ibn Masʿūd ibn Maḥmūd ibn Muḥammad Kāshānī) 122, 123, 124, 126–130, 132, 133, 135, 137, 139, 140
al-Kāshī, *see* Kāshānī 122
Kātib Çelebī *or* Katip Çelebi 7, 37
Katib-i Rumi, *see* Seydi Ali bin Hüseyin 1
Kawe (chocolate maker in Estonia) 162
Kennedy, Edward S. 122, 123, 126, 129, 135
khamā'irrīyat, see also Muhammad b. Fattūh al-Khamā'irī 108
Kharkiv (now in Ukraine) 157
khashabāt, see also stakes 25, 29
Khulāṣat al-ḥisāb (The Very Essence of Arithmetic, by by Bahā' al-dīn al-ʿĀmilī) 24
Kibre, Pearl 104
Kilitbahir, Turkey 59
Kirman (now part of Iran) 110
Kitāb-ı bahriye (Book of Navigation, by Piri Reis) 13, 37
Kitabü'l-Muhit fi İlm el-Eflak ve'l-Ebhur (Book of the Ocean *or* the Sciences of the Spheres and the Seas), *or el-Muhit* 2, 3, 4, 6, 11, 14, 15, 55(n1)
Kléber, General Jean-Baptiste 215, 225
Knight of Malta, (Jean-Auguste d'Angos) 213
Koenig, Rudolph (acoustical instrument maker) 180
Kohl, Max (scientific instrument maker) 181
Kohler & Volckmar (scientific instrument makers) 182
Kolkata, India 148
Konya, Turkey 110
Košice, Slovakia 232, 233, 239
Krätz, Otto 193
Kundt's tube 186
al-kūniyā (triangular levelling instrument) 21, 23
al-kūniyā al-hawā'iya (aerial *kūniyā*) 21, 23, 24, 25, 37
Kuo Shou-Ching (Chinese astronomer) 96, 97
Kurth, Mr. 239
Kütahya, Turkey 41, 42
Kuwait 40
Kuyucaklızade Muhammed Atıf 40
Kyriazidis, Professor 176

La Décade égyptienne 206, 224
La Hire, Philippe de 49(n17), 52
La Paz, Bolivia 165
Labassère, Hautes Pyrénées, France 248
laboratory 158, 159, 168, 169, 172, 173, 174, 176, 177, 178, 184, 186, 209, 217
Lacaille *or* La Caille, Nicolas Louis de 63, 64, 69
Lacoine, Émile (father of Maria Lacoine) 253
Lacoine, Maria (wife of Jean Verdoux) 253
Lalande, Jérôme 63, 64, 70, 71, 74, 213, 214
Laleli Mosque, Istanbul 43, 72
Lamy, E 209
Lane, Edward William 120
La Plata, Argentina 165
L'art de jetter les bombes (by François Blondel) 19
Laschott, Julius 239
Latin 68, 80, 86, 90, 91, 92, 94, 95, 97, 98, 99, 100, 102, 103, 118, 197

latitude 7, 8, 13, 70, 71, 78, 84, 94, 95, 113, 114, 216
cometary latitude 92
lunar latitude 127
planetary latitude 87, 91, 126, 127, 129, 130
stellar latitude 87, 91
Lavoisier, Antoine 214
lawyer 173
Le Corbusier 192
Le Monnier, Pierre Charles 63
Le Patriote (transport ship) 224
Le Père, Jacques-Marie 206, 222, 223
Lebanon 189
Lecoq Pasha, Joseph 251
Lefèvre-Gineau, Louis 220
Légion d'honneur, Chevallier de la (Verdoux) 252
Leipzig 17
Lemaire (clockmaker) 211
Lenoir, Étienne (scientific instrument maker) 210, 211, 214, 216
Lenoir, Paul-Étienne-Marie (son of Etienne Lenoir), Lenoir Jr. 211, 221, 222
lens 66, 142, 146, 210, 214, 230, 249
Lepère, Jean-Baptiste 220
Leroy, Jean-Baptiste 220(n48)
Les tables astronomiques (by Jacques Cassini) 72
Les tables portatives de logarithms (by Jean François Callet) 72
Les travaux de Mars ou l'art de la guerre (by Allain Manesson Mallet) 53
level (instrument) 21, 23, 25, 37, 67, 68, 188, 195, 200, 201, 202, 203, 219, 223, 247
bubble level 207, 211, 216, 219
levels (surveying), levelling 219
Levitski, Grigori 157, 158, 167
Leyden jar 179, 184
Lianopoulos, Georgios 177
Libra, first point of 135
library 61, 208, 209
Archaeology Museum Library, Istanbul 41, 42
Atatürk Library, Istanbul 41, 42
British Library, London 124
Central Library of Tehran University 123
Chester Beatty Library, Dublin 123
India Office Library, London 123
Istanbul University Rare Books Library 41, 42, 59, 60
Kandilli Observatory Library, Istanbul 41, 42, 96(n33)
Library of the Imperial School of Engineering, Istanbul 39, 54, 58
Majlis Library of Iran 124
Malik National Library, Tehran 123
Maulana Azad Library, Aligarh Muslim University, India 147
Millet Library, Istanbul 41, 42
Mulla Firuz Library, Mumbai 124
National Library of Iran, Tehran 124
Süleymaniye Library, Istanbul 41, 42, 60
Topkapı Palace Museum Library, Istanbul 41, 42, 60
Vahidpaşa Public Library, Kutahaya, Turkey 41, 42, 46
Vatican Library 143
Yıldız Palace Library, Istanbul 60, 61
Libya 189, 198, 199
Lignères, John 97(n35)
Ligue maritime française 256
Lill, Johan (baker) 163
limelight 229, 230, 231, 236
Lithuania 162
logarithms 48, 64, 65, 70, 71, 72, 73, 74, 77
London, England 67, 68, 81, 115, 123, 124, 149, 161, 182, 229, 230, 231, 233, 234, 244, 245
longitude 214, 216
celestial 84, 92, 94
geographical *or* terrestrial 123, 139, 140
lunar 127
mean 127, 130, 133, 134, 136, 137, 139
planetary 87, 91, 122, 123, 126, 127, 128, 130, 131, 133, 136, 140
solar 70, 71, 133, 136
stellar 87, 91, 133
true 123, 130, 133, 133(n33), 134, 137, 139, 141
Lorch, Richard 91, 94, 95, 96, 103
Lowell lectures 165
Lubāb Iskandarī (by Kāshānī) 132(n29)
Lugat-ı Tarihiyye ve Coğrafiyye (Dictionary of History and Geography) 54
lunar month 12, 13
Lunkenheimer, Dr. 165

Lycée Impérial Ottoman de Galata-Sérai, Istanbul 253

Macelwane, James B. 165
machine 35, 89
  Atwood 180
  calculating (pinwheel) 232, 233, 234
  centrifugal 180
  electrical 180, 181, 182, 183, 184, 246, 254
  hydraulic 224
  industrial 209, 221, 223, 224, 225
madrasa 43, 45, 58, 102, 109, 110
Madytos, Greece 179
Magdeburg hemispheres 180
magician 230, 231
magic lantern 184, 229, 230, 233, 234, 235, 238, 239
  magic lantern slides 184, 230, 231, 233–239
magnet, magnetic, magnetism 35, 36, 159, 186, 229, 241
  magnetic compass, *see* compass
  magnetic declination 10, 11, 11(n40), 14, 15, 95, 117
  magnetic needle 35, 37, 95, 118, 120
  thermoelectric magnet 186
Mahmud II, Sultan 43, 60(n21)
Maimua Shamsi (Solar System by Mualavi Abol-Kheyr) 147, 148
Mallet, Allain Manesson 53
Malta 213, 215, 220
Mamluk, Mameluke 115, 206, 212, 222, 227
Mandeleney, André 180
manufacturer 184, 240, 244, 247, 255
map, mapping, *see also* cartography 118, 120, 163, 207, 213, 214, 216, 217, 219, 227, 247
market 14, 56, 114, 159, 161, 162, 163, 196, 198, 230, 243, 244, 251, 252, 255
  marketing strategies 154, 162, 167
Margarita philosophica nova (by Gregor Reisch) 81
mariners (sailors, seamen) 1, 3, 6, 14, 15, 55,71, 77, 78, 252
Marinovich, A 243, 246
Maritime Arsenal, Istanbul 57, 59
Marrakesh 115
al-Marrākushī (Abū ʿAlī al-Hasan ibn ʿAlī ibn ʿUmar al-Marrākushī 23, 109
Mars 123, 130, 133, 134, 138, 140, 151
Marseilles 215, 244
Masing, Hugo 154, 155, 156, 159, 160, 161, 162, 163, 164, 165, 166, 167
material culture 23, 80, 192, 240
mathematical 19, 24, 33, 54, 64, 65, 73, 109, 110
  astronomy 88, 120
  knowledge 59, 65
  practice, practitioner 110, 115, 188, 203, 204
  sciences 2, 43, 53, 221
  symbols 65, 71, 73
  treatise 16, 18
  tables 59, 69
  texts *or* works 40, 63, 69, 71
mathematician 20, 40, 59, 63, 110, 122, 146, 147, 213
mathematics 29, 39, 40, 48, 52, 53, 56, 59, 64, 71, 72, 77, 109, 110, 111, 171, 177, 204, 247
Matthias Corvinus, King of Hungary 91
Maulavi Abol-Kheyr 147
Mazo, Eli, of Paris 184
mean motions, table of 127, 130, 132, 133, 136, 137, 139
measurement, units and standards of 20, 192, 196, 197, 211, 214, 223
  Christian and Islamic 20
Mecca 55, 118, 119, 198
Méchain, André 213
Méchain, Jérôme-Isaac (son of André Méchain) 213
mechanics, mechanical 56, 157 159 160, 167, 177, 180, 207, 211, 222
  mechanical workshop 154, 159, 167, 207, 220, 221, 223, 225, 226
Mecmuʿa-i Resail-i Saʿidiyye 42
Mecmuaʿ-i Ulum-ı Riyaziye (Compendium of Mathematical Sciences, by Hoca İshak Efendi) 53
medical 142, 212, 247, 249, 252, 253
Mediterranean 1, 4, 11, 204, 219, 227, 237
  Eastern Mediterranean 187, 189, 192, 203
Mediterranean Atlas (by Ali ibn Ahmad ibn Muhammad al-Sharfi) 6
Mehadia, Romania 234
Mehmed Ali Bey, Prince 238
Mehmed Ali Pasha, Mushir of Tophane 235

Mehmed Emin Vahid Efendi 46
Mehmed Sadık bin Mehmed Emin Cihangiri 41, 47
Mehmed/Muhammed Sadık Cihangiri (astrologer, d. 1812) 47(n14)
Mehmed Said Efendi (manuscript author, d. 1767) 40, 41, 42, 43, 44, 52(n18)
Mehmed Said Efendi (ambassador to France, d. 1761) 52, 52(n18)
Mehmet II, Sultan 1, 100, 110, 114
Mehmet III, Sultan 115
Melloni's apparatus 180
Memoire della Guerra (by Raimondo Montecuccoli) 53
Menou, General Jacques-François de 225, 226, 228
merchant 17, 102, 168, 169, 173, 208, 210
mercury (metal) 67, 69, 220, 224
Mercury (planet) 127, 130, 137, 147
meridian 8, 95, 97, 120, 210, 214
  Paris Meridian 213
Merrick, James Lyman 150, 151, 152
Messageries Maritimes 244
metallurgical tests 160, 202
meteorology, meteorological 160, 175, 214, 247
metre bar, standard metre 211, 223
metrology, metrological 187, 188, 191, 192, 195, 203
  First International Congress for Historical Metrology, Zagreb 191
  history of metrology 192, 193
Metz, France 212
Meudon, France 208, 212, 224, 225
Mevlana Hamdullah Ibn Şeyh Cemaleddin 3
Michel, Henri 93, 94
micrometer (for a telescope) 210, 211, 213
microscope 184, 229, 231
mikyas, see also sector 43
military 39, 56, 59, 60, 64, 78, 102, 149, 168, 208, 210, 211, 212, 215, 216, 223, 226, 227, 240, 242, 250, 251, 253, 254
Military Academy, Istanbul 56, 250
Military Engineers, Corps of, French 208, 212, 215, 216
al-mimsaḥa al-mustaʿmila fī misāḥat al-arḍ (land surveying instrument) 21
mine (surveying, blasting, digging of, etc.) 19, 25, 29, 35, 37
mineralogy 175
minerals, collection of 169
Ministry of Military Supplies, Ottoman 254
Ministry of Posts and Telegraphs, Ottoman 253, 254
Ministry of Public Instruction, Ottoman 252
Ministry of Public Works, Ottoman 252
Ministry of War, French 208
Ministry of War, Ottoman 251
al-minqala, see also protractor 21
minute (of arc) 66, 67, 70, 73, 126, 128, 131, 141, 217
mīqāti (specialists in spherical astronomy and astronomical timekeeping) 109, 111
al-miqyās (scale or unit of measurement) 21
miqyās niṣf al-dhirāʿ (half cubit scale) 21, 22
Mir Abd al-Latif Khan Shoushtari 149
Mir David 118
Mir Mohammad Hossein Isfahani 147
mirʾā (mirror) 25, 29
Mir'at-ı Kâinat (The Mirror of the Universe, by Seydi Ali bin Hüseyin) 1, 2, 4, 5, 6, 7, 8, 10, 11, 12, 13, 14, 15
Mir'atü'l-Memalik (The Mirror of Countries, by Seydi Ali bin Hüseyin) 3, 4
Mirat-ı Alem (Mirror of the Universe, by Seyyid Ali Pasha) 74
mirror 25, 29, 216
Mirza Abdollah Isfahani 146
Mirza Abu Taleb Khan Isfahani 149
Mirza 'Azima Iksir 118
Mirza Masoud-e Ansari 150
Mirza Saleh Shirazi 149
missionary 146, 150
  Missionary Fathers convent 214
al-misṭara (straight edge) 21
misṭara al-muwāzāt, see also ruler, parallel rulers 21
model 37, 98, 110, 121, 159, 211, 212, 213
  geocentric model of the universe 74
  heliocentric model of the universe 74
modernisation 40, 64, 76, 227
Mohammad Salih Gazvini 146
Moiin Mosavvar 144

Molard, Claude-Pierre 221
Moldavia 234
Molteni, Antoine (scientific instrument maker) 184
Monge, Gaspard 208, 212, 217
monocle 249
Montecuccoli, Raimondo 53
Montreal, Canada 187, 191, 190, 195
moon 146, 147, 148, 150
  moons of Jupiter 146, 147, 217
  moons of Saturn 147
Moon, the 91, 94, 115, 127, 130, 146, 147, 150, 216
Morand, Morand's café (*Café Louvre*) 251
Morocco 115, 162
Morse telegraph 180
Morton, Nicholas 102
Morveau, Guyton de 220(n48), 223
Moscow, University of 156
Mosel, River 81
mosque 43, 72, 108, 109, 110, 114
Mossy (scientific instrument maker) 209(n10)
Mouqfar, Egypt 219
mucayyab, see also sine quadrant 55
muezzin 111, 112
Mufassal logaritma (Complete logarithms) 64
Müftizade Palabıyık Mehmed Efendi 45(n11)
Muğlalı Mehmed Efendi (copyist) 41, 45
Muhadarat, Muhadarat-ı Feyzi (Feyzi's discourse) 58, 59, 60, 61, 62, 63, 64, 65, 69, 71, 73, 77, 78
Muhammad, Prophet 64, 198
Muḥammad al-Aqkirmānī 29
Muhammad Ali Pasha 237, 238
Muhammad al-Jawharī 113
Muhammad al-Wafai, *see* Iz al-din b. Muhammad al-Wafai
Muhammad b. Fattūh al-Khamā'irī 108(n1)
Muhammad Husayn 120
Muhammad Mahdī al-Yazdī 118
Muhammad Muqīm al-Yazdī 118
Muhammad Sa'id Pasha, Prince 237, 238
Muhammad Sami 10
Mühendishane-i Berri-i Hümayun, see Imperial School of Engineering, Istanbul 39, 53, 54, 56
Muhit, el-, see *Kitabü'l-Muhit*
Mu'īn al-Munajjim Kāshānī, *see* 'Abd al-Razzāq ibn Muḥammad 124
Mukai Ahmad al-Mansūr, Sultan 115
müneccimbaşı (timekeeper/astrologer) 56
munejjim (astronomer/astrologer) 6, 10, 15
Munich, Germany 232, 233
Muscat, Oman 214
Musée des Arts et Métiers, Paris 232
Museum of Science and Industry, Chicago 161
music 230, 231, 232, 235, 238
Muslim, *see also* Islamic 20, 55, 74, 99, 102, 118, 147, 242, 248
Mustafa b. Muhammed 4
Mustafa bin Ali el-Muwaqqit 7
Mustafa IV, Sultan 76
Mustafa Reşid Pasha 237
Mustafa Sıdkı Efendi 49(n17), 52
Muuga, Jaan 159
muwaqqit (mosque astronomer/timekeeper) 6, 10, 43, 108, 109, 111, 112
muwaqqitkhane (timekeeping office) 55

Nahda (renaissance, in Egypt) 228
Nahide Işık Demirakın 249
Napoleon in Egypt. Al-Jabarti's Chronicle of the French Occupation 1798 (by Abd al-Raḥmān Jabartī) 217(n40)
Napoleon, Napoleonic, *see also* Bonaparte 207, 227, 228
Nasīr al-Dīn al-Ṭūsī 93
National Museum of American History, Smithsonian Institution, Washington 190
National Museum of Science and Technology, Ottawa, *see also* Ingenium 187, 188, 191, 194
National Theatre, Budapest 232
Natural History (by Pliny the Elder) 100
natural philosophy 88, 149, 169
Nauru Island (Central Pacific) 165
Navadir al-Olum (by Mohammad Salih Gazvini) 146
Naval Academy, Nikolajevsk 155, 158
Naval School, Imperial (Ottoman) 252
Navarino, battle of 169
Neumann, Reverend Manuel 165

navigation 1, 3, 4, 6, 11, 13, 37, 48, 117
  celestial *or* astronomical 57, 65
navigator 2, 11, 13, 14, 35, 36, 68
navy 57, 61, 74, 79, 169, 238, 252
  French navy 210, 212, 213
  Ministry of the Navy, France 212
  Minister of the Navy, Ottoman 251
Needham, Joseph 96, 97
needle, *see* compass needle
Nelson, Admiral Horatio 225
Nequam, Alexander 112
Netherlands, The 162
Neuer Cursus Mathematicus (by Bernard Forest de Bélido) 18
Neugard, Evald 159
Nicholas de Heybech 97, 98
Nicholas of Cusa 81, 82, 97, 98
Nile, River 219, 239
Nilometer (*mekyas*) 219
niṣf al-dāʾira (half circle) 25, 28, 29
niṣf dāʾira li-ramī al-khumbara (half circle for the throwing of bombs) 30
Noir & Forster (scientific instrument makers) 182
Nörlund 161
Norry, Charles 217
Notre Dame cathedral 220(n48)
Nouet, Nicolas-Antoine 213, 215, 216, 217, 218, 220
Nouveau cours de mathématique (by Bernard Forest de Bélidor) 18
numerals
  abjad numerals 8, 10
  Arabic 72, 114, 197, 198
  European *or* Latin 197, 198
Nuremberg, Germany 81, 97, 98, 114
Nuzha al-Ḥadāʾiq (Excursion to the Gardens, by Kāshānī) 122, 123, 126, 127, 128, 129, 132, 133, 136, 139, 140

obliquity of the ecliptic, *see* ecliptic
observatory 149, 214, 217, 227
  Cairo Observatory 216, 217
  Imperial Chinese observatory 96
  Istanbul Observatory 7
  Kharkiv Observatory, Ukraine 157
  Malta Observatory 213, 215
  Maragha Observatory, Iran 93, 94, 95(n31)
  Naval Observatory, Marseilles 215
  Observatorio Astronómico, La Plata, Argentina 165
  Observatory of San Calixto, La Pas, Bolivia 165
  observatory of Ulugh Beg 122
  Osservatorio Ximeniano, Florence 164
  Paris Observatory 208, 213
  Samarkand Observatory 122
  University of Tartu Observatory 154, 157, 158, 159
octant 57
Odessa 158, 234, 236
office materials 209
officer (army, naval, etc.) 52, 60, 78, 207, 212, 213, 216, 217, 220, 250, 251
Old World 187
Oman, Sultanate of 214
opera, Italian 235
opera glasses 252
opposition, astronomical 97, 217
optician, *opticien* 186, 229, 231, 241, 242, 247, 248, 249, 251, 252
optics 175, 180, 210
orbit 132, 137
  of the Moon, lunar orbit 127
  of Mercury 127
organ pipe 181
Orlov, Aleksandr 158, 167
orrery 149
oscillation (period of, axis of) 157, 159
oscilloscope 186
Osman Efendi 16–38
Ottawa 187, 191, 194
Ottoman 6, 26, 17, 26, 39, 40, 51, 52, 54, 60, 61, 62, 75, 78, 96, 100, 115, 121, 123, 195, 199, 226, 227, 240, 241, 249, 250, 251, 255
  Archives 249, 253
  army 58, 60, 61, 76, 253
  court 114, 240
  Empire, domains, lands, regions, territories 7, 16, 25, 29, 37, 52, 54, 55, 64, 102, 111, 142, 168, 170, 175, 187, 192, 195, 197, 198, 201, 203, 204, 206, 214, 227, 228, 229, 235, 236, 237, 240, 241, 242, 243, 244, 248, 252, 255
  engineer 55, 73
  fleet 1, 3
  instrument 14, 189, 190, 195, 201

literature, manuscripts, texts, treatises 6, 7, 14, 54, 61, 71, 73
market 14, 114, 251, 255
mathematician, mathematics 54, 71, 77
modernisation 76
nautical astronomy 2, 12
navy 57, 78. 169
period, times 40, 54, 114, 189, 197, 202
sailors. seamen, mariners 1, 3, 14, 15, 55
science 16, 240
state, statesman 52, 115, 248, 251
Sultan 61, 110, 115, 168
Turkey 39, 51, 52, 53, 54, 56, 77

Ottomans 14, 36, 37, 39, 43, 54, 55, 57, 70, 74, 170, 189, 199, 226, 240, 241, 242, 244, 245, 249

Oxford University, England 149

painter 61, 207, 231, 232, 233, 237

painting, painted 81, 113, 144, 218, 231, 233, 234, 235, 236, 238, 239

paintings 100, 143, 194, 195, 237

Palamas, Grigorios 172

Palatinate 213

Pallados School for girls 171

Panthéon 220(n48)

pantograph 211

Papin digester 181

parabolic reflector 179

parallax (in astronomy) 69, 74, 214

parallel ruler 21, 37

Paris, France 72, 80, 90, 97, 100, 102, 103, 181–184, 195, 207, 208, 209, 211, 212, 213, 215, 220, 226, 228, 232, 233, 238, 244, 245, 249, 254

Pate, Master 116

patent 68, 232

Patriarchate 168, 169, 173, 175

Peluse, Egypt 219

pendulum 157, 159
compound pendulum 209
horizontal pendulum 157, 158
Rebeur-Paschewitz pendulum 157
seconds pendulum 180
Stückrath heavy pendulum 157
Zöllner heavy pendulum 157
Zöllner pendulum 157, 164

Pera, Istanbul 173, 245, 246, 248, 249, 251
Pera Museum 200

perch (*qaḍīb*, a proposed unit of length) 20

pergar-ı nisbe, pergar-ı nisbiyy, pergarü'n-nisbe (*see also* sector) 39, 41, 42, 43, 45, 53

Pergar-ı nisbetin aʿmali ve bazı bürhanı beyanındadır (On the Uses and Some Proofs of the Sector) 41, 42

periodic table 185

Persia, Persian 115, 117, 119, 120, 128, 162

Persian (language) 6, 11, 14, 26, 46, 122, 123, 128, 129, 143, 145, 146, 147, 149, 150

Peter of Limoges, canon of Evreux 92

Petrovic, George, *original name* Djordje Petrović, 187–205
A Vision of the Future (exhibition of Petrovic's paintings) 188, 194
publications of 193

Phanar (a district of Istanbul, modern Fener) 173

Phanar Greek Orthodox College, Istanbul 170, 171, 172, 178–181, 182, 185

Phantasmagoria 236

pharmacy, pharmacist 173, 180, 209, 247

Philological Society, Greek, of Constantinople 169, 171

philosophy, philosopher 35, 81, 147, 149, 229
natural philosophy 88, 149, 169

phonograph 183, 254

photograph, photographic 157, 163, 165, 239

physics 4, 155, 158, 159, 160, 161, 169, 170, 171, 172, 175, 176, 177, 178, 184, 185, 241, 247, 252
cabinet of physics 180, 209, 217
experimental physics 171, 172, 173, 176, 186
physics laboratory 158, 173, 174, 177

physiology 160

pi (π) 122

Piccolomini, Enea Silvio, Pope Pius II 100

Pikkand, Eduard 159

Piri Reis 1, 13, 37

plane table 26, 29, 37, 78, 210, 211, 216, 217, 223

planet, planetary 74, 87, 91, 92, 122, 123, 126–140 141, 146, 147, 148, 150

planetarium 179

Planetary Equatorium of Jamshīd Ghīyāth al-Dīn al-Kāshī, The (by E. S. Kennedy) 123

Pliny the Elder 100
Plovdiv, Bulgaria 202
plumb line, plumb bob 8, 23, 31, 34, 84, 188, 189, 195, 199, 200, 201, 202, 203
pneumotachometer (for studying respiration) 160
poet, poetry 118, 142
Poirier, L. (French optician in Istanbul) 248
Poland 23
pole (north, south, or celestial) 8, 118
Poltava (Ukraine) 158
Pompey's Pillar, Alexandria 223
Pomponius Mela 100
Pope, The 100
portrait, 101, 144, 237
  of Feyzi Bey 58, 60, 61, 62, 65, 66, 67, 68
Portugal, Portuguese 1, 11
Poulle, Emmanuel 88, 99
Praetorius, Johann 81(n3)
prayer, 45, 46, 120
  times of 55, 64, 71, 78, 109, 112, 113, 114, 218
precession 74
precision 67, 78, 93, 159, 167, 211, 221–224, 247, 252
Prešov, Slovakia 233
press (newspapers) 231, 233
priest 146
printing, printed 23, 37, 72, 81, 111, 114, 148, 207, 238, 247
  printing material 208, 209
  printing press 72, 100, 165, 208, 217
privatdozent 155
Prizren, Kosova (in the former Yugoslvia) 196
Prony, Gaspard Clair François Marie Riche de 213, 225
proportion, theories of 192
protractor 21, 211
Ptolemy 3, 94, 130, 143
  Ptolemaic ruler 94, 97(n35)
  Ptolemaic system, planetary theory 74, 127(n16), 130, 131, 132, 133, 140, 141
public bath (*ḥammām*) 33, 35
Pulkovo (St Petersburg) 157, 158
pulse time recorder 160
pyramid 220(n49)
  Great Pyramid of Giza 217, 219, 220
pyrometer 224
Pythagoras theorem 29

al-Qalqashandi 109
Qansuh al-Ghurī, Sultan 115
Qawshajī, ʿAlī *see* ʿAlī bin Muhammad al-Qawshajī
qibla (direction of Mecca) 10, 11, 13, 55, 113, 118, 119, 120, 218
qiblanuma (*qibla* indicator) 11, 117, 118, 119, 120, 131
quadrant 55, 56, 78, 94, 114, 120, 131, 214, 216
  astrolabe quadrant 2, 4, 6, 7, 12, 13, 14
  azimuth quadrant 210
  gunner's quadrant 37, 38
  quadrant of altitude 8
  sine quadrant 2, 3, 4, 5, 6, 7, 12, 13, 14, 25, 29, 30, 31, 37, 43, 55
  wooden quadrant, Ottoman 1, 7, 55, 56, 67, 74, 78
Quénot, François-Marie 213, 215, 216
Quṭb al-dīn al-Shirāzī 23

Rafael, *see also* Du-Mans, Raphael 146
Rami Bey 251
Ramsden, Jesse 67, 68, 183
Ras El Tin palace 237
rectangulus (of Richard of Wallingford) 92
Red Sea 219
reflecting circle 210, 214, 216
repeating circle 215
reforms, military 64, 168, 169
refraction, atmospheric 64, 69, 70, 73, 74, 219
Regiomontanus 80, 86, 89, 90, 90(n22), 91, 92, 93, 95, 98, 107
registration (galvanometric, mechanical, photographic) 157
regula latitudinum 84, 88
Reichstag, German 100
Reisch, Gregor 81
religion, religious 58, 109, 112, 168, 206, 229, 236
repair, repairer, repairing (of scientific instruments) 56, 76, 170, 207, 211, 220, 223, 243, 246, 247
Repsold (of Hamburg, scientific instrument maker) 157
Resale-I dar Esbat-e Hey'at-e Jadid (Treatise on the Proof of Modern Astronomy, by Abu Taleb ibn Hassan Hosseini Safavi) 147

resonators, Helmholtz 180
retailer 183, 240, 241, 242, 243, 245, 246, 252, 255
retailers of instruments in Istanbul 240, 255
Adjarian, H. and Ghazarossian, A. 245, 246
Armao, P. J. 245, 246
Caracas, L. P. 245, 247
Copello, J. B. 245, 247
Marinovich, A. 243, 246
Schapira, I. 245
Schapira, J. 245, 247
Urgliavich, Antoine *or* Urglievich, Anton 242, 243, 245, 246
Urgliavich, Madame 243, 246
Vitalis, E, et Cie. 246, 247
Verdoux, J. 186, 240, 246–256
Révolutions de Constantinople (by Antoine Juchereau) 76
Reza Abbasi 143, 144
Rhoda (island in the River Nile near Cairo) 219
Riaz ol-Olama (by Mirza Abdollah Isfahani) 146
Richard of Wallingford, Abbot of St. Albans 92
Richer, Jean-François (scientific instrument maker) 209(n10), 211
Ridwan al-Falaki 207(n7)
Rifā'a al-Tahtāwī 228
riot, rioter 206, 207, 217, 221
Risāla fī maʿrifat al-abʿād, (Treatise on the knowledge of distances, by Muḥammad al-Aqkirmānī) 29
Risale-i Ceyb, (by Seydi Ali bin Hüseyin) 5
Risale-i Da'ire-i Muaddil (Treatise on the equatorial circle, by Seydi Ali bin Hüseyin) 4, 10
Risale-i irtifa (Treatise for Altitude) 60
Risale-i İstiʿmal-i Pergarü'n-Nisbet (Treatise on the Use of the Sector) 41, 49
Risale-i Mukantarat, (by Seydi Ali bin Hüseyin) 4, 5
Risale-i Pergarü'n-Nisbet (Treatise on the Sector) 40, 41, 46, 47, 49
Risale-i Rub-ı Müceyyeb (Treatise on the sine quadrant, by Seydi Ali bin Hüseyin) 4, 5
Risale-i Şerh-i Pergarü'n-Nisbe (Commentary on the Sector) 41, 45
Risale-i Usturlab (Treatise on the Astrolabe, by Seydi Ali bin Hüseyin) 4
Risale-i Zatü'l-Kürsi (Treatise on the Celestial Globe, by Seydi Ali bin Hüseyin) 4, 5
Risale-ye Elm ol-Nojum (A Treatise on Astronomical Science, by James Lyman Merrick) 150, 151, 152
Rochette (scientific instrument supplier) 209(n10)
Rohrbeck's Nachfolger, W. J. (scientific instrument makers) 181, 182
Romania
Academy of Sciences of Romania 166
Société de Physique de Roumanie 166
Rome 146, 208
Rorweger, of London 245
Rosetta, Egypt 216, 221, 225
Roth, David 232
Roth, Imrich Emmanuel 229–239
Royal Polytechnic Institute, London 230, 231
Royal Theatre, Budapest 233
al-rubʿ al-mujayyab (sine quadrant), *see also* quadrant 25, 29
Ruhmkorff coil 183
ruler, rule, 21, 22, 29, 32, 91, 124ff, 188, 192, 195, 196, 197, 198, 199, 202, 223, 252
arshin (a Serbian type of rule) 196, 197, 198
endaze (a Serbian type of rule) 196, 197
parallel ruler 21, 37
weaver's rule 196, 197, 203
Russia, Russian 142, 149, 157, 158, 162, 227, 234, 244
Russian revolution 160
Russo-Ottoman treaty 227
Rüstem Paşa (a *madrasa*) 45

al-Sā'igh, Hasan 110
Sacred Congregation for the Propagation of the Faith, Rome 208
Sadi, also Saʿdi (translator) 41, 45
Safavid 117, 119, 142, 143
saffārīyat, see also Ibn al-Sāffar 108
Saint George of Kremnos, Monastery of, Halki, Turkey 177
Saint Louis University, Missouri, Department of Geophysics 165
Saint-Priest, Comte de 57(n7)

Salih Murad Özdilek 71(n44)
Salih Zeki 71(n44)
sailors, *see* mariners
Samarkand 110, 111(n12), 122, 124
sandūq or sandūq al-yawāqīt, (a form of sundial) 112, 113
Sarton, George 93
Saturn 123, 130, 133, 147, 150, 152
Saulfarar Efendi 237
Sautter, Harlé & Cie 254, 255
savant, (Oriental, Egyptian) 207, 228
Savart, Nicolas 211
Savoy, France 213
Savoyard 229
Saxony, Germany 23
Sayili, Aydın 143
scale (measuring instrument) 21, 22, 37, 66, 67, 68, 114, 119, 120, 195, 211
Schapira, I. 245, 247
Schapira, J. 245
Scholarios, Patriarch Gennadios 170
school 159, 168, 170, 178, 182, 186, 247, 255
  Arsakeion Girls' School, Athens 174
  Commercial School of Halki 171
  elementary school 169, 173, 174
  girls' school 169, 171, 173, 174, 175
  Greek Lyceum of Chatzihristou 171
  Greek school 180, 182, 185, 186
  high school (gymnasium) 169, 170, 173, 175, 181
  Imperial School of Engineering, Istanbul 39, 43, 53, 54, 56, 58, 59, 63, 64, 66, 67, 72, 74, 76, 77
  military high school 56
  modern school 142, 150
  Pallados School for girls 171
  Phanar Greek Orthodox College, (Great School of the Nation), Istanbul 170, 171, 172, 178–181, 185
  School of Geometry, Maritime Arsenal, Istanbul 57
  School of the Virgin Mary, Istanbul, later Zografeion high school 173
  school regulations 175, 176
  technical school 212
  Theological School of Halki, Turkey 170, 175–178, 184, 186
  Zappeion School (for girls), Istanbul 170, 171, 173, 174, 182, 183, 184, 186, 246
  Zografeion School, Istanbul 170, 173, 174, 181–184, 186, 252
Schulze, Bernhard 157
science
  history of 142, 195
  modern 16, 142, 147, 149, 150, 153, 169
  natural 169, 170, 171, 229
  science of the stars (*ilm-i nücum*) 2, 3, 5, 6, 11, 14
  popular 239
  Western, European 143, 147, 228
Science and Invention (magazine) 164
Science Museum, London 161
scientific equipment *or* product *see also* scientific instrument 229, 242
scientist 142, 160, 166, 167, 207, 208, 212, 215, 222
scriptures 111
sea 3, 4, 11, 68, 71, 77, 117, 118, 214, 216, 219, 237
seal 60, 61, 62
seamen, *see* mariners
sector (mathematical instrument), *see also Pergar-ı Nisbe* 39, 40, 42–47, 49, 51, 53, 54
  Line of Chords, Metals, Planes, Polygons, Solids (on a sector) 43, 48, 49, 51
Sefaretname-yi Rifa'at Bey, (by Rifāʿa al-Tahtāwī) 228
seismograph, seismographic 154, 155, 157, 159, 160, 161, 163, 164, 165, 166, 167
  Anderson and Wood torsion seismograph 161
  Bosch seismograph 162
  Cambridge and Paul seismograph 161
  Golitsyn seismograph 157, 158, 161, 166
  Golitsyn-Vilip seismograph 154, 155, 156, 158, 161, 162, 163, 166, 167
  horizontal seismograph 158
  Milne-Shaw seismograph 162
  seismographic commission (of the Russian Academy of Sciences) 157
  vertical seismograph 158, 159
  Wiechert seismograph 162
seismology, seismologist 154, 155, 157, 158, 164, 167

Cartuja (Granada) seismological station 165
central seismological committee (of the Russian Imperial Academy of Science) 158
Fordham University seismological station 165
Kharkiv Observatory seismological station 157
Pulkovo seismological station 158
seismological conference, Manchester 158
Seismological Observatory of Georgetown University, Washington, D.C. 161
seismological station 157, 161, 163
Tartu seismological station 157
United States Seismological Society East Wing 165
Şekerzade Feyzullah 71, 73
Selim III, Sultan 58, 59, 60, 61, 62, 63, 64, 66, 67, 76, 77, 227(n79)
Seljuk Turks 101, 102
semis (half circle) 84, 88
Serbia, Serbian 189, 192, 199, 200, 241
architecture, house 191, 192, 195, 204
Serbian language 187, 191, 193
Serenissima Signoria of Venice 115
Sermet Muhtar Alus 248, 249, 250, 251
Serres, Macedonia 111
Seven Years' War 229
Seville, Spain 80, 90, 108(n1)
sextant 55–61, 63, 65–71, 73, 77, 78
Seydi Ali bin Hüseyin, *also known as* Katib-i Rumi, Galatali, Seydi Ali Reis, Seydi Ali Kapudan *or* Seydi Ali Çelebi 1–15, 55(n1), 96
Seyyid Abdürrahim Muhibb Efendi 226
Seyyid Ali Pasha 74
Seyyid Mustafa 64
shadow square (for trigonometric calculations) 7
Shah ʿAbbas I (the Great) 117
Shams al-Dīn Muhammad ibn Abi al-Fath al-Sūfī 113
Sharḥ Nukhbat al-Tuffāḥa fī ʿilm al-misāḥa (by ʿAbd al-laṭīf b. Aḥmad al-Dimashqī) 24, 29
Shaykh the imam, *see* ʿAlāʾ al-Dīn b. al-Shātir 113
shipbuilding 227
ship, shipping 57, 78, 99, 116, 209, 210, 214, 224, 234, 239, 244, 254
Shiraz, Iran 139, 140
shop 52, 159, 240, 241, 242, 243, 245, 246, 247, 248, 249, 251, 252, 255
showmen, travelling, 229, 236, 239
Sibiu, Romania 234
Sienese 100
Siilbaum, Paul 159
sine 30, 71, 122
sine tables (*see also* mathematical tables) 29, 30, 31
silver 10, 115, 116, 121, 214, 219
siren, Helmholtz double 180
slide, magic lantern 184, 230, 231, 233, 235–239
Slovakia 232, 233
smith (metalworker) 222
Smyrna 168
Société de Géographie 256
Société de Physique de Roumanie 166
Société des lunetiers Muneaux, Videpied & Cie, Paris 244
Société générale des téléphones et des usines Rattier 255
Société internationale des électriciens 246, 256
social 178, 187, 192, 195, 204, 206, 228, 229, 236, 242, 249, 250, 251
society 109, 114, 143(n3), 228, 234, 239
sociological 241
solar parallax (correction to sextant observation) 69, 74
solar radius (correction to sextant observation) 69, 71, 74
sonometer 180, 183
Spain 23, 103, 213
Spanish (language) 24, 116
Spatharis, Professor Andreas 171, 172, 173, 180
spectacles, *see also* eyeglasses 144, 229, 241, 244, 246, 249, 251, 255
spectrometer, spectroscope 181, 186
sphere, celestial 84
Spindler & Hoyer, Göttingen 164
Split, Croatia 189, 199, 200
square, set square (mathematical instrument) 223, 252

St Cloud, William 87(n15)
St James's Theatre, London 231
St Nikolaus-Hospital, Bernkastel-Kues, Germany 81
St Petersburg 157, 158, 160
St Xavier College, Cincinnati, USA 165
stakes (*khashabāt* for surveying) 25, 29
star, stellar (in astronomy) 3, 4, 5, 6, 11, 13, 14, 64, 77, 87, 91, 92, 146, 214, 217
star (symbol) 61, 62, 63
Stassinopoulos, Stylianos 177
stationery 246, 247, 253
Stavropoleos, Metropolitan Bishop of (Konstantinos Tipaldos) 176
Stone, Edmund 40(n2), 51
Strasbourg 81
  University of Strasbourg 155
student 57, 64, 89, 142, 143, 149, 155, 158, 171, 172, 173, 175, 177, 179, 211, 213, 215, 228
Stylianidis, Kleonimos 172
Sublime Porte 168
Suez 1, 206, 219
Suleiman I, Sultan 3, 4
Süleyman Nutki Bey 78
Sultan 59, 61, 62, 112, 115, 168, 203, 227, 240, 253, 255
  Valide Sultan (Sultan's mother) 235
Sun, the 10, 13, 14, 55, 57, 64, 65, 68, 69, 70, 77, 78, 109, 115,126, 127, 129, 130, 132, 133, 138, 139, 216, 216, 218, 237
  solar 12, 13, 14, 55, 69, 70, 71, 74, 90, 126, 129, 132, 133, 134, 136, 137, 138, 147, 184, 198
sundial 96, 108–121, 131, 218
sunspots 146, 147, 150
Surat, India 1
surveying 19, 21, 25, 29, 39, 43, 47, 49, 68, 203, 206, 207, 210, 211, 214, 216, 217, 219, 223, 247
surveyor 20, 25, 219
  surveyor's chain 21
  surveyor's half circle 37
  surveyor's pole 219
Swann, Peter, President of the Association of Cultural Executives in Montreal 190
Swinger, L. H. W. 165
Syria 102, 223
syzygy 97
Szeged, Hungary 234
al-ṭabla (table, plane table) 25, 26, 27
Tabriz, Iran 110
tabula aequinoctialis 84, 88
tabula horizontalis 84, 88
tabula orbis signorum, (celestial longitude, *see* longitude) 84
Talleyrand (Charles Maurice de Talleyrand-Périgord) 227(n79)
Takhlīs al-ibrīz fi talkhīs Bārīz (The fineness of Gold: a synopsis of Paris, by Rifāʿa al-Tahtāwī) 228
Tanzimat 168, 169, 228
Taqi al-Din 7, 47(n14), 56(n6)
Tarjome-ye Heya`t-e Jadid (translation of 'Treatise on Modern Astronomy') 150
Tartu, Estonia 154–160, 162, 167
  Tartu Telephone Company 162
  Tartu University 155, 156, 158, 160, 166, 167
  Tartu University Press 165
Tata, Hungary 233
Taton, René 193
teacher, teaching 40, 52, 110, 169, 170, 171, 172, 173, 174, 175, 185, 250, 251, 253
teaching (of sciences) 43, 112, 160, 169, 170, 171, 172, 173, 175, 176, 177, 178, 180, 184, 186, 191, 227, 228
technical 40, 59, 63, 88, 98, 117, 120, 149, 150, 187, 203, 212, 242, 253
technician 40, 52, 248, 251
technology 115, 116, 147, 149, 190, 207, 228, 253
Tehran 123, 124
Tekeli, Sevim 96
telegraph, telegraphy 210, 225, 226, 244, 253, 254
  aerial telegraph 225
  Breguet and Betancourt's telegraph 225
  Chappe telegraph 225
  Morse telegraph 180
telephone 162, 246, 254, 255
telescope 57(n10), 66, 67, 68, 69, 120, 142, 143, 146, 147, 148, 149, 150, 151, 181, 183, 207, 209, 210, 213, 214, 215, 216, 217, 219, 221, 225, 226, 229, 241, 242, 252
  achromatic telescope 209, 214, 215, 216
  Dollond telescope 210, 211, 215
  Herschel telescope 149
  Newton's telescope 147

tellurium 182, 183
temperature 158, 159, 223, 224
terazi, havai terazi 24, 200, 201, 202
Tesla apparatus 186
Tessier, Alexandre-Henri 220(n48)
theatre 231–236
theodolite 78, 252
theology, theological 168, 175, 177
  Theological School of Halki, Turkey 170, 175–178, 184, 186
thermometer 176, 183, 186, 209, 247, 248, 252
Thessaloniki, Greece 196
Thomson, Elihu, electromagnetic device of 186
Thorndike, Lynn 86, 104
Thyssagetae (ancient people north of Anatolia) 100
time 10, 12, 13, 55, 70, 71, 111, 78, 109, 123 126, 128, 130, 132, 133, 134, 135, 136, 137, 213, 218, 224
timekeeper, *see also muwaqqit* 6, 11, 40, 43, 55, 56, 68, 72, 78
timekeeping, time measurement 6, 13, 15, 47, 55, 56, 64, 71, 74, 77, 78, 109, 111, 112, 113, 115, 117, 213, 224
Timişoara, Romania 234
tinplate 223, 225
Tipaldos, Konstantinos, Metropolitan Bishop of Stavropoleos 176
Titl, Anton Emil (composer of music) 231
Toderini, Abbé Giambattista 57
Tohfat al-Alam (by Mir Abd al-Latif Khan Shoushtari) 149
toise (old unit of length) 20, 31
Toisé des dépenses (list of expenses), General Caffarelli's 208, 215
Toledan Tables 86 (n11), 86(n12), 87, 88, 90
Tondorf, Reverend 161
tools 202, 203, 209, 216, 221, 222, 223
Tophane, Istanbul 198, 199, 235
Tophāne-i Āmire (the Imperial Foundry in Istanbul) 198
Topkapı Palace, Istanbul 41, 42, 60, 67, 235
topographer, topography, topographical 78, 179, 206, 207, 208, 209, 210, 215, 216, 217, 219, 222
Toronto, Canada 195
torquetum, *see* turketum
Torricelli 30
Tott, Baron de 57
Toulon, France 210, 220
Toulouse Tables 87
trade, trading 100, 117, 162, 163, 164, 242, 243, 245, 247, 255
  trade catalogues 163, 167, 182, 184
trader, tradesman 52, 243, 251, 255
Traité de la construction et des principaux usages des instrumens de mathématique (by Nicolas Bion) 22, 26, 40, 43, 45(n9), 47, 48, 49(n17), 51, 52, 53, 54
Traité élémentaire de Physique (by A. Ganot) 169
translator 18, 40, 41, 45, 51, 53, 241
traveller 24, 52, 117, 142, 146, 149
travelogue, travel book 143, 149, 150(n20)
Treatise on Modern Astronomy (translated as *Tarjome-ye Heya`t-e Jadid*) 150
Treatise on the Sector (*Risale-i Pergarü'n-Nisbet*) 40, 41, 45, 47, 49
Trebizond, Turkey (modern Trabzon) 214
trigonometry, trigonometric 5, 7, 12, 13, 25, 31, 43, 55, 64, 71, 73, 220
tripod 26, 29
Tripoli 115
trisect an angle 69, 69(n42)
Trouvé, Gustave 255
tughra (monogram or signature of a Sultan) 61
Tünel Square, Istanbul 247, 248, 249
tuning fork, electrically maintained 186
Tunisia 189
Turcae, Tyrcae (ancient people living north of Anatolia) 100, 101
Türkengerät (instrument of the Turks) 98, 99
turketum, *also written as* torquetum, truqueti, trunqueti, *or* tarqueti (with variant endings) 80–107
Turkey 39, 40, 41, 42, 52, 54, 55, 56, 72, 73, 76, 97, 99, 101, 111, 162, 170, 185, 189, 200, 201, 214, 236, 241
  Ottoman Turkey 39, 51, 52, 53, 54, 56, 77
Turkish 80, 97, 98, 100, 101, 102, 168, 178, 241
  Turkish notables, Turkish lords 228, 241, 242
  Turkish Republic 60

Turkish (language) 3, 6, 11, 14, 17, 24, 26, 36, 39, 40, 41, 42, 43, 46, 49, 51, 52, 56, 58, 71, 72, 73, 75, 78, 96, 98, 197, 201, 228
Turkmenistan 99
Turks 80, 97, 98, 99, 100, 101, 102
Turkomans, 142
Turner, Anthony J. 84
Tychonic system 74, 143
Tynan, John 164, 165

Ukraine 157, 158
Ulugh Beg *or* Ulugh Bey 74, 110, 111(n12), 120, 122
unit, of measurement 20, 21, 192, 196, 197, 198
United States of America 157, 161, 162, 163, 165
universe, model of 74
university, *see under the name of the city*
Urgliavich, Antoine *or* Urglievich, Anton 242, 243, 245, 246
Urgliavich, Madame (widow of Antoine Urgliavich) 243, 246
ustunc takımı (drawing instruments, *see* mathematical instruments) 54
al-usṭurlāb al-basīṭ, (simple astrolabe) 29
Usul-i Hendese (translation of Elements of Geometry by John Bonnycastle) 73
Uzun Hasan, Sultan 110, 142

vacuum pump, *see also* air pump 35
Valle, Pietro della 143, 145, 146
Vancouver, Canada 195
Vandermonde, Alexandre-Théophile 212
Vanek, Joseph 236
varnish 221, 223, 224
Vasfi, Mehmet 184, 185
Veliyyüddin (copyist) 45
Venice 100, 114, 115, 142
Venus 56(n6), 123, 127, 130, 133, 139, 140, 147, 140, 151
Verdoux, Jean 186, 240, 246–256
Verdoux's optical shop 247–251
vernier scale 66, 67
Versailles, France 222
veterinarian 247
Vicar General, Baghdad 214
Viceroy (of Egypt) 237, 238
Vienna, Viennese 17, 91, 91(n24), 182, 231, 232, 241, 244, 245, 247
Vilip, Johan 155, 158, 159, 160, 161, 165, 166, 167
Vitalis, E. 246, 247
Vitez, János (Archbishop of Esztergom) 89
Volta's electric lighter 184

al-Wafā'ī, *see* 'Abd al-'Aziz ibn Muhammad al-Wafā'ī al-Miqātī
war, warfare 6, 53, 76(n64), 149, 208, 221, 229, 251, 253
Anglo-Turkish War 76, 149
Greco-Turkish War 168
World War I 160, 162
World War II 185
watch (timepiece) 56, 114, 115, 117, 240, 246, 249
marine watch 209, 213, 214, 215, 216, 221
weapons 171, 222, 223
weaver's rule (Serbian) 196, 197, 203
Weiller, Lazare 255
West, the, *see also* Europe 19, 37, 80, 95, 142, 187, 239
Western, *see also* European 16, 23, 36, 37, 102, 143, 147, 149, 150, 162, 228, 242
wheel 99, 222
windmill 222, 223
Winnipeg, Canada 195
Wolf, Rudolf 93
Wolff, Christian Freiherr von 19, 22, 25, 26, 29, 30
woodwork, *see* carpentry
workshop 52, 56, 154, 159, 160, 167, 175, 207, 209, 212, 220, 221, 222, 223, 225, 226, 240
workshop equipment and tools (bellows, forges, lathes, hammers etc.) 222, 255
Wright, Edward 115, 116, 117
writer 95, 99, 241, 242, 248

yard (a type of Serbian rule and unit of length) 197
Yedikule, Turkey 254
Yemen, Yemenite 102, 112
Yıldız Palace, Istanbul 60, 250, 254
Yugoslavia 187, 189, 192, 195, 196, 200

Zagreb 191
Zaječar, Serbia 191
Zappas, Constantine 173, 174
Zappeion School, Istanbul 170, 171, 173, 174, 182, 183, 184, 186, 247
Zappios Science Contest, Athens 171
Zarifis, Georgios 173
zatü'l-kürsi (celestial globe, *see* globe) 4, 5, 7, 8, 12, 14
zenith 8, 216
Zeynal-Din Lari 143, 146
zij, *see also* astronomical tables 10, 55, 74, 126, 127, 128, 130, 141
  Zīj-i Khāqānī 132, 139, 140
  zīj al-jadīd (by Ali Qushji) 110
  zij of Ulugh Beg (*or* Ulugh Bey) 74, 120
Zinner, Ernst 98, 99
zodiac 8, 69, 70, 126, 135, 126, 218
Zografeion School, Istanbul 170, 173, 174, 181–184, 186, 252
Zografos, Christakis 173
zoology 175

www.ingramcontent.com/pod-product-compliance
Lightning Source LLC
LaVergne TN
LVHW010641300926
849161LV00005B/1009